Central Eurasia in Global Politics

International Studies in Sociology and Social Anthropology

VOLUME 92

Central Eurasia in Global Politics

Conflict, Security, and Development

Edited by

Mehdi Parvizi Amineh

Henk Houweling

SECOND EDITION

BRILL

LEIDEN · BOSTON

2005

This book was published with financial support from the International Institute for Asian Studies (IIAS), Leiden, the Netherlands.

This book is printed on acid-free paper.

Library of Congress Cataloging-in-Publication Data

Central Eurasia in global politics : conflict, security, and development / edited by Mehdi Parvizi Amineh, Henk Houweling.
p. cm. — (International studies in sociology and social anthropology, ISSN 0074-8684 ; v. 92)
Includes bibliographical references and index.
ISBN 90-04-14439-0 (pbk. : alk. paper) 1. Asia, Central—Politics and government—1991- 2. Post-communism—Asia, Central. 3. Geopolitics—Asia, Central. 4. Asia, Central—Foreign relations—1991- 5. Asia, Central—Social conditions—1991- 6. Asia, Central—Economic conditions—1991- 7. Petroleum industry and trade—Asia, Central. I. Amineh, Mehdi Parvizi. II. Houweling, Henk. III. Title. IV. Series.

JQ1080.C46 2005
327.58—dc22

2005042153

ISSN 0074-8684
ISBN 90 04 14439 0

Cover design by Sandra van Merode

Cover illustration by Mehdi Parvizi Amineh

PRINTED IN THE NETHERLANDS

CONTENTS

Part Three

Interactions between Outsiders, Neighbors and Central Eurasian Republics

Part Four

Local Conflicts

PREFACE TO THE SECOND EDITION

The first edition of this book was largely put together during the invasion of Iraq by the Anglo-Saxon powers in the spring of 2003. The editors were not tempted by the war to chronicle the unfolding events leading to that war and its subsequent progression. Instead, they were working on a book about US power projection in the post-Cold War era and the fate of Theory in International Relations after the disintegration of the Soviet Union. The creation of eight independent states in Central Eurasia (South Caucasus and Central Asia) changed the geopolitical landscape of the Soviet era. However, due to the justifications put forward by the American government and its allies for invading Iraq, we could not neglect the war. These justifications did contradict the geopolitical hypothesis on US power projection that we were working on at that time. If the Bush Administration, and the governments allied with it, had not lied about their motives for committing international aggression, our geopolitical analysis would lose credibility. We therefore tested the motive statements for their behavioral implications. We found that US and allied behavior on the ground was very different from what one would expect it to be if the invading powers had spoken the truth. On January 12, 2005, the Bush Administration, after having spent hundreds of millions of dollars on finding these weapons, quietly acknowledged that Iraq did not in fact have such weapons. On an earlier occasion, the US also acknowledged that its intelligence had failed to find any evidence of a connection between the September 11th terrorists and the regime in Iraq. We therefore feel more confident in our geopolitical hypothesis: the invasion of Iraq is part of the process of constructing a new leg in America's Cold War 'defense perimeter.' By studying the history of American power projection from the early 19th century to the present, or its 'conjuncture,' as Braudel used that term in chapter two of his book, ***Écrits sur l'histoire***, (Paris: Flammarion 1969), we hoped to contribute to a better understanding of American foreign policy on the Eurasian landmass, particularly in Central Eurasia and the Middle East since the end of the Cold War.

The approach we have brought to bear in the work is called critical geopolitics. We argue that this approach is particularly relevant for studying the foreign policies of projecting power beyond borders in the era of sequential industrialization. A process of power projection by a state-making elite during several generations is 'anonymous history.'

A partially industrialized world is characterized by vast inequalities in wealth and power.

In a multi-state system characterized by sequential capitalist industrialization, power projection is inevitably a competitive undertaking.

Competitive power projection by western countries since the beginning of the industrial era has 'phased out' of existence in the name of 'progress' small autarchic and mutually isolated societies. However, until the beginning of the industrial era, empires were successful in fending off penetration by Western powers. This began to change during the late 18th century. Historical roles between expanding Islam and Europe under threat were reversed. As early as 1798, Napoleon invaded Ottoman territory. In the colonial era, the Muslim empires disintegrated, ending in colonization by Western powers due to the force of modern firearms, mustard gas and air power.

Policies of power projection beyond legal borders by industrialized and industrializing societies are therefore the driving force in the continuous process of transition from a world composed of small-scale societies, with domestic orders untouched by one another, and agro-aristocratic empires in Asia, to a single interdependent world society in a global capitalist economy.

In the constraining bipolar military order of the Cold War, US-engineered regime change in sovereign countries was legitimized as a contribution to the struggle against the global threat of communist dictatorship. However, US rationales for unilateral action abroad differ, whereas American behavior reveals a persistent pattern. Between 1798 and the outbreak of World War I, the US initiated 135 military operations abroad, including 14 in China alone, not to mention Sumatra and elsewhere in East Asia, without having been attacked first. At that time, the rationale given by the US was probably best described by President Wilson as the "army following the flag of commerce" and of "uplifting the host from barbarity into civilization." In the less constraining unipolar military order of the post-Cold War era, military force again implements regime change in resource-bearing areas.

The critical geopolitical framework we bring to bear in this work aims at better understanding the position of the Central Eurasian region in the global order and the socioeconomic and ideological forces that penetrate the region from the outside. In that part of the world, the foreign policies of competitive power projection bring together China, Japan, Russia, the US and the European Union. The US is the only major power that since World War II invaded and occupied an oil-rich country. In that role, it succeeds Great Britain, the true creator of the state of Iraq.

The operating assumption on which we studied American power projection underway in Iraq and the Caspian region is twofold. Firstly, America's policy-making elite brings into its foreign policy projects directed at the region the functional requirements of the domestic socioeconomic order, part of which is to maintain its domestic position, further enrich it and create honor, including the spread of true religion, for itself. Secondly, to acquire the ability to set conditions for potential rivals, such as China, India, Japan, the European Union, to tap the fossil fuel sources on the Eurasian continent. This project will not be complete by inserting Anglo-Saxon oil companies into the Iraq oil stream. We anticipate that Russia's state-controlled oil and gas industry will be a future target.

However, in the current global system of instant communications, such operations are acted out in an ever more tightly interconnected system. Acting in a closed system, increasingly interconnected into one responding whole by means of instant communication, prevents policy makers from controlling outcomes. Lies do not matter as long as no one takes notice and the host being visited can be crushed without having fired a shot. Things are different when the elites of major powers, with the self-declared mission of civilizing the world and getting rich by it, are condemned by a world public as liars and torturers, while the dismal fate of the host is brought to the attention of a global audience. The political culture of power projection by a particular country, or its ideological form, depends on its prior history, on current power relations in the interstate system and on responses by host societies.

We argue in this book that the energy needs of expanding industrial economies in East Asia, Russia and Continental Europe are transforming the scattered industries on the Eurasian continent into one industrial system. That level of integration, however, is opposed by the Anglo-Saxon maritime powers.

The focus of the book is on Central Eurasia in today's global political, economic, military and sociogeographic global system.

However, the powers bordering on that region are part of the story as well. This applies in particular to Russia and its relations with the EU and the US, China, Iran and Japan. Under President Putin, the Russian state is reasserting control over regions in order to reestablish central control over the country's wealth in natural resources. Russia is therefore doing something that is apparently unacceptable to the US foreign policy elite: to impose conditions for Anglo-Saxon companies to access Russian natural resources, particularly energy resources. As long as the Russian state exists, extends from the Ukrainian-Belarus borders to the Pacific and is equipped with a second strike capability, the Anglo-Saxon powers will have a hard time changing that fact, missile shield or not. A disintegrating

Russia is no doubt in the interest of the maritime powers. 'Democracy' is the weapon of choice in present circumstances. Disintegration of the Russian state into a loose confederation without legitimate governments at the member-state and union levels, would deprive Japan, the EU and China from a potential counterweight against the US, that is prevent them from creating an integrated industrial system on the Eurasian landmass and linking that system to the fossil fuel sources of Russia, the Middle East and the Caspian region. The Orange Revolution in the Ukraine came too late to be incorporated into the book. We expect, however, that the democratic revolution in the Ukraine, in which the US is reported to have invested $65 million, will open up a new chapter in US-Russian and EU relations in the new post Cold-War order.

Mehdi Parvizi Amineh
Henk Houweling

Amsterdam, January 2005

ACKNOWLEDGEMENTS TO THE SECOND EDITION

Several people have contributed to the second edition of ***Central Eurasia in Global Politics***.

First of all we thank our contributors for their willingness to go over their texts again, removing textual imperfections that slipped through the nets of correctors of the first edition.

We thank Eva Rakel again for her patience in editing and in managing contacts between editors and book contributors.

The second edition is an improved reprint of the first edition. Several chapters have been partially rewritten. However, we could not substantially expand the number of pages. The task of strengthening the theoretical framework and of integrating chapter contributions in it has to wait for a revised and expanded third edition.

Mehdi Parvizi Amineh and Henk Houweling, Amsterdam

CONTRIBUTORS

Pınar Akçalı was born on October 19, 1965, in Ankara. She graduated from the Department of Political Science and Public Administration of the Middle East Technical University (METU), Ankara, in 1987. Following her graduation, she went to the Bologna Center of the Johns Hopkins University in Italy for a "one-year diploma program" in international relations. Upon completing the program, she started her master's degree at the Middle East Technical University's Department of Political Science and Public Administration. While in the program, she passed the exams for research assistantship and started to work as a research assistant at the department. She completed the master's program at METU with the acceptance of her thesis, titled "Polish Solidarity Movement." In August 1991, she went to the United States for her Ph.D. degree, with a full scholarship from the Council of Higher Education (Yüksek Öğretim Kurumu). She completed her master's degree and her Ph.D. degree at the Department of Political Science of Miami University, in Oxford, Ohio, in August 1998. The title of her Ph.D. dissertation was "Identity Politics and Political Mobilization in Central Asia: The Case of the Islamic Renaissance Party." Since December 1998, she has been working as an assistant professor in the Department of Political Science and Public Administration at the Middle East Technical University. She has publications in Turkish and in English on Central Asian identity issues.

Shirin Akiner is Lecturer in Central Asian Studies at the School of Oriental and African Studies, University of London, where she convenes and teaches courses on Central Asian history, politics, and religion. She is also an Associate Fellow of the Royal Institute of International Affairs, London. She has extensive firsthand experience of the region. Authored publications include: *Islamic Peoples of the Soviet Union* (London: KPI, 1983, 1987); *Central Asia: New Arc of Conflict?* (London: Royal United Services Institute, 1993); *The Formation of Kazakh Identity* (London: Royal Institute of International Affairs, 1995); *Minorities in a Time of Change: Prospects for Conflict, Stability and Development in Central Asia* (London: Minority Rights Group, 1997); *Tajikistan: Disintegration or Reconciliation?* (London: Royal Institute of International Affairs, 2001); and numerous scholarly articles on such topics as Islam, ethnicity, political change, and security challenges in Central Asia.

Mehdi Parvizi Amineh is a Senior Research Fellow in post-Soviet Central Asia, Caucasus, and West Asia at the International Institute for Asian Studies (IIAS), Leiden, the Netherlands; Research Fellow in post-Soviet geopolitics of energy security in Eurasia and the Caspian Region at the Clingendael International Energy Programme (CIEP), Den Haag, the Netherlands; and Research Fellow at the Amsterdam School for Social Science Research (ASSR), University of

Amsterdam, the Netherlands. He also teaches at the Webster University, Leiden, the Netherlands, and at the International School for Humanities and Social Sciences (ISHSS), University of Amsterdam, Netherlands.

His main fields of study are International Political Economy and Social History. His specific areas of study are Central Eurasia, the Caspian Region, and the Middle East. His research interests include globalisation, and regionalism/regional integration, geopolitics, politicized Islam and global energy security. Recent publications include *Globalisation, Geopolitics and Energy Security in Central Eurasia and the Caspian Region* (Den Haag: CIEP, 2003), *Towards the Control of Oil Resources in the Caspian Region* (New York: St. Martin's Press, 1999/2000); *Die Globale Kapitalistische Expansion und Iran (1500-1980): Eine Studie der Iranischen Politischen Ökonomie* (Lit Verlag: Muenster, Hamburg, 1999); and *Globalisation and Islam: The Rise and Decline of Islam as Political Ideology* (forthcoming).

Robert M. Cutler is Research Fellow at the Institute of European and Russian Studies, Carleton University, Canada. Educated at the Massachusetts Institute of Technology. With an earned doctorate from University of Michigan, he has held fellowships and teaching and research positions at major universities in the United States, France, Switzerland and Russia. He is a member, elected under open nominations, of the Executive Committee, Central Eurasian Studies Society. His publications include articles in such journals as *Global Governance, Journal of International Affairs, International Organization, and International Review of Social History*; and an extensive online presence through current-affairs commentaries (Central Asia/Caucasus Analyst, Asia Times OnLine, and elsewhere). He has a presence, likewise, in the printed and broadcast mass media, and consults widely with international institutions, governments, nongovernmental organizations, and industry. He was an External Collaborator of the Caucasus Working Group, Centre for European Policy Studies, and maintains his own website at *http://www.robertcutler.org*

Ayça Ergun is a lecturer in the Department of Sociology at Middle East Technical University, Ankara, Turkey. She works on state-society relations, democratization, nation-building, state-building, civil society formation and political elite in Southern Caucasus, particularly in Azerbaijan.

Henk Houweling is a senior lecturer in International Relations at the Department of Political Science of the University of Amsterdam. His fields of interests are history and theory of International Relations. Co-author of articles on the causes of war, he has published his work in several journals such as the *Journal of Conflict Resolution, Journal of Peace Research, Journal of Arms Control and Disarmament* and *International Interactions*. Recent publications include: "Destabilizing Consequences of Sequential Development," in L. van de Goor, Kumar Rupesinghe, Paul Sciarone, eds., *Between Development and Destruction: An Enquiry into the Causes of Conflict in Post-Colonial States* (London: Macmillan, 1996, pp. 143-169); "Industrialization in East Asia," in Holger Henke and Ian Boxill, eds.,

The end of the Asian model? (Amsterdam: John Benjamins Publishing Company, 1999, pp. 1-50); "Dutch-Chinese Bilateral Relations: Constant Elements in a Changing World System," in Europe, *China and the Two SARs: Towards a New Era* (London: Macmillan, 2000, pp. 103-137); "China's Transition to Industrial Capitalism," in Alex Jiberto Fernandez, ed., *Good Governance in the Era of Global Neoliberalism* (Routledge Publishers, 2003 [in press]).

Armine Ishkanian is a Postdoctoral Research Fellow at the University of California at Berkeley Institute of Slavic, East European, and Eurasian Studies. Her research interests include NGOs, civil society, development, and gender in post-Soviet Armenia, the Caucasus, and Central Asia. She is conducting research on Armenia's NGO sector developments. Her recent publications include: *"Importing Civil Society? The Emergence of Armenia's NGO Sector and the Impact of Western Aid on Its Development," Armenian Forum: A Journal of Contemporary Affairs*, vol. 3(1) 2003; *"Post-Soviet Women Encountering Transition"* (Woodrow Wilson Center Press) 2003; *"Central Asia and the Caucasus: Transnationalism and Diaspora"* (Routledge) 2004.

Michael Kaser is an Honorary Professor in the Institute for German Studies at the University of Birmingham and Reader Emeritus in Economics at the University of Oxford, where he taught between 1963 and 1993. He first travelled in the Caucasus in 1949 when he was Second Secretary in the British Embassy, Moscow. He was on mission in Central Asia (1957) when in the Secretariat of the United Nations, and has often returned to the region. He is chairman of the Advisory Board on Central Asia and the Caucasus of the Royal Institute of International Affairs, London, and author of many publications on the region, including 'The Economies of Kazakhstan and Uzbekistan' (RIIA, 1997); he is contributing two chapters on Central Asian economic history to 'The Cambridge History of Inner Asia' (CUP, forthcoming).

Anatoly Krutov is Operations Officer for Water and Environment at the World Bank office in Tashkent, Uzbekistan. He has extensive experience in areas such as environmental protection, water management, irrigation and drainage in the Aral Sea Basin, and has a background in hydraulic engineering and environmental management. Before joining the World Bank, he was Head of Department in the Central Asian Scientific Institute of Irrigation (SANIIRI) in Tashkent, and worked in a comparable position in the Syr Darya Water Basin Authority. Apart from coordinating and contributing to various feasibility studies and project reports, he published chapters in two recent books on Central Asian environmental problems: Bos, M. (ed.), *The Inter-Relationship Between Irrigation, Drainage and the Environment in the Aral Sea Basin* (Dordrecht/Boston/London: Kluwer Academic Publishers, 1996); and Glantz, M. (ed.), *Creeping Environmental Problems and Sustainable Development in the Aral Sea Basin* (London/New York: Cambridge University Press, 1999).

Hooman Peimani works as an independent consultant with international organizations in Geneva, Switzerland, and does research in International Relations. His research has centered on the Persian Gulf, the Caucasus, Central Asia, and the Indian subcontinent. Apart from various journal and newspaper articles, his recent publications include: *Falling Terrorism and Rising Conflicts: The Afghan "Contribution" to Polarization and Confrontation in West and South Asia* (Praeger) 2003; *Failed Transition, Bleak Future: War and Instability in Central Asia and the Caucasus* (Praeger) 2002; *The Caspian Pipeline Dilemma: Political Games and Economic Losses* (Praeger) 2001; *Nuclear Proliferation in the Indian Subcontinent: The Self-Exhausting "Superpowers" and Emerging Alliances* (Praeger) 2000; *Iran and the United States: The Rise of the West Asian Regional Grouping* (Praeger) 1999; and *Regional Security and the Future of Central Asia: The Competition of Iran, Turkey, and Russia* (Praeger) 1998.

Kurt W. Radtke with a Ph.D. in Chinese from the Australian National University, has held academic positions in the fields of Japanese and Chinese studies in New Zealand and the Netherlands before moving to his current position as professor as Chinese and Japanese studies at the Institute of Asia-Pacific Studies, Waseda University, Tokyo. He has written books on Chinese poetry and China's post-war relations with Japan, and co-edited several books on aspects of East Asia in the age of globalization. In addition, he has published numerous articles in English, Japanese, and Chinese on politics (international), security, and society in China and Japan. His recent publications are: *Competing for Integration. Japan, Europe, Latin America, and Their Strategic Partners* (New York: M.E. Sharpe, 2002, co-edited with M.L. Wiesebron); *Comprehensive Security in Asia* (Leiden: BrillPublishers, 2000, co-edited with R. Feddema); *A Dutch Spy in China* (Leiden: BrillPublishers, 1999, co-edited with Teiler).

Eva Rakel is a Ph.D. student at the Humboldt Universtät Berlin, Germany and University of Amsterdam, The Netherlands. Her Ph.D. deals with the possibilities and impediments for political elite change in contemporary Iran. She has published several reviews and presented at various international conferences on the Near and Middle East.

Max Spoor is Associate Professor of Transition Economics and Coordinator of the Centre for the Study of Transition and Development (CESTRAD) at the Institute of Social Studies (ISS) in The Hague. He has published widely on agricultural policies, environment and rural economies of developing and transition countries, with articles in international journals such as *World Development*, the *Journal of Agrarian Change*, the *Journal of International Development*, the *Journal of Development Studies*, *Food Policy*, and *Europe-Asia Studies* (formerly *Soviet Studies*). In the past decade his research has focused in particular on the transformation of the rural economy and environmental problems in the Post-Soviet Central Asian states. His most recent books are: *The 'Market Panacea': Agrarian Transformation in Developing Countries and Former Socialist Economies* (London: Intermediate Technology Publications, 1997); *Beyond Transition: Ten Years after the Fall of the Berlin Wall* (New York: UNDP/Regional Bureau for Europe and the CIS, 2000); and *Transition, Institutions and the Rural Sector* (Lexington Books, 2003). He has worked as a development consultant and economist for many international agencies.

ABBREVIATIONS

AI	Amnesty International
ADB	Asian Development Bank
b.	Born
BBcm	billion cubic meters
bbl	barrels
bcf	billion cubic feet
BP	British Petroleum
BVO	Water Basin Association
CAS	Central Asian States
CEA	Central Eurasia
CEPS	Centre for European Policy Studies
CESTRAD	Centre for the Study of Transition and Development
CIS	Commonwealth of Independent States
CPAz	Communist Party of Azerbaijan
CPI	Consumer Price Index
EBRD	European Bank for Reconstruction and Development
ECO	Economic Cooperation Organization
EIA	Energy Information Administration
EU	European Union
FDI	Foreign Direct Investment
FH	Freedom House
FOG	Friends of the UN Secretary-General for Georgia
GDP	Gross Domestic Product
GID	Gender in Development
GNP	Gross National Product
HDI	Human Development Index
HRW	Human Rights Watch
ICAS	Interstate Council for the Aral Sea
ICG	International Crisis Group
ICWC	Interstate Commission for Water Coordination
IDEE	Institute for Democracy in Eastern Europe
IFI	International Financial Tnstitution
IGO	International Governmental Organization
IISS	International Institute for Strategic Studies
IMF	International Monetary Fund
INGO	International nongovernmental organization
IR	International Relations
JCC	Joint Control Commission (Tskhinvali/South Ossetia)
Mbbl/d	thousand barrels per day
MEP	Main Export Pipeline
MMbbl/d	million barrels per day
NATO	North Atlantic Treaty Organization
NGO	Non-Governmental Organization

OIC	Organization of Islamic Conference
OPEC	Organization of Petroleum Exporting Countries
OSCE	Organization for Security and Cooperation in Europe
PFA	People's Front of Azerbaijan
PFG	People's Front Government of Azerbaijan
PPP	Purchasing Power Parity
SDC	Sustainable Development Commission
SSR	Soviet Socialist Republic
TAR	Trans-Asian-Railway
Tcf	Trillion Cubic Feet
TCGP	Trans-Caspian-Gas-Pipeline
TI	Transparency International
UAE	United Arab Emirates
UN	United Nations
UNDP	United Nations Development Programme
UNECE	United Nations Economic Committee
UNOMIG	United Nations Observer Mission in Georgia
US	United States
USAID	United States Agency for International Development
US DoJ	US Department of Justice
WHO	World Health Organization
WID	Women in Development

Introduction: The Crisis in IR-Theory: Towards a Critical Geopolitics Approach

Mehdi Parvizi Amineh
and Henk Houweling

This anthology brings together studies of post-colonial, post-Cold War, interactions between state and non-state actors regarding CEA. This part of the world is in transition from Soviet institutions to independent statehood, nation building and the release of market forces. The intent of governments of the new states is to develop state and society. The theoretical framework of the study is called "critical geopolitics." This approach merges elements borrowed from political realism, from two domestic-society schools and from constructivism. Critical geopolitics enriches that mixture with spatial variables. Realist theory leaves unexplained why a militarily capable contender for world power such as the former Soviet Union was "defeated" without a shot fired between Cold War adversaries. Liberal and Marxist approaches are on opposite sides in the class dimension. However, they share the effort to construct theory in IR on *a universal domestic society* model of an *immanent* process of development. These domestic society schools claim universal validity for the actor's intent to bring under control of its designated agency the immanent process of development. Constructivism in IR originates in hermeneutics. It is the late nineteenth century "anti-method" for positivists and scientific realists alike. The assumption of an immanent process of development asserts the primacy of structure over the agent. Constructivists assume the primacy of the actors in relation to their social and ecological context. Constructivist scholars reject the assumption of the existence of an immanent process of development. This school therefore rejects the assumption that truth and reality exist outside the minds and will of agents. Accordingly, patterns or regular behavior exist only as rule-guided behavior. This approach extracts ideas and images of "self" and "other" from actor reports, studies actor conceptions of appropriate behavior, focuses on interactions and explains behavior on the basis of materials on self-reporting. The search for regularity without motivation is rejected as a misguided "naturalistic" turn in theory construction.

The Failure of Theory in International Relations

The unanticipated collapse of the Soviet Union initiated a crisis in the theory of International Relations (IR).[1] The Cold War is rapidly moving into history. Some scholars believe that the post-Cold War interregnum came to an end on September 11th, 2001. We share that view. What contribution does "theory" make to the comprehension of incisive change in international affairs where it failed to anticipate the peaceful exit from the international system by a contender for world power? How can the recent rift in Atlantic relations and the new pattern of alliances that is emerging be understood without sound theory? The obvious failure of IR-theory cannot be overlooked by students of state-societies that achieved independent statehood as a result of the collapse of the Soviet Union. Accordingly, at a time in which understanding change is most needed—if for no other reason than to anticipate its unanticipated consequences—the theoretical framework required is not available.

This anthology studies change underway in one particular region of the world: Central Eurasia (CEA). That part of the world comprises the newly independent states of the Caucasus (Armenia, Azerbaijan, Georgia) and

[1] Several social scientists and diplomats, however, did anticipate the collapse. George Kennan, the founder of containment by economic rebuilding of war devastated societies, being one of these (Kennan 1947: 580). Scholars working in the tradition of Max Weber's classification of forms of political authority and their potential for economic development, did very well in comparison with other schools of social thought. Ithiel de Sola Pool announced the collapse in 1965 and had the timing of the event almost right. Ken Jowitt worked in the 1970s on the sociology of Rumania's state-society complex and later extended that analysis to the Soviet system. He anticipated the failure of the Soviet system at a time when government-hired social scientists-alarmists loudly proclaimed the competitive advantage of the Soviet system over their western counterpart. Randall Collins applied the bench vice analogy, which is invented by the 'Great-Germany' school of thought that criticized Bismarck's work, to the Soviet Union after the Second World War. He found a build-up of simultaneous crisis in the Soviet state-society complex and wrote in the mid-1980s about the impending implosion of the Soviet system.

of the former Russian Central Asia (Kazakhstan, Kyrgyzstan, Tajikistan, Turkmenistan, and Uzbekistan). The objective of the work is to provide greater comprehension of the nature of the post-colonial "Great Game" underway in that part of the world and to relate that game to the transformation underway in CEA.

With the end of Soviet control over the CEA and Caspian region, many players are trying to get access to the oil and gas reserves in this part of the world. Regional neighbors, such as Iran, Russia, Turkey, China, but also Afghanistan and Pakistan, meet each other directly in this region in the absence of the former dominating power of the Soviet Union. Governments of high-income countries thrive on imported energy. Without uninterrupted access to imported energy at affordable prices, the states, governments, and households in the high-income part of the world would utterly cease functioning. These countries provide head-quarters to the world's most active and advanced oil companies. The distribution of military capabilities among these countries, however, is highly skewed. Military might is concentrated in just one of them: the United States (US).

Late-industrializing countries such as China and India, whose populations dwarf those of the high-income countries, try to close the productivity-power gap with countries that industrialized first. With colonization and hostile penetration by the western powers fresh in mind, these countries face the choice of renewed humiliation or to catch-up in power and wealth with their former rulers and enemies—as well as with their neighbors who have the same objective.

New non-state actors, such as ethnoreligious groups and external faith-based organizations, cross-border traders in people, cocaine and weapons, are active in the region. How can we understand this mixture of actors, their aims, means and the social forces they generate? One of the most interesting questions is how conflicts and social struggles will be articulated and who will gain the upper hand, if any. Will multilevel conflicts and cooperation in the region be controlled by the US, as the now dominant military actor, in cooperation with local governments? How will Russia and China respond to the insertion of US military power in the region? Will transnational battles among non-state actors force the major powers involved in the region to cooperate and collectively impose a local peace order?

The governments of the newly independent states in the region inherited territorial borders before they achieved the capability to penetrate society by administrative means. These governments, unlike the pioneer states of the western world, did not need to mobilize societal capital and manpower to establish borders and to defend them against outsiders. Western states are shaped by war, trade, and overseas expansion: the constitution of these states, the legitimacy of their governments, and their administrative capacity to penetrate society were put into place by the interaction between warfare, overseas expansion, and market forces. The governments of the newly independent states, by contrast, are highly dependent on outside friends to stay in power in domestic society. These governments are vulnerable to trans-national non-state actors. Such governments have to negotiate with power groups in domestic society in ways not very different from international bargaining between governments. In these circumstances, IR-scholars have to go beyond established wisdom. Many scholars have indeed attempted to develop new approaches to anticipate the future direction of international affairs.

Francis Fukuyama's *End of History and the Last Man* (1992) envisions a post-Cold War world order in which communism has proved to be nonviable. Governments in the post-Cold War era are therefore *universally and irrevocably* turning towards a state-society model that favors democracy and markets in a private-enterprise system. The future of IR, and of domestic orders that compete within the international system, is therefore the extension of these western institutions to the rest of the world.

For two reasons, this is not the theoretical avenue we intend to go into in this book. The first one is the absence of a natural ecology capable to carry on such a world system of mass production/mass extraction and mass pollution indefinitely. The other is Fukuyama's neglect of the polarizing effect of sequential industrialization in the capitalist system.

As we see it, Fukuyama follows the trail of utopia builders that anticipate a future without societal change. He has taken a page from Hegel's book and plays a trick on the Marxist action-credo that the history of all and any class-divided social formations will come to an end when the laboring class of industrial capitalism has prevailed for good over the capitalist class. Fukuyama's liberal-democratic utopia shares with all other utopias the inability to explain why it is impossible that anything new could ever be invented after liberal victory has been proclaimed everywhere. The connection between industrial capitalism, representative democracy, and respect for human rights is, in our view, too recent and too vulnerable to change as to serve for trend projection. Capitalism is very flexible as it comes to forms of political order in which that system is capable to grow. In the history of the first-industrialized countries, representative democracy came to fruition at the time in which industrial mass-production systems were producing largely for the home market and conscription mobilized the manpower in society for mass armies, which got trained for mass slaughter for the sake of national greatness. That system launched national governments in the center of society as mobilizing and mediation forces.

Accordingly, the claim to equality of status of all citizens found support from the economic system of industrial capitalism and the military system of international politics. Industrial capitalism, however, is based on the opposite of equality and, unless corrected by government intervention, contributes to further inequality.

The explicit recognition by western governments that further economic development of society requires the world market has created in the last quarter century a new, "outward-reach" form of *industrial* capitalism. Technology-intensive fighting is for specialists, not for mass armies. Accordingly, the domestic context in which representative democracy emerged has changed. We do not yet know enough about the effect of that change on its functioning. What we do know is that governments turn to entrepreneurs established in their jurisdiction and search for external opportunities for them (Harvey 1995: Chapter 3). Consequently, cross-border operating private actors create pressure to shift political agendas to international policymaking networks whose actors are not under the legal obligation of public representation. When policymaking beyond borders becomes more important, citizenship as a legal status tends to get separated from citizenship as substantive opportunity for participation.[2] Fukuyama is therefore too weak a reed to rely on in anticipating the future of the region under study in this anthology.

Samuel Huntington does anticipate indeed a future after liberal victory. In his book *The Clash of Civilizations* (1996), Huntington studies the dimension of cultural and institutional heterogeneity between societies omitted by both liberals and Marxists. He thus does not neglect the obvious fact that societies are not all organized "naturally" to generate long-term growth in GDP-per capita. He also takes into account the expansion of Europe and the response from host societies. The core idea of his new theory is that "civilizations" are the units of analysis in International Relations. His concept of a "clash of civilizations" originates in the work of Bernard Lewis (1990), a Princeton University historian on the Middle East. Huntington brings his borrowed construct to the

[2] The era of consolidation of representative democracy and its extension into Southern Europe in the 1950-1970s coincides with consensus among political elites about the welfare state and the rebuilding of predominantly nationally oriented industries. Not left wing, but market radicals took over power at the end of the decade, first in the UK, to be followed by the US. In Europe, governments first transferred the authority to define the scope of the market to Brussels and subsequently declare themselves compelled to reduce welfare spending and to break down the institutions of tri-partite macro-economic management in the new European-global economy. The problem is, says Birnbaum that high-income societies stumble into a new epoch, whereas large numbers of citizens think in terms of the labor market of the past (Birnbaum 2001: 204).

global level by arguing that humanity is divided in apparently rather internally homogeneous civilizations. He anticipates a twenty first century in which the revolutionary impact of globalization induces irrational violence along axis of religious values on which the *Orientalism* of Lewis is based. Huntington shares his opinion that religious values are at the heart of human "civilizations" and gives this insight universal application. Huntington is part of a group of authors whose works together have been properly designated as "global chaos-theory" (Sadowski 1998: Chapter 2). Other scholars working in the school of global chaos-theory see, instead of principled violence, cases of anomie coming from the destitution produced by population growth, resource scarcity and environmental decay.

In his book *Jihad versus McWorld* (1995), Benjamin Barber combines both Fukuyama's and Huntington's ideas, stating that while there is an increasing cultural fragmentation taking place around the globe, the emergence of "jihad" is part and parcel of the process of globalization. He sees a dialectical relation between globalization and jihad, in which the first will, in the end, overcome the latter.

Findings from empirical research contradict hypotheses extracted from works of global chaos theorists. Instead of moving into global chaos of anomic, respectively of religious-based violence, *all* measures of warfare, thus irrespective of the data set used, aggregated to the global level, are down since the mid-1990s, with some regions lagging behind in pacification and development. If one takes into account world population growth and the increase in the number of state actors, global society is definitely pacifying. Conceptual ambiguity, the lack of internal consistency, and selection of cases on the dependent variable [3] have further compromised the explanatory power of global chaos theories. We therefore do not accept the theoretical framework of this school to organize the anthology, either. In the region under study in this work, the "clash of civilizations" turns out to be fratricide between neighbors and co-religionists.

The editors of this collection of studies on CEA have undertaken the adventurous task of contributing to the development of a new approach, one hopefully better capable of contributing to an understanding of what is going on in CEA and West Asia.

Section 2 of this Introduction intends to create that alternative framework. It is called "Critical Geopolitics" (Agnew and Corbridge 1995: Chapter 3). It is an uneasy as well as incomplete mixture of hypotheses and

[3] For example, the findings of the World Values Survey Project, which samples world society, indicate that Americans feel substantially closer to God in their lives than populations in Western Europe. For Islamic societies the score is 9 on a scale from 0-10, Americans score 8.5, which questions the homogeneity of civilizations as construed by Huntington (cf. Díez-Nicolás 2003: 245).

propositions from political realism, from two domestic-society schools and from constructivism. Critical geopolitics enriches that mixture with spatial variables.

Critical Geopolitics

Where does critical geopolitics stand in relation to classical theories of IR?[4] In this section, we position our approach in relation to structural realism, to two domestic-society schools, and to constructivism.

Scholars working in the tradition of structural realism could not explain at the interstate system level the unanticipated collapse of a military superpower. A contender for world power got "defeated" without a shot fired between adversaries. Instead of being defeated in war, Soviet leaders were compelled to opt out of the bipolar confrontation by domestic weakness. Domestic weakness, brought to the international level by Soviet foreign and military policies, induced a power struggle between the President of the Russian Federation and Soviet leader Gorbachev. Russian nationalism played havoc on Soviet Man. Critical geopolitics takes off from the failure of structural realism by considering self-constructed identity as a social force impacting on behavior. It is a learning effect of the disintegration of the Soviet Union whose contested identity induced Machiavellism in domestic politics unconstrained by agreement about the definition of Russia's political community.

Liberal and Marxist approaches are on opposite sides in the class dimension. However, authors working in these traditions share the aim of constructing theory in IR on *a universal domestic society* model of an *immanent* process of development. Liberals and Marxists differ on how to conceptualize the immanent process of development. Both schools claim participation of their actors in such a process. That claim implies for liberals the voluntary ambition to spread development outwards, which they consider to be progress. Marxists agree that such a spreading out of the immanent process of development does occur, but conceive it as compulsive and destructive. Compulsion and destruction, however, are steps on the way to bring that process under deliberate control of the Marxist agency: the workers class.

In liberal and in Marxist theory of IR, *domestic society* actors and private interests are brought by foreign policy of governments into the domain of interstate relations. The role of the state itself is secondary and derivative of interests of sub-state actors. In the Liberal School, external behavior of

---

[4] For an elaboration on the concept of critical geopolitics see also Amineh, Mehdi Parvizi, *Globalisation, Geopolitics and Energy Security in Central Eurasia and the Caspian Region* (Den Haag: CIEP, 2003), ch. 1.

states depends on relations to citizens and their interests. The absence of a global government therefore does not compel state behavior in external relations to be governed by security. In classical Marxist thought, classes and class interests are the driving forces of the immanent process of development. States in their external behavior bring the interests of one class into IR. The absence of a global government, therefore, does not compel state behavior in external relations to be governed by security.

Critical geopolitics shares with domestic society schools the explanation of foreign policy by inputs from domestic society. However, they reject the secondary, derivative role of the state in international politics. It shares with structural realists the selective function of the international system. But the object of selection is different: instead of states, state-society complexes get selected. The core of a state-society complex consists of institutionalized state-business-military relations within states.

Liberals and Marxist scholars in IR operate on a *universal domestic society* model. These schools have no place for cultural diversity and national identities among their actors. For liberal scholars, the progress of learning, wealth, and industry will overcome superstition and fear related to frontiers and to people living on the other side of the border. For Marx, members of the labor class have neither local roots nor national identities. In his conception of the actors, the laborer and the capitalist are as homogenous as individual items in mass-produced manufactures. Critical geopolitics as an approach to the study of IR considers the missing variable of identity to be the fatal weakness of both *universal* domestic society schools.

Constructivism in IR originates in hermeneutics. It is the late ninteenth century "anti-method" for positivism and scientific realism. The assumption of the existence of an immanent process of development asserts the primacy of structure over its agents. Constructivists assume the primacy of the actor in relation to his social and ecological context. Constructivist scholars reject therefore the assumption of the existence of an immanent process of development. Scholars working in this tradition reject by implication the idea that truth and reality exist outside the minds and will of agents. The argument is that social reality, be it a game of chess or an international system, does not exist outside human consciousness. Such conception as truth and reality "exist" only as social concepts constructed by human awareness.

Constructivists are therefore capable of incorporating into their theory the obvious fact that human groups that get enmeshed in enduring interactions at the group level do set themselves apart from each other in the dimension of awareness, which gets expressed as self-identification. This approach elevates ideas and images of "self" and "other" into proper objects of study. It extracts from these materials actor conceptions of appropriate behavior; it focuses on interactions and construes explanations of behavior on the basis of self-reporting and social rules of conduct.[5] Accordingly, patterns, or regular behavior, exist only as rule-guided behavior. The search for regularity without implicit or explicit actor motivation, or project of action, is rejected as a misguided "naturalistic" turn in theory construction. Constructivists refuse to go beyond shared, linguistically construed social reality and, therefore, cannot answer the question where these constructions come from. Accordingly, why Soviet Man eventually opposed Liberal Man and why that first construct subsequently disappeared from the system cannot be answered in this tradition by analyzing social reality external to the actors that created these models of man.

In critical geopolitics, human constructions of identity are considered to be intermediate *variables* between social reality and actor behavior. Let us further explicate what we have in mind. As an intermediate variable, self-constructed identities may assumed to be *fully endogenous* to the immanent process of development. On that assumption, the impact of political and economic factors on actor behavior works out partially through their effect on ethnic consciousness, which itself is an effect of political and economic causes of behavior. There is no feedback effect of action on actor consciousness and on political and economic causes of actor behavior. Thus, scientific realists may bring into political realist theory the variable of ethnic awareness. For constructivists, awareness or self-consciousness stands at the beginning of action. As measured variables, ethnic constructions are assumed to be fully *exogenous* to actor behavior and to its political and social causes. In constructivism, there is no feedback effect of behavior and its political and economic causes on self-created identity. In critical geo-politics, expressions of awareness of group identity are assumed to be *partially endogenous*. That is, political and economic factors impact on expressions of identity and on other constructions of self awareness, which however have a feedback effect[6] on political and economic factors that explain actor behavior. Expressions of self-awareness have an independent impact on actor behavior; however, actor behavior has feedback effects

[5] Constructivism and its counterpart, scientific realism, (which should be distinguished from the school of political realism in IR) differ in philosophy of science, methodology and in model of explanation. IR schools of thought reflect these wider divisions among social scientists and historians. IR students focus on a problem area without agreed upon boundaries and without a philosophy of science, methodology and model of explanation unique to that field.

[6] The substantive reason for taking this position is that action undertaken on the basis of identity tends to persist beyond the break-even point of interest-based cost-benefit calculations, suicidal terrorism being an extreme example of this.

on subjective awareness. The upshot of these distinctions is that critical geopolitics avoids the constructivist fallacy of confusing self-understandings of actors with the objective social processes that constitute actors and the international system in which they operate.

Let us relate these distinctions to the materials presented in Chapter 1 and see where realists, domestic society schools, and constructivists part ways. European immigrants, by moving to the Americas, created a "boundary of civilization" between themselves and Amerindian societies. By creating the collective "we," of being the civilized portion of mankind, individuals who participated in it turned into an objective force of social change. That change is summarized by the frontier moving up to the Pacific, liquidating native societies it met in that process. For realists, social change has causes and consequences. Realists could refer to the difference in population density between pre-Columbian Western Europe and the Americas. The construct of a boundary of civilization is to be explained by political and economic factors that caused people to move from Europe to overseas and impact the immigrant behavior that they found there. That construct, however, is neither impacted by behavior itself, nor is it changed by the causes of behavior after it is in place. Consequently, the boundary concept is part of the immanent process of development of social reality and does not change its course.

Constructivists study changes from the perspective of participants in the collective we. For participants in the creation of the collective we, the language of social causes and consequences is alien to their experience. Participants in the creation of the collective we experience themselves as self-conscious, autonomous, actors in the pursuit of their interests. For them the thought of being constituted as actors and self-conscious subjects by an immanent process of development falls beyond imagination. As actors in the pursuit of their interests, they establish control over the social and natural environments in which their interests are situated. Accordingly, actors stand at the beginning of social change and are its deliberate creators. That is why President Bush could say in his Inaugural Address, "Ours is a great story and we need to tell it to the world."

In critical geopolitics, self-constructed identity is a partially endogenous intermediating variable between social reality and actor behavior. Such constructions are dependent on political and economic causes, but have feedback effects on social reality. These constructions help to explain actor behavior, which has feedback effects on self-constructed identities. Critical geopolitics conceives its domain of study, therefore, as "complex." In complex reality, outcomes do not necessary follow from intentions, and intentions change under the impact of outcome of behavior. Methodological implications of this stand have recently been clarified in Jervis (1997: 73ff).

Schools of thought in IR differentiate on what they take to be the unit of analysis and the level at which that unit is being studied. In critical geopolitics, units of analysis are state-society complexes of self-identifying groups that are in continuous interaction with other self-identifying groups. In the industrial age, state-society complexes at their core are institutionalized state-business-military relations. State-business-military relations are part of growth promoting or growth restraining institutions of societies. Self-identifying groups of humans subsist on natural resource systems in their reach of mobility. It is here that the variable "space" and control over space comes in.

In critical geopolitics, state-society complexes and the natural resource systems in which societies survive, or vanish, are taken as units of analysis. State-society complexes interact at the system level. Foreign policy of state and non-state actors brings these social units into interaction. These interactions create a system level of social order.

Since the mid-nineteenth century, the system-level social order is characterized by *sequential industrialization of self-identifying*, state-incorporated, societies. In critical geopolitics, actors that operate in state-society complexes engage in cross-border activity to get access to resources beyond legal borders. Cross-border activity connects domestic society and its institutions to the external world. Such connecting activity is called "power projection." Power projectors have therefore *spatial representations* of the external world and of their own position in it (Agnew and Corbridge 1995: Chapter 3).

Critical geopolitics is therefore, first of all, *a theory of action*. It should be distinguished from classical geopolitics. The latter is an approach to the study of *world order*. The editors have the ambition to contribute to the geopolitics as a theory of action and to illustrate its relevance by applying it to the study of the transition process underway in CEA. Accordingly, we conceive critical geopolitics as the study of spatio-temporal aspects of action beyond legally or otherwise recognized borders by actors that manage state-society complexes and use the natural resource base of the ecological niche in which society is located.

We operationalize power projection activity of state and non-state actors by the *dimensions of control sought beyond borders*. Dimensions of control refer to the timing of power projecting activity, to actors in named locations, and to situations in target societies the power projector aims at bringing under its control. Objectives of power projectors are inferred from the *timing* and *spacing* of activity, from *the resources being allocated* to it, and from the *target actor*, or *situation*, power projectors seek to bring under their control.

Actors that use material capabilities to project power beyond legal borders, operate in multicultural environments. Multicultural environments change under the impact of power projection activity into self-conscious "we" versus "they" groups. These social constructions have feedback effects on actors' behavior. In critical geopolitics, the material interests that are pursued by power projectors have for this reason a distinct ideological imprint. In critical geopolitics, that imprint has an independent impact on external behavior of power projectors. In the collisions between material forces released by power-projecting activity and host society responses, cultural-historical experiences of home and host change. In host societies, encounters with power projectors set off a process of change called 'hybridization,' or 'intermingling' of identities and institutions. The messages delivered by power projectors to domestic and foreign audiences, to bystanders, respectively to allies and opponents, are formed by previous acts of power projection. Observers studying power-projecting activity meet therefore in that activity the past historical experiences of home and host societies.

As referred to above, classical geopolitics in IR studies world orders. World-order students conceptualize and measure world orders by variables that are defined at the world-order level. Such global, analytic and structural[7] characteristics (Lazarsfeld and Menzel 1961: 422-439) of the international system refer to systemic features. Examples of systemic features of the international system used in research include the number of major powers; measures of stratification, respectively polarization among them; measures of upward and downward mobility of actors; alliance patterns; and the distribution of control in networks of production, labor allocation, trade and plunder in that system. Geopolitical studies of world order map potential capability positions of state and non-state actors in various locations as well as the redistribution of capabilities through conquest and through cross-border market forces. Classical geopolitical theorists, Mahan and Mackinder in particular, aimed at constructing a geopolitical theory of change of world orders. At the time of their writings, Germany and the United States were catching up and overtaking Great Britain in industrial capability and military power.

Mackinder (1904: 421-437) anticipated a future in which the spread of industrialization into Eurasia had brought about an overland transport revolution. He predicted an end to the "Colombian age" of sea powers dominating land powers. More specifically, he feared that an alliance between rail-connected Russia and industrialized Germany would turn to sea as naval powers, using the resources of Eurasia, and would be able to conquer the world. He therefore recommended France, Italy, Egypt,

[7] The distinction between properties of an individual member of a collective and of collectives defines different levels of analysis in IR (cf. Lazarsfeld and Menzel 1961: 422-439).

India, and Korea to ally with naval powers and thus divert the resources of the land powers to fight along the periphery of Eurasia. Mahan, on the other hand, offered in the 65 pages preceding his detailed description of seven great maritime wars fought between 1660 and the American Revolution, a general hypothesis of the influence of sea power on recent world orders.[8] As a practical man, he belonged to the US advocates of imperial and commercial maritime expansion. However, commerce in his view did not pacify the world.

State and non-state actors that undertake to project power beyond their previously recognized borders will be guided by some conception of the world order as well as by the aspired position in it. As noted above, this is a crucial variable for constructivist theorists and an intermediate variable in critical geo-politics. Part of such a conception of the wider system will be geographic in nature; another component refers to resources in specific locations; the third dimension is cultural, such as boundary of civilization that separate actor and target. Whatever the precise nature of these conceptions may be, they are shaped by the past and frame perceived policy projects *at home* and *abroad.*

These statements may be highlighted by the current stage of US power projection. In Chapter 1, we argue that the post-Cold War interregnum is brought to an end by America's effort to create a new extension to its World War and Cold War "defense perimeter." The new leg to America's military border extends from Southern Europe, onto the Caucasus (Georgia) into CEA, separating that region from Russia, from integrating Europe, and from the People's Republic of China. Power-projection activity underway by the US is bringing its military forces up to the western borders of China. Part of power-projection activity is its verbalization of the action and of actor's self-awareness. The US pursues its ambitions under the general category of, first, securing the blessings of America's conception of western civilization for home society; second, extending that civilization to societies that still live beyond that civilization; and third, removing obstacles to that extension, such as "rogue states," from the system. These folks are believed to live on the other side of the good-evil distinction. In the capacity of the good side, the American leadership construes itself as acting subject in IR in the role of *redeemer* at

[8] George Modelski has put Mahan's core idea of a succession of leading naval powers developing into world leaders into social science jargon of Parson's systems theory. He anticipates a new century of US dominance based on control over sea-lanes and land-based resource areas required for sustaining maritime (and aerospace) hegemony (Modelski and Thompson 1988). For a critical debate on the scientific merits of functionalist theory of world wars see (Houweling and Siccama 1993: 387-408; 1994: 223-226).

the world level. The realist will assert, "Yes, this is what powerful actors do in the pursuit of national interests."

In the critical geopolitics approach, such a role as "redeemer" may bring about actor behavior that hurts, or even is destructive of national interests, which will change social reality. Realists and constructivists make different assumptions about the status in theory of constructions of awareness. Let us use an example from the material presented in this work. The Anglo-Saxon war against Iraq has brought into the open a rift between the major state actors of the "Western World" about methods and objectives of America's undertaking. That rift denotes change in concepts expressing self-identity on each side of the Atlantic. The schism extends into the ranks of European partners that are involved in the creation of an "ever closer union."

The disunity in NATO and within Europe about the integration project is verbalized in terms of cultural constructs such as "the West" and "Europe" in relation to "brutes," who still live on the other side of the boundary of western civilization. These cultural constructs functioned in the Cold War era as identity markers of both the members of the NATO-Alliance *and* of participants in the process of European integration. The dispute between America and some of its allies, as well as among the integration partners in Europe, cannot be fully comprehended by the clash of power and of material interests alone. We, therefore, argue in the first Chapter that critical geopolitics is better equipped for this task. In terms of realist interests, Europeans are more dependent on imported oil than the US. Without demand or supply-induced scarcity of oil around the corner, Europeans share with the US the interest in low energy prices. They should therefore give full political support to America's preventive war in a region where 60% of the world's oil stock is found, but leave fighting an evil regime, which nationalized the oil industry and turned into the capital of the producer cartel OPEC, to the US—this is what the Dutch government did.

In critical geopolitics, power-projection activity by state and non-state actors changes the world order. The Anglo-Saxon war against Iraq opens the door for the US to create a long-term military presence in West- and CEA. Whether or not the US succeeds in creating a long-term military presence in the region, its effort induces responses from other actors. The US activity in West- and CEA is a cause of change in alliance patterns and in the identities of the American "self" as well as in the self-definitions of America's allies. American success in creating a long-term military presence in the region opens the door to enterprises headquartered in the US, to America's faith-based and non-faith-based non-governmental organizations to become established in the region. Getting a firm foothold in the region gives the US the capacity to shape host societies as well as to set conditions for outsiders to access the oil and gas wealth of the region. Accordingly, the leadership will get confirmation of its self-created role of being the redeemer of the world. Success changes reality; it gives the US government indirect control over the economic and technological development of potential contenders as the EU, China, India, and Russia. Failure would undermine America's standing to be the sole remaining superpower and the redeemer role of the country in IR and thus provoke compensating action to regain its lost status, which would change social reality.

Power-projection activities are policy inputs into the world order. World order changes feedback into foreign policymaking at later points in time. In that sense, the world order "returns" to the actors as rewards and punishments. Both are consequences of power-projecting activity by actors that are in charge of state-society complexes. The function of the world order thus is to select over time among the diversity of state-society complexes that their policymakers bring to the world level by foreign policy, leading to change in the world order itself. In political realism, state actors get selected. In critical geo-politics, domestic social orders get selected. Accordingly, one aspect of change in the world order comes down to the changing frequency over time in its constituent domestic orders, including their ways of living. To quote anthropologist Marshall Sahlins (1976) on this aspect: "Western civilization does not hesitate to destroy any other form of humanity whose difference from us consists in having discovered not merely other codes of existence but ways of achieving an end that still eludes us: the mastery by society of society's mastery over nature" (P. 221). We thus find an overtime feedback process at work between motivation, interests, action undertaken to shape domestic order, change of the international system and its impact on domestic orders that collectively constitute the international system. Critical geopolitics focuses on the input side of forces of change in the international system.

Internationally operating actors in state-society complexes form conceptions of what is *possible* for them to achieve in the world beyond borders and what is *required from them* to undertake in the *external domain* in order to sustain the *domestic social* system. These conceptions are political constructions of self-awareness intermediating between the domestic and international level. As noted above, power-projection activities are inputs into the world order. Such inputs originate in the domestic level organization of externally operating actors equipped with ideas about what is required from others beyond borders to sustain domestic society.

Our stance implies three things. The *first is* "subjectivity" of action: actor conceptions of what is possible and what should be accomplished abroad in order to sustain a domestic order, cannot be conceptualized as being *only* effects of impersonal social or economic structures at home and abroad. *Second*, if subjective-actor conceptions of what is *possible* in external relations and *required to achieve in the external domain* did not have any connection whatsoever to externally existing social reality, power-projecting actors would find themselves in an *ego-centric box*. In such a box, actors are unable to learn from past encounters with reality and thus are not capable of developing new conceptions about it. Actors in an egocentric box cannot arrive at conceptions of the immanent process they participate in and how to bring it under their control. Accordingly, acts of power projection could never reach out and get in touch with social reality. Critical geopolitics therefore accepts the existence of a social reality external to the will. Cognitive frames, preferences, intentions or beliefs, or reasoning of actors originate in social reality but experience an independent impact when action projects are implemented. If a project of power projection backfires, the international order will be different from before. The same is true in case such a project succeeds in achieving its intended objectives. It should be further noted that intended effects and unintended effects are distributed over time. It is simply impossible for an actor to do only *one thing* at *one time* in relation to *only one* target. Third, the subjective act of power projection in the geopolitics of foreign policy of a country or the policy of a non-state actor becomes "absorbed" by the reality, which is the international order.[9]

In the conception of "reality" of the structuralist, social research aims at extending knowledge of social reality that antedates its actors and projects of action. Reality "waits" as an object of discovery for researchers. The structuralist position implies that actors are *captives in a social cage*. For the structural realist, therefore, students of critical geo-politics approach the topic at the wrong level of analysis. However, human actors may, and do, *re*define their conception of social order by the experience of moving out into the world. In one way or another, actors do have the ability to venture with speculative reasons beyond the limits of what they experience. This implies the possibility of *social order change* induced by subjective action. For constructivists, social reality is produced by human cognition, beliefs, and reasoning. Accordingly, social phenomena such as money, commodified oil, or axis of evil, do not really exist outside the human minds that invented and named them. This is to state the obvious: no one will credibly

[9] Kuhn's work helped to release the debate among theorists of IR on "paradigms." His notion of paradigm shifts is a crucial step from studying the unchanging reality of the international system of anarchy towards actors changing their self-conception by moving out in the international system and its consequences.

assert that, were the human species to go extinct, entities such as states, axis of evil, rogue regimes, and flows of oil would survive. Mountains and rivers would survive such an event. Constructivists, however, do go beyond platitude. For the constructivist, social reality is a construction in the minds of men mediated by language. Accordingly, the naming of action and interests gets the quality of subjectivity beyond which no reality does exist. The implication of this position is that actors reaching out beyond previously recognized borders do not get in touch with externally existing social reality.

Constructivists confuse two ways in which rule-guided behavior mediated by language and other symbols "exists." In the game of chess, the activity of the players has no independent existence outside the rules according to which the game is played. Outsiders, not knowing that game at all, will fail to figure out the rules according to which the players move the pieces on the chessboard. Outsiders cannot participate in the game without getting access to and comprehending the rules and calculations in the minds of the players. Accordingly, in chess, the rules of the game define *the existence* of the activity; that activity has no existence outside those rules and its players. This is the proper domain of the constructivist theory.

Power projection policy differs from the game of chess in that it is not called into being by socially construed rules according to which actors operate. "Power politics" is not *constituted* by some set of agreed upon rules according to which it should be played. Actors involved in projects of power projection no doubt appeal to such rules. However, it is mistaken to believe that rules so addressed *constitute* the action and the actor. Of course, the actors play the game of power politics under constant reference to rules under which they believe to operate. However, these rules may, at best and under well-specified circumstances, *regulate* an activity that *precedes* the rules that should regulate it. The activity itself is not called into existence by the regulative frameworks to which actors refer. For realists, regulative frameworks are part of the play of material power and interests in relationships among actors. In critical geopolitics, world order is the meeting points where state-society formations and subjective actor concepts relating such formations to the external world confront each other and *get selected for survival.*

The Organization of the Work

In Part I, we study US power projection in the framework of critical geopolitics and apply that framework to US power-projection activity in CEA. In Chapter 1, we argue that power projection activity by he US has been in the past, and still is, construed by policymakers as a world-redeeming undertaking. Borrowing conceptually from post-

Medieval interpretation of the law of nature, the distinction between "self" and "other" has been, and still is, conceptualized as a "boundary of civilization." The natural right claimed by the Bush government for the US as a sovereign state to wage preventive war originates in early modern political thought. Classical political theorists as Grotius, Hobbes, and Vattel reflected on the new, post-Medieval, world order created by the double process of state formation and overseas expansion. These authors agreed on the natural rights of states to wage preventive war, but coined that right as "pre-emptive" war. America's war against "rogue states" reinvents the barbarian that America's founding fathers created conceptually when immigrant society expanded from the Atlantic to the Pacific coast. Chapter 2 singles out for special attention the oil and gas resources of West and CEA. Fossil energy creates an ecological niche in which industrialized societies subsist. Since the 1850s first-industrializing countries have adapted the internal "metabolism" of society to only one sort of "food," and put their "excrement" into the global atmosphere of industrialized and not-yet industrialized societies. America and its allies moved to the top of the world's hierarchy of power and wealth by getting access to this niche and exploiting it before others did. However, the spread of industrialization into India, China, and Indonesia brings vast populations to the area where the world's oil stock is largely concentrated.

Part II brings together articles on various aspects of nation-state building, economic development and household, respectively, and gender relations in society in the newly independent states of CEA. These are the host societies of power-projecting activity.

Pinar Akçali studies, in Chapter 3, the process of state formation and nation building. She finds that the carving out of new polities began in the Soviet era. Before the Soviet era, peoples in Central Asia existed as horse mobile ethnic groups in steppe land under clan leadership. A system of territorial-based rule was imposed on the region by Stalin's delimitation of "nationalities," which were intended as a transition to the creation of one Soviet people. Instead of Soviet Man, a process of cultural and institutional intermingling occurred, leading to hybridization. The official discourse of the state-making elites has to take off from the failure of Soviet Man to materialize. They therefore appeal to past glory, re-creating the history of newly invented peoples in that process. Her conclusion is that the new republics are establishing themselves but that the process of nation-state building is far from complete.

Shirin Akiner in Chapter 4 analyzes in more detail one aspect of this large process, that is, the creation of a viable social order and the means employed by governments to achieve it. Colonial powers of the West used peoples overseas to enrich the motherland. Soviet colonization depended on the transfer of resources from the motherland to colonies. Akiner finds that the newly independent governments fail to mobilize resources of society required to maintain the level of education and health care produced in the Soviet era. She fears the consequences of pauperization and the bitter competition the gap between haves and have-nots is unleashing. In Chapter 5, Michael Kaser looks at the impact of the transition process on economic development and external commercial policy in the five Central Asian republics and in Azerbaijan. During the Soviet period the economies of the countries of Central Asia depended mainly on the production of primary goods and produced mainly for the Soviet domestic market. When they gained independence, they thus depended on a market in deep recession. While Kazakhstan and Azerbaijan show some positive signs of economic recovery, the main obstacles to economic development in all countries of Central Asia and Azerbaijan are a "shadow economy," as well as authoritarian and corrupt political regimes. Armine Ishkanian, in Chapter 6, explores the impact of political and socioeconomic transition in post-Soviet CEA on women. She asserts that political and economic developments of transition have had a negative effect on the lives of women in CEA. However, women have also found similar coping and survival strategies. For example, NGOs have given women the possibility to gain the knowledge, skills, and social connections needed to promote progressive developments in their countries. Ishkanian stresses that women are not simply the victims of developments in post-Soviet CEA but are agents of change.

Part III, covering Chapters 7, 8, 9, looks at the interests, conflict, and cooperation between the various actors striving for influence in CEA. In the seventh Chapter, Kurt Radtke explores the impact of the strategic and security discourse between China and India on CEA. China sees itself as an emerging world power in IR. US' global-reach military power approaches China's landborders in its "Far west." Countries in Southeast and Central Asia tend to adopt policies of diversification by strengthening their links with all major global powers. Radtke concludes that a policy of "comprehensive security" could be a solution for policymakers in IR of today. Mehdi Parvizi Amineh and Henk Houweling in Chapter 8 analyze US and EU interests in CEA and look at US relations with the major regional powers active in CEA. The main challenge for US policymakers is to balance commercial and security interests in the CEA region, and foreign policy goals in CEA. A major American objective is to break Russia's control over oil and gas resources and transport routes, as well as to prevent Iran from extending its influence in the region. Iran is cooperation-partner of the Shanghai-Five group lead by China. The US encourages NATO activities in CEA, aims at deepening its a strategic alliance with Turkey and cooperates with the countries in CEA in the economic and military field on a bilateral basis. However, as Amineh and Houweling conclude, the US cannot act on its own. It also depends on a partial approval of Russia for its activities in CEA and lately has been confronted with another major competitor in the region: China. The EU sees CEA as part of its security environment and is interested in importing its oil and gas resources. Until now, however, it has not yet developed a comprehensive policy for CEA. In Chapter 9, Eva Rakel looks at the interests and activities of Iran in CEA. Iran has historically close links to CEA and hopes to be able to use these to get a greater influence in the region. Politically, good relations with the countries of CEA could strengthen Iran's position in the greater Middle East after Sunni-rule in Iraq has been destroyed by the US. Economic cooperation with the region, could help Iran to overcome its severe economic crisis. However, internal political and economic problems as well as its bad relations with the US have hampered Iran so far to become more active in CEA.

Part IV has three chapters on local conflicts in CEA that not only threaten security in the region but also pose a threat to global security. In the tenth Chapter, Hooman Peimani studies the effects the absence of a legal regime of the Caspian Sea could have on CEA. The persistence of this dispute creates ground for conflicts, crises, and wars in the Caspian region. There is an increasing development towards militarization, and possible military conflicts, not least because of Turkey's efforts to deny Iran political and economic gains in the Caspian region, the growing American military presence in Eurasia, and the expanding American-Azeri military ties since September 11th, 2001. Max Spoor and Anatoly Krutov in Chapter 11 study the ecological crisis in the Aral Sea. Since the 1960s, the Aral Sea has shrunk rapidly in surface area and in volume of water. Because of increased demand for water for irrigation and hydroelectric power by the competing newly independent states they see a potential for resource-based conflicts or even a "water war." In Chapter 12, Robert M. Cutler examines conflicts in the South Caucasus with a view towards means for their interdependent resolution. Particularly, he focuses on the potential for nongovernmental actors to create potential transgovernmental and transsocietal sociopolitical coalitions, that is, an institution such as a transnational Assembly for Regions and Peoples of the South Caucasus. Issues of institutional design are considered and assessed on the basis of existing comparative work on international parliamentary formations. Finally, Ayça Ergun in Chapter 13 studies the impact of international governmental and nongovernmental organizations on the state-society relations in Azerbaijan. She argues that post-Soviet transition is not only a question of internal politics but is also affected by the redefinition of its relationships with the outside world.

She concludes that the international element in state-society relations in Azerbaijan is far from democratizing and/or promoting democratization when the domestic actors, particularly governments, are resistant to it.

References

AMINEH, M.P.
2003 *Globalisation, Geopolitics and Energy Security in Central Eurasia and the Caspian Region.* Den Haag: CIEP.

AGNEW, J. AND S. CORBRIDGE
1995 *Mastering Space: Hegemony, Territory and International Political Economy.* London: Routledge.

BARBER, B.R.
1995 *Jihad vs. McWorld: How Globalism and Tribalism Are Reshaping the World.* New York: Times Books.

BIRNBAUM, N.
2001 *After Progress: American Social Reform and European Socialism in the Twentieth Century.* Oxford: Oxford University Press.

DÍEZ-NICOLÁS, J.
2003 "Two Contradictory Hypotheses on Globalization: Societal Convergence or Civilization Differentiation and Clash?" in Inglehart, R. (ed.), *Human Values and Social Change: Findings From the Values Surveys.* Leiden: Brill.

FUKUYAMA, F.
1992 *The End of History and the Last Man.* New York: Free Press.

HARVEY, R.
1995 *The Return of the Strong: The Drift to Global Disorder.* London: Macmillan.

HOUWELING, H. AND J.G. SICCAMA
1994 "Long Cycle Theory: A Further Discussion." *International Interactions* 20: 223-226.
1993 "The Neo-Functionalist Explanation of World Wars: A Critique and an Alternative." *International Interactions* 18: 387-408.

HUNTINGTON, S.P.
1996 *The Clash of Civilizations: Remaking of World Order.* New York: Touchstone.

JERVIS, R.
1997 *System Effects: Complexity in Political and Social Life.* Princeton: Princeton University Press.

KENNAN, G.
1947 "The Sources of Soviet Conduct." *Foreign Affairs* 25(4).

LAZARSFELD, P. AND H. MENZEL
1961 "On the Relation Between Individual and Collective Properties." Pp. 422-439 in Etzioni, A. (ed.), *Complex Organization: A Sociological Reader.* New York: Rinehart and Winston.

LEWIS, B.
1990 "The Roots of Muslim Rage." *Atlantic Monthly*, September.

MODELSKI, G. AND W. THOMPSON
1988 *Seapower in Global Politics, 1494-1993.* Seattle: University of Washington Press.

SADOWSKI, Y.
1998 *The Myth of Global Chaos.* Washington: Brookings Institution Press.

SAHLINS, M.
1976 *Culture and Practical Reason.* Chicago: Chicago University Press.

PART ONE

THE POLICY OF PROJECTING POWER INTO VACUUM AREAS

I. The Geopolitics of Power Projection in US Foreign Policy: From Colonization to Globalization

MEHDI PARVIZI AMINEH AND HENK HOUWELING

ABSTRACT

This Chapter studies continuity and innovation in the geopolitics of America in projecting power beyond legally recognized borders. Exporting cultural symbols expressing what America has on offer plays as crucial a role in the opening of societies beyond borders as commodity exports, the activities of the CIA and the US Air Force do. The historical part summarizes early experience and aims at uncovering continuity in the foreign policy of getting America offshore. The hypothesis is that the US objective of inserting power and influence in West and CEA is to deny to a single state, other than the US itself, or coalition of powers not including the US, the capability to set conditions for accessing the energy resources of West and CEA. Our argument is that such a dominating coalition of actors not including the US, would arise from the creation of overland energy and other transportation links among the industries of Western Europe, Russia, Turkey, Northeast Asia, and China, leading to economic unification of Eurasia. Economic unification by creating overland energy and transport links of much of Eurasia would deprive the US navy of its power to interdict supplies of oil and food to core industrial areas of Eurasia and Japan. The reassertion of Russian power in the Caucasus and Central Asia should therefore be prevented. The EU and Japan should be prevented from developing autonomous military power and be kept dependent on maritime transported energy and food supplies. China should not host pipelines connecting energy resources of West Asia and CEA with the industries of Japan

> and Korea, whose unification and economic and strategic merger with China should be prevented. Iraq, Iran, and the Saudi Kingdom should be reformed into powers friendly to the US. Energy unification by overland transport systems, leading to economic unification between industries of these entities, would give major powers of the Asian landmass the potential for setting conditions for the US state and non-state actors to access the resources on the largest of world's islands. Such a power shift between the world's continents would reduce the Western Hemisphere to a rather dependent offshore island between the Atlantic and Pacific Oceans.

> Geopolitically, America is an island off the shores of the large landmass of Eurasia, whose resources and population far exceed those of the United States. The domination by a single power of either Eurasia's two principal spheres—Europe or Asia—remains a good definition of strategic danger for America, Cold War or no Cold War. For such a grouping would have the capacity to outstrip America economically and, in the end, militarily. That danger would have to be resisted even were the dominant power apparently benevolent, for if the intentions ever changed, America would find itself with a grossly diminished capacity for effective resistance and a growing inability to shape events. (Kissinger 1994: 813)

> Once the image of a hostile Soviet Union has been destroyed.... America will be expelled from Eurasia. (Kissinger 1988: 4)

> ...a dominant consideration [in US defense strategy should be] to prevent any hostile power from dominating a region whose resources would, under consolidated control, be sufficient to generate global power. Those regions ... are Western Europe, East Asia, the territory of the former Soviet Union and Southwest Asia. (Wolfowitz 2000)

> We have 50% of the world's wealth but only 6.3% of its population. In this situation, our real job in the coming period ... is to maintain this position of disparity. To do so, we have to dispense with all sentimentality ... we should cease thinking about human rights, the raising of living standards and democratization. (Pilger 2001: 98)

Overview of the History of the Geopolitics of the US and Its Legacy Today

To clarify our approach to the geopolitics of power projection (see Introduction), one may contrast mental maps of top policy makers in the US [1] on the one hand and Imperial China or of Yi Korea on the other hand. These empires were not institutionally organized for spatial growth,

[1] "The US" stands for executive branch dominance in foreign policy. Following classical political theory of Hobbes and Locke, decision-making in foreign policy was right from the beginning concentrated in the executive branch. George Washington's Proclamation

that is, to exercise *direct* control over actors or markets beyond borders. [2] The temporal dimension of imperial authority consisted of the dynastic chronology, not of stages of development and territorial expansion. In foreign policy, court bureaucracy and its pomp and circumstance set the stage for dealing with foreigners, including external traders. In terms of dimensions of control, imperial maps are in theory "total." The land and its people were conceived as property of the emperor. Securing state-society survival is not predicated on domestic institutions that require for their proper functioning expansion beyond borders, thus, on power projection. Accordingly, "anarchy," which is the pillar of realist and neorealist theory in IR, is unable to explain the foreign-policy orientation of members of this class of state actors. [3]

In The US, mental maps were dynamic right from the beginning. [4] George Washington anticipated a future of the US as a new form of Empire, as in the words of Van Alstyne (1960), "...a domination, state or sovereignty that would expand in population and territory, and increase in strength and power" (P. 1). From the earliest colonial charters onwards, US domestic and foreign policy aimed at populating all the land between the Atlantic and the Pacific coasts, and beyond. In the policy of expanding towards the Pacific, depriving Mexico of approximately two-third of its territory on the way, "North Americans will spread far beyond their present

---

of Neutrality of 1793 settled the issue. Lincoln's Secretary of State Seward is sometimes quoted for the oxymoron that American democracy elects a king for four years, and gives him absolute power within certain limits, which after all he can interpret for himself.

[2] The Qin, who unified the Zhou Kingdoms in 221 BC under central administration, and Han (206BC-221AC) Empires unified China proper. This was the area ruled by the Ming until 1644. However, following the Manchu conquest of China, Russia expanded across Siberia, moving to the Pacific through northern Manchuria. Russian expansion exposed China to the risk of an alliance between Russians and Mongols. China became involved into Mongolian politics and into the affairs of East Turkestan and Tibet. Foreign relations were managed by the system of tribute trade, the construction of walls and through court ceremony instead of territorial conquest (Borthwick, *Pacific Century. The Emergence of Modern Pacific Asia*, Boulder: Westview Press, 1992, p. 21).

[3] These systems are studied in Kautsky (1982). Political scientists disagree on the classification of political systems. McNeill, neglecting social provision, emphasizes the exploitative nature of aristocratic empires (McNeill: 1982, 7ff). Polanyi emphasizes the production of collective goods by the court; he classifies agro-empires as "redistributive" (Polanyi 1944/1957: 47ff). Classifications are eye-openers as well as blinders. Taking the redistributive side of imperial systems serious, Davis studied the impact of the rampage of the Western powers in China at the end of the ninteenth century on the ability of the court to provide relief during El Nino years. He finds famine of horrendous proportions (Davis 2001: 277ff).

[4] See, for power projection for the 125 years before Wilson, the first four chapters of Hogan (2000).

bounds.... New territories will be planted, declare their independence and be annexed. We have New Mexico and California. We will have Old Mexico and Cuba.... Have not results in Mexico taught the invincibility of American arms? The Anglo-Saxons have been sweeping everything before them on the North American continent ... and establishing an empire which is felt, respected and feared in every corner of the globe" (see Bermann 1986:14). US diplomacy of expansion could profit from quarrels among the European powers (principle of the "tertius gaudens").[5] In the minds of the founding fathers of the Republic, a free holding peasantry would settle the territory. To understand the creation of a liberal social order amidst the dispossession of the Amerindians, one has to go back to the political theory of the founding fathers of the Westphalian state system on the one hand and to the modern handgun on the other. The handgun is the weapon of choice of the individual settler: its use does not require large-scale military organization.

The Union created the legal basis of property rights in land (see Linklater 2003).[6] Expansion therefore cemented the Union by making it more important in material sense as well as for creating a national consciousness coming from opportunities for development of settler society created by the dispossession of Indians and the creation of legally protected property rights for immigrants. Through expansion, people from "old Europe," transformed themselves into loyal "Americans." A distinct "Yankee" nationalism developed out of British eighteenth-century civic nationalism, for which John Locke stands model.

Loyalty to the Union created the boundary between those who could participate in elections and outsiders at home and abroad.[7] Agreement on who is insider and who is outsider is the origin of US idealism of seeing itself as blessed by the will of heavens to civilize the rest of the world. Or, in other words, settler expansion, made possible by warfare, the handgun, alcohol, plague-infested blankets, and the system of legally protected property rights created by the state, are tools of power projection aimed at liquidating native society in the name of progress. In the mental map of its founding fathers, the US had the preemptive right, if not a duty,

[5] Examples are given in Houweling and Siccama (1985: 641-663). See, for the US taking over UK positions in Central American and the Caribbean at the turn of the century, Beale (1956/1984: 101, 106, 114, and 131).

[6] The US government engineered the largest transfer in history of real estate and commodified it. That process opened the door for European emigration, creating an Atlantic economy based on labor moving out of Europe's still agrarian economy, on investment capital for the building of rail, harbors and steamers for transporting food to industrializing Western Europe. The "Atlantic explosion" is its outcome.

[7] Such boundaries are not agreed upon by elections, which is a logical impossibility.

to promote the progress of humanity, which is to conquer the area and to populate it.[8]

The Reception in the US of a Boundary of Civilization and Its Use in US Foreign Policy

The legitimizing doctrine for "removing the Indians" is of older vintage. Political theorists in sixteenth and seventeenth century Europe, partitioned the world order, which was unfolding before their eyes by the double process of state formation in Europe and overseas expansion, into two parts. In their conception, the state of nature had been abolished within European countries. Out of the "anarchy" of the state of nature, a legal order had been created. However, relations between European monarchs and the lives of peoples dwelling overseas without a state continued to be in its grips.

The reasoning is that for those who live in a state of nature, natural law prescribes self-preservation by whatever means to be the preeminent natural right. That right overrules obligations towards others. Europeans moving overseas dwelled among people living in a state of nature. The legal regime applicable to relations between them and people they met overseas is the one of natural law. Accordingly, the law of nations governing relations between migrants and natives is the law of nature. As quoted by Grotius (cited in Tuck 1999) on the two fundamental laws of nations governing relations in a state of nature:

> ...first, that it shall be permissible to defend one's life and to shun what which threatens to prove injurious; secondly that it shall be permissible to acquire for oneself, and to retain, those things which are useful for life. The latter precept, indeed, we shall interpret with Cicero as an admission that each individual may, without violating the precepts of nature, prefer to see acquired for himself rather than for another, that which is important for the conduct of life.
>
> ...the order of presentation of the first set of laws and those following immediately thereafter has indicated that one's own good takes precedence over the good of another person—or let us say that by nature's ordinance each individual should be desirous of his own good fortune in preference over that of another... (pp. 85-86)

For Grotius and those who follow in his footsteps, the rights of individuals in the state of nature are modeled on the rights of the sovereign state:

> It is evident that private persons held the right of chastisement before it was held by the state. The following argument, too, has great force in this

[8] The language of American state-builders about "Indians" has several similarities to the language used by early Zionists on "Arabs."

> connection: the state inflicts punishment for wrong against itself, not only upon its own subjects but also upon foreigners; yet it derives no power over the latter from civil law, which is binding upon citizens only because they have given their consent; and therefore, the law of nature, or law of nations, is the source from which the state receives the power in question.

The sovereign state has no rights which individuals in nature did not possess first:

> Kings, and those who are invested with a Power equal to Kings, have a Right to exact Punishments, not only for the Injuries committed against themselves, or their Subjects, but likewise, for those which do not particularly concern them, but which are, in any Persons whatsoever grievous Violations of the Law of Nature or Nations.... Punishments, which at first, as we have seen, was in every particular Person, does now, since Civil societies, and Courts of Justice, have been instituted, reside in those who are possessed of the supreme power. And upon this account it is, that Hercules is so highly extolled by the Antients, for having freed the Earth of Antaeus, Busiris, Diomedes and such like Tyrants, Whose Countries, says Seneca of him he passed over not with an ambitious Design of gaining them for himself, but for the Sake of vindicating the Cause of the Oppressed.

In his struggle with the Scholastics, Grotius defended the right to wage pre-emptive war in the interest of expanding civilization as case of just war. To quote from De Iure Belli ac Pacis:

> For the same reason we make no Doubt, but War may be justly undertaken against those ... who kill Strangers that come to dwell amongst them; against those who eat human Flesh; and against those who practice Piracy. And so far we follow Innocentius, and others, who hold that War is lawful against those who offend against Nature; which is contrary to the opinion of Victoria and others who seem to require, towards making a war just, that he who undertakes it be injured himself, or in his State, or that he has some Jurisdiction over the Person against whom the War is made. For they assert, that the Power of Punishing is properly an Effect of Civil Jurisdiction; whereas our Opinion is, that it proceeds from the Law of Nature... (Grotius cited in Tuck 1999: 102-03)

Gentili, referring to the now fashionable concept of "world community," justified Spanish conquests in the same wordings as Grotius had used: "I approve the more decidedly the opinion of those who say that the cause of the Spaniards is just when they make war upon the Indians, who practiced abominable lewdness even with beasts, and who ate human flesh, slaying men for that purpose. For such sins are contrary to human nature, and the same is true of other sins recognized as such by all except by brutes and brutish men."

Today's international community, which is the tiny fraction of world population that pretends to speak for all, thus does not wage war for the sake of interests of its elite members. The aim is no less grandiose than to expand civilization beyond the self-created boundary between "we" and those living beyond it in darkness. The common currency of western public diplomacy thus has a long pedigree. Lawyers of the Church of Rome at the time of "discoveries" could not accept the arrogance of the moderns. Scholars in the tradition of Christian Medieval natural law explicitly rejected the right of preemptive warfare against peoples overseas, or to conquer their lands, let alone the right to exterminate such peoples. To quote the Dominican Cajetan, criticizing the lawyers that served the interests of overseas commerce:

> Some infidels do not fall under the temporal jurisdiction of Christian princes either on law or in fact. Take as an example the case of pagans who were never subjects of the Roman Empire, and who dwell in land where the term 'Christian' was never heard. For surely the rulers of such persons are legitimate rulers, despite the fact that they are infidels and regardless of whether the government in question is a monarchical regime or a commonwealth; nor are they to be deprived of dominion over their own peoples on the ground of the lack of faith. No king, no emperor, not even the Church of Rome, is empowered to undertake war against them for the purpose of seizing their lands or reduce them to temporal subjection. Such an attempt would be based on no just cause of war. (Cited in Tuck 1999: 70)

The doctrine of the right of civilized peoples to exterminate—or "punish," as the more polite Grotius would have put it—primitives living in the anarchy of the state of nature reveals the rupture in the European social and normative order during the creation of the Westphalian interstate system. American state-nation builders drew upon the legal doctrines of the moderns as the justifying belief system to get rid of the primitives they found on the continent.[9] Emmerich De Vattel, who is likely the most well-known international lawyer of the Westphalian order and who had been a lifelong diplomat in the newly created interstate system noted the following:

> But let us repeat again here that ... the savage tribes of North America had no right to keep to themselves the whole of that vast continent.... The cultivation of the soil is ... an obligation imposed upon man by nature. Every nation is therefore bound by natural law to cultivate the land, which has fallen to its share.... Those peoples, such as the ancient Germans and certain modern Tartars, who, though dwelling in fertile countries, disdain the cultivation of the soil and prefer to live by plunder, fail in their duty to themselves, injure their neighbors, and deserve to be exterminated like wild beasts of prey... Thus while the conquest of the civilized empires of Peru and Mexico was a notorious usurpation, the establishment of various colonies upon the continent of North America might, if done within just

[9] He has to be "civilized off the face of the earth," as Charles Dickens commented on a group of Zulus on display in London in 1853.

> limits, have been entirely lawful. The people of those vast tracts of land rather roamed over them than inhabited them. (Cited in Tuck 1999: 195)

Vattel is one of the first authors on international law to clearly conceptualize the distinction between the state as a public institution and the state as a tool used to serve private interests of its rulers. This is, of course, the very foundation of representative democracy—and of state-organized genocide. The "raison of state," therefore, does not concern the private interest of the Prince. The US government accordingly acted as a public agency. State-created legal titles in land, which could be traded or used as collateral for loans, "founded" the state, the political community, and the market. In the Introduction, we described units of analysis in critical geopolitics as "state-society complexes in their natural resource niche." The expansion of the US frontier is a pre-industrial example of such a complex. Vattel's *Le Droit des Gens; ou Principes de la Loi Naturelle appliqués à la conduite et aux Affaires des Nations et des Souverains* was on the shelves of sovereigns and their diplomats well after a century it was published (probably the latest reprint has been published in Paris by Editions A. Pedone in 1998). This is the time period in Western Europe in which sovereignty moved from the Prince to the state. Vattel was widely read and quoted by nation state builders in the US. He wrote for practical men in charge of a state active in colonizing ever-larger parts of the world. Practical men are, of course, not held back by words of law-pronouncing clerics. Practical men are powerful enough to sway the normative debate to their advantage and to get away with the benefits.

Carr (1979/1939: 79) gives the key realist insight into the foundation of the morality of the members of the international community: "Theories of social morality are always the product of a dominant group which identifies itself with the community as a whole, and which possess facilities denied to subordinate groups of individuals for imposing its view of life on the community."

The following example helps are to better understand what practical men accomplish. Infuriated by the Cherokee declaring sovereignty and adopting a constitution modeled after the US document, President Andrew Jackson submitted to Congress on December 7, 1829, the final version of the bill to remove the Indians. The Senate passed it on April 26th, 1830. On May 28th, 1830 the President signed the Indian Removal Act (reprinted in Steel Commager 1958: 259ff). The passing of the bill by both houses of Congress reflects the development of democracy in the US. Earlier presidents, who were among the authors of the first truly democratic constitution, expelled Indians by executive order. In subsequent decades, white bison hunters destroyed the prairie ecosystem in which prairie Indian societies subsisted.[10] How easy it is for the stronger party to twist the rights of man into a device for ousting a rival group. Why is this rather old and dry stuff relevant for today's diplomatic practice?

Current US war policy of "pre-emptive strikes" as a tool to extend civilization draws *direct* inspiration from the early modern period and nineteenth-century US practice. The concept of pre-emption in *The National Security Strategy of the United States* is identical in its intended meaning to the use of that concept by the founding fathers of post-Medieval natural law doctrine. The concept of the "natural right to pre-emptive warfare" obfuscates the distinction between "preventive" war and a "pre-emptive" mode of attack in war planning. That distinction is one of the more clarifying conceptual contributions to the vocabulary of strategic interaction made by strategists of the nuclear era of bilateral deterrence. In current US doctrine, nuclear weapons are enlisted into the arsenal of *preventive* warfare against the enemies of civilization as defined by the US. The US leadership has decided that states it has so designated by itself, should never gain access to such weapons, irrespective of whether target governments intend to use them for deterring the US, for defense against western aggression, or respectively, for retaliation in kind in case the US attacks them first or for staging an attack against the US homeland. It is ironic that due to the spread of industrialization, latecomers may confront their enemies from civilization with weapons systems that were once deemed to be essential for the defense of civilization: "At a time when our nation may be called upon to fight a war to protect Americans from chemical and biological terrorism, it is tragic to learn that four decades earlier, some American soldiers and sailors were unwittingly participants in tests using live chemical and biological agents." The Report, prepared by the Pentagon, details tests conducted from 1962 to 1971, and included, among others, "exercises in Alaska, Hawaii and Maryland, and that a mild biological agent was used in Florida" (*International Herald Tribune*, October 20, 2002).[11]

[10] Skins of the animal got the same function in early US industrialization as colonial rubber from the Congo and Malaysia in early European industrialization. The demographic impact is comparable (Isenberg 2000: 156ff). Population losses in Belgian Congo between 1885-1908 are estimated to be about 10 million (Hochschild 1999: 233); see also Lindqvist (1996), quoting Herbert Spencer, founder of the functionalist theory in classical sociology, on "extermination" as a requirement of achieving progress. Accordingly, *the universal* language of progress is the language of *national elites* who cast the pursuit of their interests in universal terms, expressing the global ambition of their undertakings. On January 19, 1864, The Anthropological Society arranged a conference on the theme "The extinction of lower races."

[11] In its Memorandum on biological warfare of September 21, 1951, the Joint Chiefs of Staff arrived at the conclusion that biological weapons possess a great potential as a weapon

To summarize, a boundary of civilization had been created in the early modern period. It separated the legal order of state-organized societies of the interstate system in Europe from the outer world still living in lawless darkness of anarchy. By moving overseas, European settlers in the Americas, in Australasia, and New Zealand argued that they found themselves thrown back into the anarchy of the state of nature and thus had the legal right to behave accordingly.[12] Today, the "boundary of civilization" continues to shape the foreign policy of the civilized. The geopolitical map proclaimed by the founding fathers of the first free Republic was brought to completion by the victory of the US army at Wounded Knee in 1890. In 1910 the Bureau of Census made its first serious head count of surviving natives. That office identified 266.000 Indians. In 1492, estimates of their number ranged from between 2 to 6 million and above. A *Liberal Democratic* US is thus founded on a holocaust. Theodore Roosevelt did not see any trouble with that: "I do not go so far as to think that the only good Indians are dead Indians, but I believe nine out of ten are, and I shouldn't inquire too closely the case of the tenth. The most vicious cowboy has more moral principle than the average Indian" (cited in Dower 1986:15). President Bush in his State of the Union Address of 2001 could not agree more with his great predecessor: "Ours is a great story and we need to tell it to the world."

It is the great story of the redeemer nation, legitimized by a *gospel* of "progress." In 1847, US Secretary of Treasury, referred in his annual report to "a higher than an earthly power" which "still guards and directs our destiny, impels us onward, and has selected our great and happy country as a model and ultimate center of attraction for all nations of the world." The rhetoric of George Bancroft in his Memorial Address of 1866 on Lincoln tells the story of the America as the gospel moving out into the world: "Thousands of years have passed away before the child of the ages could be born. From whatever was of good in the systems of former centuries, America drew her nourishment; the wrecks of the past were her warnings.... The fame of this only daughter of freedom went out in all the lands of the earth; from here, the human race drew hope." Even before the Pacific had been reached, Seward, Secretary of State under Lincoln and Johnson, thought beyond the purchase of Alaska: "Give me fifty forty, thirty years of life, he told his Boston audience in 1867, and I will engage

---

of war. Accordingly, national security demands that the US acquires a strong "offensive" BW capability without delay.

[12] Robert Manne considers British policy towards natives as "genocidal" (cf. Manne 2001). The 2002 film *Rabbit Proof Fence*, produced by Philip Noyce, traces the track of three "half casts" all the way back home after their escape from civilization.

to give you the possession of the American continent and the control of the world" (cited in Chace, November 21, 2002: 33).

The US Goes to the Global Level

At the end of the nineteenth century, the US had succeeded in catching up with and then overtook its European competitors. Closing the productivity gap and overtaking European competitors originate in the continental size of the country. In Europe, markets at that time were divided by borders and oriented to colonies overseas. The number of people living in the new world created the potential for a mass market—and for mass production, finding the natural resource base at home. The entrepreneurial system Americans created to exploit the opportunities for development protected infant industries [13] by high tariff walls and profited from 'pirating foreign technology.' [14] In the first decades of the twentieth century, US enterprises forged ahead on European enterprises in technology, enterprise organization, and product innovation of manufactured products. Accordingly, in the decade of the 1890s, the powers of Europe saw themselves as being caught in a bench vice. A continental-sized US, which joined the competition for colonies, being one part of the device, and Russia, which resumed its westward expansion after its defeat in the Japanese war, being the other. In that decade, the US initiated a new, transoceanic stage of power projection. The dimension of control sought, however, was more oriented towards compelling others to admit US traders to imperial holdings than to acquire territory for itself. The Hawaiian Islands, "even if they were populated by a low race of savages, even if they were desert rocks, would still be important to this country from their position.... The main thing is that those islands lie there in the heart of the Pacific, the controlling point of commerce of that great ocean," argued Senator Lodge (quoted in Zimmermann 2002: 151). The US thus kept aloof from direct colonization, but did demand its fair share in the commercial benefits of it.

The Japanese leadership managed to get onto the path of the civilized, but failed to get accepted in their ranks. Today, Japan is not a part of the West. Japanese equally failed to secure a Monroe-zone of dominance in East Asia. Due to that failure, the civilized agree, to this day, that it

[13] Between the Morill Tariff of 1861 and the Underwood tariff of 1913, US rates vary between 40% and 50%; the Smoot-Hawley Tariff Act brought the rate close to 60%. US history as a selected free trader begins after US enterprise had achieved competitive dominance over its European competitor. Since that time, "free trade" is considered to be a *moral* obligation whose territorial sphere of validity is universal.

[14] Words used in the *International Herald Tribune*, October 16, 2002, summarizing a British report on intellectual property rights.

is their duty to civilization to hold Japan in check. Latecomer Germany found itself trapped on the European continent. However, at the end of the First World War, German military men, due to their success on the Eastern Front, began to imagine a blockade-proof and economic autarkic German Eastern Empire.[15] Outside Europe, following upon the British withdrawal in 1904-05 of fleet squadrons from the Caribbean, the US stepped into Britain's positions. The dispossession of Colombia from the Panama Canal Zone in the name of liberation and the building of the Canal itself turned from a joint Anglo-Saxon project into a purely US undertaking. In 1912, the British Isles, coming under threat from rising Germany, Britain recalled home the Malta squadron. The US soon got a strategic interest in controlling that waterway. The US sided with France and England at the Algeciras Conference of 1906 that settled the Moroccan crisis.

In Asia, the US took the opportunity to trade Korean independence for Japanese recognition of its conquests of the Philippines and Hawaii in the Taft-Katsuara Memorandum. Before the great wars of the twentieth century had broken out, the US thus had begun to act as participant in disputes between European major powers on each side of the Eurasian landmass.

The growth of a worldwide system out of the limited European system confronted the colonial powers of Western Europe with the problem of resource allocation; none of them proved able to solve the problem by policy choice. Accordingly, the powers on the European continent were humiliated by defeat and preemptive surrender. Germany in the East humiliated itself by being defeated in a racial war of extermination fought out with industrial weapons with the objective to cleanse large stretches of land from racial inferiors. European settlers in North America and Australasia had preceded them in such an objective.

The Geo-politics of Managing an Industrial Market Economy in the Multipolar Major Power System of Rival Imperialisms

In 1890, Jefferson's free Republic of white farmers had been transformed into an industrial economy of the first rank, despite its racially segregated labor market and high tariffs. In the year of Wounded Knee, the US had passed Britain in iron and steel production, in energy consumption,

[15] The food blockade imposed by Britain onto Germany between November 11th, 1918 and June 28th, 1919, bringing widespread starvation to these modern "Huns" is another input in Hitler's "Lebensraum" by the Eastern Empire. Social science methodologies designed to capture delayed effects systematically are assembled in McCleary and Hay (1980); see also Jervis (1997: Parts Two and Three).

and the two were even in total industrial potential (Kennedy 1987: 200-201). Accordingly, the capability to project power beyond borders no longer depended on the handgun, on brandy, and on plague-contaminated blankets.[16] The closing of the frontier, and inspired by Mahan's work on the rise of the British Empire, President Theodore Roosevelt had to go to war with Spain in order to create a navy for the new world role of the US. To impress US potential on Germany in Europe and on Japan in North East Asia, Roosevelt ordered in 1907 the newly acquired fleet to sail around the world. When the fleet returned on February 22, 1909, Roosevelt greeted the sailors with the words "As a war machine, the fleet comes back in better shape than it went out. In addition, you, the officers and men of this formidable fighting force, have shown yourselves the best of all possible ambassadors and heralds of peace"[17] (quoted in Zimmermann 2002: 4).

At the turn of the century, two visions of how to expand the role of the US in world politics had crystallized into two recognizable programs of expansion. Their spokesmen were equally impressed by US power and shared the belief in the blessings it would bring to mankind. But they disagreed on how to spread these benefits beyond the Western Hemisphere. Albert Beveridge, a vocal supporter of imperialist expansion, anticipated the future in which the US "will be the sought-for arbitrator of the disputes of nations, the justice of whose decrees every people will admit, and whose power to enforce them none will dare resist" (Osgood 1953: 87). Speakers supporting colonial expansion favored continuing the mercantilist trade policy of the US first recommended by Hamilton and Carey, arguing that "the measure of protection should always at least equal the difference in the cost or production at home and abroad."[18] William Jennings Bryan, the democratic nominee for the Presidency in 1900 and anti-imperialist activist, saw a no less powerful and blessed US unfolding before his eyes than his challenger. The mission of the US President in the world, however, was beyond solving disputes. Instead, its calling was

---

[16] See, for a fully operationalized and tested model of *competitive* power projection in the era of heavy industry in the multipolar system of Rival Imperialism of 1870-1914, Choucri and North (1975).

[17] As quoted in Warren Zimmermann, *First Great Triumph: How Five Americans Made Their Country a World Power*, New York, Farrar and Giroux, 2002, p. 4. The Noble Peace Prize selection commission seems to have preference for practical men such as Teddy Roosevelt and Henry Kissinger. Recently, President Bush, stepped in the shoes of his great predecessor, declaring victory over the brutes in Iraq.

[18] The Fordney-McCumber Act of 1922 aimed at destroying the basis of international trade—cost differences between home and foreign producers—by requiring import duties to the level of price difference between foreign-made and made products.

nothing less than being "the supreme moral factor in the world's progress": "Behold a republic increasing in population, in wealth, in strength and in influence, solving the problems of civilization and hastening the coming of universal brotherhood—a republic which shakes thrones and dissolves aristocracies by its silent example and gives light and inspiration to those who sit in darkness. Behold a republic gradually but surely becoming the supreme moral factor in the world's progress and the accepted arbiter of the world's disputes..." The Democratic platform in the nomination battle of 1912 countered the imperialist-mercantilists that "American wages are established by competitive conditions and not by the tariff" (Eckes 1995: 43). In 1912, the Democrats nominated the tariff reformer Woodrow Wilson as their candidate. Wilson indeed looked forward to the day in which "all shall know that she [America] puts human rights above all other rights, and that her flag is not only the flag of America but of humanity" (Osgood 1953: 178).[19] If prosperity in the US is not to be checked, Wilson believed, its government had to break down the tariff barriers that had protected its producers against previously more productive enterprises in Europe.

Having become the most productive economy in the world, the US now sought trade and investment access to late-industrializing societies—in the name of universal *moral* progress and as a matter of right in natural law. Consequently, protection for late-industrializing countries is damaging universal morality as well as the interest of the US and of the country that was protecting itself. The US had therefore the moral duty to pierce open these countries by whatever means. Indeed, in the *National Security Strategy of the United States (2002)*, free trade is defended first of all as a *moral* issue.

The shift away from creating an empire by colonization towards creating an empire by trade and overseas investment, supported by small wars required for trade opening, debt collection, regime change, outright plunder, and commerce protection,[20] impacts on three aspects of power projection in US foreign policy after the Second World War. *First*, Free Trade liberalism, instead of seeking perfection of domestic society, requires from rulers beyond borders to "open-up" to integrate into the liberal trading system on conditions set by the US—the most productive economy[21] of the world. *Second*, the European colonialists at the turn of the

[19] This is also the message of "*The Security Strategy of the United States*," published in September 2002.

[20] Between 1800 and 1934 US marines crossed legal borders 180 occasions, in addition to numerous army, navy and air force operations against cannibal kingdoms and other primitives (cf. Max Boot 2002).

[21] That is, per working man/women, per hour of work; Japan, Germany, among others, are more productive.

century could no longer avoid taking some responsibility for the well-being of colonial societies. The trade and investment liberalism of open markets avoids such responsibility. In the dimension of control sought overseas, the US undertakes putting into place local rulers conducive to integrating into the international system before the economy and society have been able to close the productivity-power gap. Projecting power by market expansion, with the air force held in reserve,[22] was practiced first in Central and Latin America after the US had turned away from colonial conquest under the presidencies of Taft and Wilson. *Third*, the US under Taft and Wilson began to envision a universal time path of social change for *all* societies, whether its rulers liked it not, called development. Wilson noted in 1900:

> The East is to be opened and transformed whether we will or not; the standards of the West are to be imposed upon it; nations and peoples which have stood still the centuries through ... will be made part of the universal world of commerce and of ideas. It is our peculiar duty ... to moderate the process in the interests of liberty. This is we shall do by giving them, in the spirit of service, a government and rule which shall moralize them by being itself moral.

> Since ... the manufacturer, he noted in 1907, insists on having the world as a market, the flag of this nation must follow him, and the doors of nations which are closed against him must be battered down. Concessions obtained by financiers must be safeguarded by ministers of state even if the sovereignty of unwilling nations be outraged in he process.

Wilson intervened in Haïti in 1915; in the Dominican Republic 1916; in the Mexicai Revolution, 1914 by shipping arms to Huerta, bombed and invaded port city of Vera Cruz. Continuity of US policy objective of elite enrichment and securing it by transforming the world outside the US by trade and violence is reflected by Thomas Friedman, a US nationalist commentator, in 1999 that

> We want enlargement of both our values and our Pizza Huts. We want the world to follow our lead and become democratic and capitalistic, with a website in every spot, a Pepsi on every lip, a Microsoft Windows on every computer and with everyone, everywhere, pumping their own gas.

> (All quotes from Andrew Bacevich, *American empire. The realities and consequences of American empire.* Harvard: CUP, 2002, chapter 4.)

Today, development has resulted in about 15% of the world population controlling about 75% of its productive resources. The US had traveled

---

[22] See, for early trials with the use of airpower on rebellious peasants in Nicaragua during the Sandino War, Bermann (1986: 183f). In that war the US had deployed Marine garrisons throughout the country, "supported by numerous airstrips."

that path first, had run ran ahead of it, and was connected to the mission of paving the way for others to follow suit. The new policy initiated by Taft and Wilson is known as Dollar Diplomacy: the use of bank loans, of US financial advisors, backed up by marines, regime change and fiscal reforms facilitating debt servicing. Between the early 1900s and the financial chaos of the 1930s, which, due to US-based international finance, spread around the world, dollar diplomacy was limited to Latin America and the Caribbean (cf. Rosenberg 2000).[23]

The depression decade and the experience of the Second World War[24] fshifted the dimension of control sought in US foreign policy. First, the shift was towards setting the rules of the game in the *world* economy to the advantage of the most productive participant. Carr's (1978: 124) description has become classic: "The use of the economic weapon as an instrument of national policy, i.e. its use to acquire power and influence abroad, has been fully recognized... It takes two principal forms: (a) the export of capital, and (b) the control of foreign markets. ... Free trade, championed by England in the 19th century, was the cause of the stronger in pure commercial competition. The 'sphere of influence' with its special rights as the objective of states which sought to compensate for the weakness in such competition by the direct application of military power." Second, controlling the rules of conduct in the global economy to provide the ability to shape domestic institutions of countries that integrate into the global economy became a significant shift. In a globally integrated economy, market-friendly governments should rule the national state components of the international system. This shift in the dimension of control turned regime change to the top of America's external policy agenda. The question why some regimes and populations have to be bombed into progress, instead of being eager to receive its blessings, are left for those at the receiving end to answer. The new world war confirmed in the mind of US war leaders the Wilsonian lesson that an integrated, hierarchical global economy with the US at the top should be policed by the major powers through a world organization, in which the US would be the most powerful actor. It is not a paradox therefore for the US to

[23] The work is instructive for a better understanding of the present worldwide practicing of dollar diplomacy cum armed intervention.

[24] The purpose of this chapter prevents reference to three controversies between historians about this new stage of US power projection: (i) the outbreak of war between Japan and the US in the Pacific; (ii) Truman's resort to nuclear weapons against defeated Japan; (iii) the onset of the Cold War in Europe and its subsequent worldwide spread. See, for shifting interpretations on the outbreak of the Pacific War, Stinnett (2001). However, no serious historian will argue today that US behavior in any of these cases was just a brave and essential response of free men to aggression against it.

be the great champion of multilateral institutions and at the same time to extricate itself from their regulatory frameworks.

For US post-war planners, the collapse of the international system in the 1930s demonstrated the incompatibility between rival imperialism at the international level and survival of the US liberal domestic system: "War restored American prosperity—something the New Deal had been unable to do.... It was not until 1943, two years after this nation took up arms that the last of the jobless finally disappeared" (Ellis 1970/1995: 533). Consequently, the alternative came down to either a state-controlled economy at home, which would constrain the power of the corporate elite over introducing new technologies in the lives of Americans, or to transforming the international system into one compatible with domestic freedoms considered by the US government as appropriate for Americans. US post-war planners intended to create an external environment in which the US system could thrive, cold war or no cold war. [25] This new dimension of control sought abroad is probably best summarized by Henry Luce in February of 1941:

> It is for America and America alone to determine whether a system of free enterprise—an economic order compatible with freedom and progress—shall or shall not prevail in this century. We know perfectly well that there is not the slightest chance of anything faintly resembling a free economic system prevailing in this country if it prevails nowhere else.... It now becomes our time to be the powerhouse from which the ideals spread throughout the world and do their mysterious work of lifting the life of mankind from the level of the beasts to what the Psalmist called a little lower than the angels. America is the dynamic centre of ever-widening spheres of enterprise, America as the training centre of the skilful servants of mankind, America as the good Samaritan, really believing that it is more blessed to give than to receive, and America as the powerhouse of the ideals of Freedom and Justice.... (cited in Hogan, 11-29)

Victory in the Second World War gave the US government the interests, the will, the self-definition, and the capability to reshape the international system with these objectives in mind.

---

[25] Robert A. Pollard gives the history of these efforts in US post-war planning. In 1945, the US was, in terms of colonial possessions, a have-not nation. Due to industrialization, its self-sufficiency no longer existed. The war itself had depleted easily accessible minerals. The post-war alarm over natural resource scarcity was inspired by US inter-war diplomacy towards Japan (cf. Pollard 1985: 198); see also Simmons (1995) and Kindleberger (1973: Chapter 14).

Power Projection in the Cold War Era: A New "Defense Perimeter" and the Notion of the West[26]

This subsection describes the process of giving military teeth to global Wilsonian idealism, leading to the spatial separation of the US legal border from its military border. Due to that separation, it is possible to commit aggression against the US without physically attacking US legal territory.

The origin of the spatial separation of US land borders from the military borders of the country is centered in the Second World War. In order to supply US and allied forces across the Atlantic and Pacific, the US created a ring of naval and air bases around the landmass of Eurasia. At the end of the Second World War, the supply line stretched in Asia from the Aleut Islands, Okinawa, and the former Japanese mandates in the Pacific to the Philippines. In the Atlantic, it ran from Greenland, Iceland, and the Atlantic Islands to Britain, with extensions under construction in Northern Africa and South Asia (Monrovia, Casablanca, Algiers, Tripoli, Cairo, Saudi Arabia, Delhi, Calcutta, Rangoon, Bangkok, Saigon, and Manila). Post-war planners decided that the war front supply line would become the new *strategic frontier*, the country's new *"defense perimeter"* (Leffler 1984). US post-war military doctrines are put into effect at the strategic border, not at the border of US legal territory.[27] Aggression against the US, therefore, is not aggression against the territory of the US. Accordingly, Wilson's idealism has military teeth on both sides of Eurasia, even before the militarization of containment in April 1950. At that time, US territorial sphere extended into effective control of the worlds' oceans and their coastal peripheries.

At the end of the war, the US had demonstrated its ability and willingness to destroy a city in defeated Japan by dropping one bomb from one plane.[28] The US decision to shift its military border away from hemispheric defense, which was implemented due to the fall of France in June 1940, to across the world's oceans around the landmass of Eurasia, should prevent hostile powers to reciprocate with such a mission

[26] Essays on issues in American foreign policy covering the period from the Pacific War up to the Cold War are brought together in Hogan (1995).

[27] Carr correctly anticipated in 1939 a Pax Americana, not a Pax Anglo-Saxonica. Britain's fear of the US dismembering the British Empire was part of the motivation of Chamberlain's appeasement policy.

[28] Just war theorists since have not been able to decide whether or not the blessing by a US armed forces chaplain of the mission of the plane that took off on August 9th, 1945, to Japan and killed, among others, about 8000 out of the 12000 Catholics living in Northern Nagasaki, who originate in Portuguese times, made any difference to the fate of these believers.

in the Western Hemisphere. The decision to shift the military border of the US across the world's oceans did not take into account whether or not the Russians or Chinese would see it as compatible with their own security.[29] Due to that shift, local powers on the wrong side of the defense perimeter are able to threaten the security of the US by going to war with their neighbors or having a revolution the US does not like. Geopolitical strategists in the US during and immediately after the war (cf. Spykman 1942)[30] anticipated Cold War US security policy in Western Europe and East Asia. In their view, Nazi power, not German power, had to be destroyed. In the Far East, Japan's militarism had to be defeated, not Japan as an ongoing society. After the war, these powers had to be rebuilt as local counterweights against Russia and China, respectively. Governments in Europe served the US to achieve its objective by continuing to divert resources to colonies overseas, asking the US to defend home territories. It should come as no surprise therefore to learn that NATO's origin is as much in Europe, particularly in the activities of Britain, as is the origin of European integration in US Cold War foreign policy. Accordingly, the role of Britain in Europe of the first Labor government under Attlee is not very different from "new labor" invented by Tony Blair. However, due to the Cold War, new identifying concepts did emerge: the "West," respectively the "Atlantic Community." The notion of being the West signals an upward shift of national identities. The new expressions create a roof over pre-war national self-conceptions on each side of the Atlantic. In other words, the notion of an "Atlantic Community" expresses consensus among the allies regarding the following:

(1) The US aim of containing the Soviet Union; it is a club good of the newly created West.
(2) The position of Germany. The German leadership agreed on the anchoring of Germany in western institutions to prevent a return of Germany, which was in the middle between major powers.
(3) Acceptance of France's interest in preventing an autonomous strategic role for Germany after its rearmament had become inevitable with the outbreak of the Korean War.

However, agreement on strategy should not be confused with sharing the same identity. US nationalism is universal. In his first public statement as Chancellor of the Federal Republic of Germany, Adenauer, however,

---

[29] In the Spaatz-Tedder Agreement of June 1946, the British government, leaving Parliament in the dark, provided access to British airfields in East Anglia for US heavy bombers capable of nuclear strikes against the Soviet Union.

[30] Spykman, a Dutch sociologist, turned to the Yale Institute of International Studies. During the Second World War, that institute paved the way for appending military power to Wilson's commercialism.

declared that: "Es besteht für uns keine Zweifel, das wir nach unserer Herrkunft und nach unserer Gesinning zur *westeuropäischen Welt* gehoren" (quoted in 1995: 174). By mutual accommodation the leaders of France and Germany agreed to constitute their countries within the US-dominated Atlantic Community *as the core of a resurgent Europe* in search of an *"ever closer union."*

For realists, the new outward shift of the US border reflects the bi-polar structure of the interstate-system. Critical geopolitics, however, looks beyond the distribution of capabilities to domestic society variables. The first domestic society variable requiring attention is the spatial construction of a modern, urban-suburban industrial US. Since the 1920s, state and society in the US have been spatially structured around the automobile. In 1908, Ford produced the first model-T, of which 15.5 million were produced in 1927 on the assembly line. In 1945, 26 million cars were in service on US legal territory; in 1950, that number had increased to 40 million. In 1999 alone, the US put 9.300.000 new passenger cars on its roads (Ruigrok and van Tulder 1993: 219). Mass-production and use of household appliances put household organization on an energy-intensive path. The White House got its first air conditioner installed in 1929. In 1928 Congress legislated to air condition the Capitol. After the Second World War, the urban US in the South of the country expanded into the deserts.[31] Air-conditioned cities thrive on fossil fuel, whose wealthy inhabitants may drive in air-conditioned Sport Utility Vehicles.[32] Without access to fossil fuel beyond its legal borders,[33] lives and vehicles would

[31] The air conditioner revived the debate between "the white man can take any temperature," and to prove his strength should therefore take it, versus the "white man cannot take it" (Ackermann 2002). Nixon sometimes put temperature down to such a low level that he could fire a log in mid-summer time to remove the artificial chill. Carter, however, describing the US as the most wasteful society in the world, prescribed by law a temperature level of 80F for offices of government and private business. His energy legislation was not helpful for his reelection.

[32] Today, SUV's take 24% of the car market in the US and produce 40% more carbon dioxide than ordinary passenger cars. Between 1990-1995 each American produced 5.3 tons of carbon, each Chinese 0,7 tons. To increase sales, energy legislation classifies the SUV as a light truck for which the fleet-wide legal standard of energy efficiency set for passenger cars do not apply. In case passenger car standards would apply, the US would save on imports one million barrels a day, which is over one-third of what it imports from the Gulf region (see *International Herald Tribune*, February 10, 2003: 8).

[33] Until 1940, the US thrived on domestic supplies. In that year, the entire Middle East produced less than 5% of world oil production. During the war even Republicans discovered that private enterprise without state involvement could not serve American interests abroad and thus called in the state to put together a US energy policy, using so-

come to a standstill. Without having control over its price, US power and wealth would become hostages to decisions taken by foreigners.

US Cold War oil diplomacy has been thoroughly analyzed; there is no need to repeat it here.[34] However, freed from the military constraint of the US-Soviet stand-off, the US is currently involved in a new project of power projection. It builds on the legacy of having created, since 1920, a distinct American way of life based on the assumption of unlimited supplies of energy at affordable prices. In today's uni-polar military order, and following upon the US attack on Iraq, the US is creating a new military corridor. It extends from Southern Europe, across the Black Sea and Caucasus (Georgia) into CEA. It ends in the border region between China, Kazakhstan, Kyrgyzstan, and Tajikistan.

Critical Geopolitics and Power Projection into West and Central East Asia

The Socioeconomic Context: the Global Business Revolution Is Changing the World Order

During the Second World War, the US took the lead in setting up multilateral institutions for managing the international security and economic development of countries in the "grand area" in which the US was dominant. Never before had a country launched such a far-reaching program of creating a rule-based international order. At the same time, however, the US managed to extricate itself from the restraining impact of these institutions either by dominating them or by creating exceptions favorable to domestic producers—such as exempting the agricultural sector from free trade or by discarding them, the best example of which is the International Trade Organization, or by declaring them irrelevant for the conduct of US foreign policy, the best example of which is the United Nations since the adoption of the *Kassebaum Amendment to the Foreign Relations Authorization Act* in August 1985. However, the US failed to bring nuclear weapons under a system of multilateral control favorable to itself: the Acheson-Lilienthal Plan aiming to create institutionalized supervision of the nuclear fuel cycle worldwide was stillborn. The US wanted to restrain others in their domestic order and in their international relations, not itself. A power able to create rules according to which other state and non-state actors play at home and abroad has a form of influence over domestic orders and foreign

---

called private oil companies to reach out to the stock beyond borders. See for Roosevelt's oil map of February 1944, Yergin (1991: 401ff).

[34] On US Cold war oil policy in Iran see e.g. Amineh, Mehdi Parvizi, *Die globale kapitalistische Expansion und Iran: Eine Studie der iranischen politischen Ökonomie 1500-1980* (Hamburg, Münster, London: Lit Verlag, 1999), ch. 6.

policies of weaker states more effective and efficient than bargaining on a case-by-case basis. It is thus not an index of US strength to have to discard that form of influence and consequently to resort to an ad-hoc approach in buying friends and punishing enemies.

Since the 1980s, the international order of managed industrial capitalism of the Cold War era is at the same time *transforming* and *expanding* beyond the grand area of the Free World. Transformation and expansion are brought together in the notion of a *global business revolution*, which entails, among others, the following:

(i) Consensus among elites in western countries that further enrichment in society requires from companies and other owners of investment funds to move to the international market. Accordingly, governments should better prepare society to offer efficient locations to affiliates of multinational companies. The US has a labor component in globalization policy: massive immigration at the bottom of the labor[35] market and systematic skimming the world labor market, particularly of third world countries, for cheap talent.[36] The US seems to be able to induce in these people an upward shift in identity. Political classes in Western Europe fail, not unlike castle owners in the era of state-formation, to create such an upward dynamic in identity and economy.

(ii) Worldwide competition between enterprises from high-income economies to get control over the most rewarding positions in the value-added curve, respectively, to push enterprises in the host country into the role of supplier of intermediate products or to the status of subcontractor. Intermediate product suppliers and subcontractors, unlike assemblers and retailers, often operate on perfect markets, and thus see profits dwindle to zero.

(iii) Worldwide competition between enterprises from high-income economies to get access to competitiveness enhancing partnerships in late industrializing countries. It is a misnomer to consider high-income economies as industrialized. The shares of the workforce in these countries allocated to the manufacturing sector decrease. However, the shares of the manufacturing sector in total export increase. This situation implies exporting expensive items and importing labor intensive manufactures for further processing and

[35] The US Bureau of Census estimates US population to grow, between 2000 and 2050, from 275.6 million to 403.9 million (mid-term projection); the EU-15 is estimated to have population decline from 377.8 million in 2000 to 350.3 million in 2050.

[36] In 2000, almost 60% of people with tertiary education from Mexico, Central and Latin America, and the Caribbean have left their country and live in the US.

for re-exporting. International trade in manufactured products is turning into intra-sectoral transactions within multinational enterprises:

(iv) Worldwide competition between companies to get access to raw material inputs in a tripolar world economy.[37]

(v) The use of diplomacy, by kings and presidents as sales agents, and aggressive unilateralism in external commercial policy.

(vi) Changes in corporate structure, in governance, and in pattern of share-holding.

(vii) International financial deregulation and the rise of global financial system.

(viii) State support for technological innovation and protection of technological assets.

(ix) In representative democracies, efforts to financially capture elective officials by business interests that finance election campaigns. The shift from *citizenship politics* to *entrepreneurial politics* at the national level feeds back into external commercial policy.

The entrepreneurial state of high-income countries seeks business opportunities beyond borders for home-based owners of investment funds and for owners of technological assets. In domestic society, governments of entrepreneurial states reduce, or try to reduce, social protection costs. The outcome is the return of economic competition, though in a new, postcolonial form, and of renewed income polarization between countries.

The new ego-centered approach to domestic and international economic order building stands in contrast to the tripartite management of mass production for the home market and to stabilize that system by the rule-based international financial order of the gold-dollar standard. The conclusion is that the western elites, after having lost their red enemy at home and abroad, also lost the disciplinary force that kept their competition at home and abroad within the bounds of reason.

In diplomacy, the global business revolution is currently transforming NATO from a standing collective defense organization into shifting ad hoc coalitions. In the new order of things, allies operate as subcontractors in the implementation of the US security policy. States prepared to assist in implementing US foreign and security policy do so with the expectation of getting an *individual* return in mind. In this set-up, it no longer makes sense

[37] In 1956, 42 out of the 50 Fortune's largest corporations by annual sales were headquartered in the US; in 1987, that number had decreased to 17. In that year, the 50 largest corporations were distributed among nine countries. The number of industries in which non-US enterprises are competing with US-based enterprises expanded from oil to electrical goods, chemicals, food, transport equipment, office tools, and computing (Bergesen and Fernandez 1999: 11-23).

for the allies to routinely speak about "the West." Interests and identity *both* play a role in this transition (see also introduction).

The war against Iraq reveals that in the post-Cold War era, the civilized West no longer speaks with a single voice to the uncivilized. The once unified West is drifting apart. The US claims the natural right of a sovereign state to wage preventive (in the Bush-terminology: pre-emptive) war against targets of its own choosing. Today, the US is civilizing the world by the *universal* application of the *moral* principles of free trade and democracy, with the US Air Force held on standby. Not all allies agree.

This use of moralizing language in public diplomacy is a stark reminder of the survival in the US of the missionary ethos created by the process of expansion from the Atlantic to the Pacific. Why does the United Kingdom under Blair join the US in that great task? Why did France and Germany stay out of it, the latter particularly in public diplomacy? By economic integration and political cooperation during the Cold War, France and Germany compensated for post-war tension arising respectively, from French's decline and from Germany's restoration as a major power on the European continent. Since the formation of the EEC, the French and German governments have committed themselves to an "ever closer union." Britain was, and is, hostile to that commitment. New Labor of Blair is not that new. In foreign policy, it operates on the program of his predecessor Attlee and Bevin, that is, to prepare Europe to receive the US as security provider and shaper of domestic institutions and policy preferences.

Unlike France, however, England was not compelled by the post-war reality of power to compensate for its decline by joining hands with the new Germany. To safeguard its colonial interest against the rising commercial realm of the US, Britain induced the US to be on guard within Europe and thus to protect the British Isles. Accordingly, the special relationship which Churchill invented as a last resort for England in the North Sea during the Second World War still has special application on the continent of Europe. The US shares with Britain the interest to prevent integrating Europe to establish itself as a major power actor capable of defending its interests outside Europe. With German unification firmly set in the framework of "ever closer union"[38] by France under Mitterrand, who clearly anticipated (Schabert 2002: 323-24) the Soviet decline, if not its impending collapse, France could accept, and in a sense even enjoy, German unification and use a united Germany to rebel against American Europe. England,

[38] Thoroughly documented by Schabert (2002). Mitterrand's rebellion against bloc-structure connects German unification with European unification; see summary in one phrase in Schabert (2002: 320).

under Thatcher, anticipating that a united Europe dominated by Germany and France, would turn into a competitor outside Europe, waged verbal war against it.[39] Blair thus continues the policy of his social-democratic predecessor and liberal Thatcher in Europe of keeping the continent divided, though in a very different sort of international system.[40] For that reason, we believe that the Blair-approach to the foreign policy of British social democracy will fail in the end.

For France, the centralized commercial and monetary policy of the EU is the beginning of the building of an *autonomous* power on the European continent. England joins the US in resisting the French design for post-Cold War Europe as an autonomous strategic actor in world politics. The US joins Britain in that effort. We suggest therefore that the British government under Blair is fighting its anti-European war in Iraq. The Anglo-Saxon attack on Iraq provides some evidence of the rise of European public opinion (BBC News, 19 January, and 16 February, 2003). It stands in marked contrast to those governments willing to support the US-led invasion of Iraq. Apparently, leaders and populations no longer agree that they share living in a strategically undivided Atlantic Community. Some governments are willing to operate as subcontractors of the US endeavor to civilize the world. France, Germany, and Russia reject to be reduced to such a role. Even in Britain, public opinion is divided on this issue. Recent past history of intra-European politics argues therefore against the suggestion that Franco-German resistance against the British-US attack on Iraq is a policy for domestic leadership survival only. Is the ad hoc coalition of France, Germany, Russian and China the beginning of a balancing act of a proto-coalition in the making?

How to comprehend the fragmentation of the West? *First*, the rise of policy disagreements within the Atlantic Alliance after the short-lived unity due to the September 11th events is consistent with the realist theory. Realists consider a uni-polar military order to be incompatible with institutionalized international cooperation. Cooperation between the most powerful and their clients should instead be based on an ad hoc "coalitions of the willing." *Second*, the Anglo-Saxon powers have been able to go on the rampage overseas *and* never lost. Realist theory is silent on who will stand where. We suggest that identity is the missing variable (see

[39] Thatcher consented to Kennan's article in the *Washington Post* of November 12, 1989, that there should never be a united Germany again. Blair's anti-European policy in the Iraq crisis is the implementation of that statement.

[40] The conservative Heath government ended the special relationship by refusing over flight rights for US airplanes during the 1973 Israeli-Egyptian war. Heath also refused to punish Soviet clients for not obstructing Soviet planes flying over their territory to bring arms to Egypt and Syria.

introduction). Freed from the bipolar constraint, the US has resumed its global mission and is waging war for the sake of expanding the zone of civilized humanity. For Britain under Blair, colonial history and European politics may be expected to impact on current decisions. In several ways, war in Iraq reminds one of the one waged by Blair's predecessor. There are similarities in the way in which Iraq gets ruled then and now. On March 17, 1919, Lt-General Frederick Stanley Maude issued the following declaration:

"I am charged with absolute and supreme control of all regions in which British troops operate; but our armies do not come into your cities and lands as conquerors of enemies, but as liberators.... It is the hope and the desire of the British people and the nations in alliance with them that the Arab race may rise once more to greatness" (cited in *Le Monde Diplomatique*, April 2003).

Today, a US commander is setting himself up as "the authority," with derived regional authority for his British assistant, which may further subcontract portions of the job to still lower-level service suppliers, as Poles, Dutchmen, and people from the Baltic States. The US as the occupying power, delivers to Iraqi's the same message as its British predecessor did: "We are the boss here, but we rule in your interests, not to our own benefit." The similarity in behavior is equally striking: natives who disagree get arrested or shot. The losers among the civilized powers are the same too: France and Germany, who seek compensation in European integration. Between the conclusion of the Sykes-Picot Treaty and British-French agreement on borders between their mandates, which put the stamp of legality on rampaging the territories of the former Ottoman Empire, France lost its zone of influence in the oil-rich part of Iraq around Mosul and Kirkuk. Latecomer Germany never completed its Berlin-to-Baghdad railroad project.

In continental Europe, elites in the twentieth century met defeat at home and humiliation in colonies. In France and Germany, memory of both world wars no longer serves a national mobilizing purpose. In Britain, world wars are remembered as victories over Germany and serve as mobilizing events against repetition, that is, against the project of a united Europe in which Germany will inevitably be the most powerful actor. To escape from one's past into the dream world of bringing civilization to brutes is more difficult for losers. Losers inevitably face their own brutal past, and thus lose the capacity to convince themselves of having the duty to carry on civilization at gunpoint. This is one source of the fragmentation of the West, particularly between the Anglo-Saxons on the one hand and the French-German dyad on the other side. The other source is in the dimension of capabilities: France and Germany have no guns to carry on the civilizing mission in oil-rich West Asia. The US does have the guns. The conclusion from this is that exit from the international system by the Soviet Union and the defeat of the reds at home, prepared the way for the primitive to reappear on the stage and of the civilized to discipline them in the name of progress, as Kagan, referring to Robert Cooper writes:

> Some Britons, not surprisingly, understand it best. Thus Robert Cooper writes of the need to address the hard truth that although "within the post-modern world there are no security threats in the traditional sense," nevertheless, throughout the rest of the world, threats abound.... The challenge to the post-modern world, Cooper argues, is to get used to the idea of double standards.... Among ourselves, we keep the law, but when we are operating in the jungle, we must also use the laws of the jungle. (Kagan, June 2002: 15)

There is probably no better concise summary of the classical theory of European expansion than the quote from Kagan.[41] The application of "double standards"—one for treatment of natives, the other one for relations among the civilized—thus continues into the present. It is not difficult to find other examples of the application of double standards by the Anglo-Saxon powers.[42] The law of the jungle, of course, exists for those who go on their own initiative to dwell among the natives living there. A detailed analysis of the use of political language used by civilized rulers during the present crisis could provide further clues about the fragmenting West and double standards recommended by Kagan.

This brings us back to the theme of critical geopolitics: western societies cannot continue functioning without expanding by state and market forces beyond legal borders. The topic of regime change is nothing new to western external policy. The new factor in the present unipolar structure is that regime change is played out by the US openly, even gets announced before the start of operations to remove a regime, kill a head of state,

[41] In the practice of bringing civilization to barbarians, "double standards" get squared. In Somalia and Rwanda, civilization withdraws. In Indonesia and Cambodia, forces of civilization gave diplomatic and financial backing to the brutish regimes.

[42] America critiques China for not having ratified the Comprehensive Test Ban Treaty. Both governments signed the treaty text on September 24, 1996. The US has informed the world it will not become a party to that treaty. China's contribution to a non-proliferation regime is criticized by the US. At the same time the US government announced its intention to develop so-called "mini-nukes," to integrate the new weapon in its preventive war doctrine and to protect itself against retaliation in kind by a missile shield. The US gives full backing to proliferator Israel. The US first "punished" proliferating India and Pakistan to embrace these countries after a few years. Probably more interesting than "double standards" is the apparent lack of control over the process. The Foreign Minister of the second proliferator, the United Kingdom, "warns" Iran not to follow the British example, though only after the US has threatened the country with military action.

or lock him up as a war criminal. In the Cold War, such events were undertaken as covert operations.

Regime change in the region under study in this book includes Afghanistan from mid-1979 onwards.

Quoting from an interview published in Le Nouvel Observateur, Paris, 15-21 January 1998, Carter's National Security Advisor Brzezinski referred to the CIA arming Mujahadeen in Afghanistan 6 months before the Soviet invasion of that country, as follows. Brzezinski:

> According to the official version of history, CIA aid to the Mujahadeen began during 1980, that is to say, after the Soviet army invaded Afghanistan, 24 Dec 1979. But the reality, secretly guarded until now, is completely otherwise. Indeed, it was July 3, 1979 that President Carter signed the first directive for secret aid to the opponents of the pro-Soviet regime in Kabul. And that very day, I wrote a note to the president in which I explained to him that in my opinion this aid was going to induce a Soviet military intervention.

"That secret operation was an excellent idea. It had the effect of drawing the Russians into the Afghan trap and you want me to regret it? The day that the Soviets officially crossed the border, I wrote to President Carter. We now have the opportunity of giving to the USSR its Vietnam war." Referring to "some stirred-up Moslems" he believed the threat to the west posed by fundamentalist to be 'nonsense.'

The US produced over 2 million direct war deaths in Vietnam, a war called by one of its main architects, McNamara, criminal as well as completely superfluous. During operation "***Rolling Thunder***," the Dutch government, apparently knowing better than McNamara, arrested demonstrators in the streets of Amsterdam. Its successor today seems to feel more at home with the motor cycle gang "Rolling Thunder" expressing their roaring support for the re-election of President Bush. The Afghan War produced 1.8 million Afghan deaths, 2.6 million refugees and 10 million unexploded landmines, all left, for the sake of civilized humanity, freely behind for the honest finder.

European governments could pretend not to know about such events. In today's coalitions, those who are willing are drawn into these events as subcontractors. Disunity among the NATO allies is the outcome. The public goods theory of alliances, however, predicts that America's partners should agree that the US should do the fighting, with all the others giving diplomatic support to that effort, as the Dutch did in the case of Iraq.

In the theory of public goods, the availability of oil at prices affordable for energy-intensive, high-income economies is a Club Good. The US allies in Europe are even more dependent on the outside flow than the US. For members of the high-income Club, consumption is, as long as the flow lasts, non-rival (there is no demand-induced scarcity for oil or gas; see also Chapter 2). Low-income countries that cannot afford prices affordable to the rich get excluded from the flow of oil by market forces. Collective goods theory of alignments would thus predict to unify western *diplomatic* support for the US effort to remove a brutish regime that no longer serves the collective interests of Club members. The opposite occurred. The US decision to invade Iraq provoked a rift in the West, which extends into European integration and into domestic politics (see Introduction on identity as an intermediate variable). Above, we noted that since the end of the Cold War, the NATO-alliance is changing from a collective defense organization into a shifting "coalition of the willing." Franco-German resistance against the Anglo-Saxon war in Iraq will no doubt reinforce that trend. An anonymous State Department official in Europe is quoted in *International Herald Tribune,* March 15-16, 2003 as follows: "We will want to make sure hat the United States never gets caught again in a diplomatic choke point in the Security Council or in NATO."

In the setting of the Cold War, the US did provide quasi-public goods for the western bloc, one of these goods being the creation of the West as common identity, which is the foundation of collective action. During intense crisis, the West even got to the brink of the nuclear abyss together.[43] In the post-Cold War order, military unipolarity sits on top of a tripolar world economy. In such a structure, the military overlord no longer faces the threat of getting destroyed by an opponent of the same military rank. Such a country is therefore no longer under the compulsion to seek consent among its allies. However, due to its military might, the US is able to create benefits for itself and thus to recruit followers who want a share of the pie. This is the global business revolution in alliance politics. That revolution transforms shared identities into ad hoc coalitions based on individual interest and competition among members. Supporters no longer care for the legality of the actions of their master.[44] Both sides operate on a market where demand for support is traded against supply of it. NATO as a coalition of the willing means that a standing organization transforms into the political equivalent of an imperfect market of demand for, and supply of, support. On the demand side is the US. It recruits friends prepared to support policy decisions made by the US alone. Many offer support in the expectation of getting a good return, and thus have to

[43] For the principled lecture of Mitterrand to several European policymakers on why flexible response should be rejected by allies in Europe, see Chapter 6 of Schabert (2002).

[44] For example, the Dutch government was one of the first among small allies to offer diplomatic support to the US irrespective of an authorizing Security Council Resolution, despite the provision in the constitution of the state about the higher-order level nature of international law (see Trouw September 6, 2002: 1).

compete for attention and favors. Allies that join in battle, or help clearing up the rubble when their overlord has declared victory, have to be bought with material rewards on a case-by-case basis.[45] This is the reason why the British, Spanish, and Polish supporters of the US will not get a grip on US foreign policy, nor receive a good return for their investment. The conclusion therefore is that, with the unipolar West gone, the US allies have lost their grip on the behavior of their former protector and that the US is turning from a collective goods provider for members of the Club into a self-centered "Machtstaat."

This brings us back to the matter of energy dependence and access to energy holding sites beyond borders. Currently, the EU dependence on imported oil amounts to 76 percent and gas dependence amounts to 40 percent. These numbers are expected to increase in the next 20-30 years, respectively, to 90 percent and to 70 percent. The EU candidate states have an oil dependence ratio of 90-94 percent and a gas dependence of 60-90 percent. The OPEC represents 45 percent of the current EU oil imports. The US too is increasingly dependent on imported fossil energy (see for details on EU policy towards CEA Chapter 8). Both the launching of the EU-Russia strategic energy partnership on November 30, 2000, in Paris and the competition for access to the vast energy potential of CEA have led to a refocusing of EU attention to the necessity to diversify its energy imports. In the world order of economic tripolarity and unipolar military control, the old law of the jungle is creeping into intra-core relations. That is one reason why we think that the unipolar military order will not survive that far into the twenty-first century.

Energy as a Commodity: Control over Energy Sources as a Strategic Asset

Liberal economists write textbooks on the assumption that markets are mechanisms for decentralized coordination of economic decisions. Accordingly, international markets are an extension of national markets. Realists in the industrial era assert that foreign trade and foreign investment are tools of power and tools for creating wealth. State actors that manage state-society complexes in foreign relations, including external commercial policy, use these tools for national, and elite advantages. A century and a half ago, Friedrich List, anticipating Schumpeter's industrial economics, formulated the basic principle of economic realism for the industrial age:

[45] Reno observed the pattern in sub-state violence and coined it as "warlord politics" (see Reno 1998: Chapter 1). The assumption of isomorphism of politics at different levels of analysis has demonstrated to be a powerful tool of model building in International Relations.

> States included "giants and dwarfs, well-formed bodies and cripples, civilized, half-civilized, and barbarian nations. ... The popular [British] school [of free trade] betrays an utter misconception of the nature of national economic conditions if it believes that such nations can promote and further their civilization, their prosperity, and especially their social progress, equally well by the exchange of agricultural products for manufactured goods, as by establishing manufacturing power of their own.... We ask, would not every sane person consider a government to be insane which, in consideration of the benefit and reasonableness of a state of universal and perpetual peace proposed to disband its armies, destroy its fleets, and demolish its fortresses? But such a government would be doing nothing different in principle from what the popular school requires from governments when, because of the advantages which would be derived from general free trade, it urges that they should abandon the advantages from protection." (cited in Haslam 2002: 157-58)

One of the first rules of economic development is that profits come from market imperfection. An important source of market imperfection is, being the first, in technological advancement. Unequal access to modern technology thus creates imperfect markets. Profits dwindle to zero under free competition. Latecomers in industrialization face the risk of getting pushed into raw material exporting or intermediate product producer roles. That is why the West agrees that free trade is good for *African* cotton farmers, for *Chinese* component suppliers and for *Third World* people who get employed in transnational companies. Profits have to finance new investment in capital and innovation. Profits come from sale at home and abroad. Countries that escaped from the follies described by List, such as Japan, South Korea, and Taiwan, are located at the bottom of the rank order of national economies according to their level of transnationalization as published by UNCTAD's annual *World Investment Report.* These countries are outliers in the systemic process of income polarization underway since the mid ninetieth century (cf. Bairoch and Levy-Leboy 1981). That process has never accelerated as fast as it did in the new era of US-pushed globalization.

Today, the calling by the US on late-industrializing countries is to disarm and open up, irrespective of their level of development, for market forces driven by western state-society complexes. The conclusion therefore is that, for those in control of societies with productive superiority, military power is a tool to expand *markets* as well as for getting control *over* markets of late-industrializing countries. Transnational companies are thus not just an extension across borders of civil society. Trade law is international public law, not private contract law. Companies working beyond borders have to go home first before they ask host governments to open up for business. This reality brings the play of power among states and intrastate society relations into the domain of international economy and vice versa.

The US attack on Iraq is a reminder that this function of military superiority is not a thing of the past. In domestic systems, democracy and markets are not separated by an impenetrable boundary. The same is true in the international system. The US victory in Iraq brings the following economic benefits for home-based enterprises. Military victory is instrumental for the transfer of property rights in the energy sector from the state of Iraq to *selected* private-sector interests; that sector will be dominated by companies headquartered in the US and Britain: "The contracts signed while Iraq was a pariah state, are so lopsidedly favorable to France that no successor regime would respect them, and a new team in Baghdad brought to power by Washington will certainly want to think again," a Bush Administration official said (*International Herald Tribune*, March 15-16, 2003). By getting control over the quantity of oil exported from the location called Iraq, the US gets a hold over market price of oil. The share of 11 OPEC-members in the proven world stock in 2002 is 80.5 percent. High-income economies will become more dependent for their energy needs on imports from OPEC-members. With an Iraqi regime dependent on the US, the US military victory will get influence over oil prices. Victory in Iraq will be useful also to US and British private sector interests in arms production and arms trade; Iraq will also get new arms suppliers. Opportunities created by America's victory also attract subcontractors. A Polish contingent goes to Iraq. Poland may subcontract a portion of the work to its friends in the Baltic States. American arms sales to Poland are reported to bring offsetting nonmilitary investments from the supplier country into the client's economy, giving enterprises from the US preferential access to investment sites outside the military sector. Notably, military espionage technology can and has been reported being used for economic and security espionage. Gentlemen *do* read each other's mail—when they are capable enough to do so: "...espionage has emerged as one of the most serious irritants to transatlantic relations... That the US eavesdrops on foreign business is not in doubt ... economic interest and national security have become so closely intertwined. At the centre of continental European anxieties is a global surveillance system, operated by the US, Britain, Canada, Australia and New Zealand" (*Financial Times*, May 31, 2002).

Oil companies headquartered in western countries want to open up the Chinese markets for refinery and retail, and thus pre-empt Chinese enterprise in this rapidly expanding sector. Due to the military predominance of US in oil-rich West and CEA, Western oil multinationals are able to threaten Chinese companies with exclusion from participating in the bidding for exploration sites if the Chinese energy market is not "opened up" enough to their taste. The Chinese are therefore deeply concerned about military power projection by the US up to the Kazakh-

Chinese border, just as the US is interested in extending the defense perimeter into the oil-rich part of the globe.

In our judgment, the current controversy within the Atlantic Alliance will prove to be the beginning of a more brutal stage of intra-core competition for unilateral advantage in wealth, power, and status. In the unipolar military reality of today, the energy-addicted US claims as a matter of right in natural law to wage pre-emptive (that is, preventive) war against countries its government considers to be uncivilized:

> "...in an age where the enemies of civilization openly and actively seek the world's most destructive technologies, the United States cannot remain idle while dangers gather."

Accordingly, the law of nature, as defined by the moderns, is the legal right of the militarily most powerful state. We therefore anticipate a future in which Japan and the EU members of NATO on the European continent will discover that US military might is turning from being a protector of common interests, respectively a creator of opportunities open to members of the Club, into a tool of strategic competition. Today, Japan, China, and EU members are becoming hostages to US policy in West and CEA without having a say in the content of that policy.

The Regional Context of the New Stage in US Power Projection: The Rise of East Asia

The Kennan restoration in East Asia of 1947 incorporated the economies of Japan, South Korea, and Taiwan into the trading zone of the Free World, which was militarily protected by the US. These countries were taken out of their geographic region and have been connected in strategy and economy to North America instead. America's allies in Northeast Asia did not face the compulsion, as France and Germany did in Europe, to accommodate. Japan, South Korea, and Taiwan are allies of the US, without being allied among themselves. Bilateral relations therefore dominate in Northeast Asian security policies. Despite being allied to the US, these countries once belonged to an undivided Asian West. Northeast Asian governments protect national identities and limit access to the economy by foreign investors. Accordingly, capital accumulation in these countries is mainly under control of nationals. The pressure to open up by their military protector failed in the catching-up stage of industrialization. Accordingly, some of Kautsky's "Asiatic hordes" [46] of the early 20th century

[46] Kautsky's observation at the beginning of the 20th century is that "The people of the East have been defeated by the Europeans so often that they thought it hopeless to resist. Europeans had the same opinion." Karl Kautsky, "Der Weg zur Macht," in Patric Goode, ed., *Karl Kautsky: Selected Political Writings*. New York: St Martin's Press, 1983, p. 76.

have become fierce competitors in home and foreign markets of enterprises located in the Atlantic economy.

The US and European-based enterprise in search of usable labor in mines and plantations no longer ship Chinese 'coolies' around the world. As the Chinese government reasserts itself in regional and global-level international relations, it has not forgotten past humiliations meted out by western governments. Chinese historians tend to consider the 1840s as the watershed between the country's ancient and modern periods. In the ancient period, the Empire got all the respect it thought it deserved. In the modern period, the Empire got overrun, its people exposed to opium dealers, its treasures plundered. Accordingly, civilization clashed with barbarians from overseas. The ambition is to regain dignity and status by successful development.[47] Economic development is therefore more than acquisition of wealth [liberals] or wealth and power [realists]. Since the reform program, China is turning into the workplace of the world for labor-intensive manufactures and the component supplier for electronic goods producers. However, its declared ambition is to imitate the developmental state model of Japan and South Korea. To succeed, Chinese enterprises have to create branded products under their own name, to sell them worldwide and to organize production in an international supply network for Chinese assemblers. China's reformers saw in this project acceptance of the normative framework of multilateral cooperation in western institutions.[48] It now sees that this approach to diplomacy being discarded by its self-appointed mentor.

The Geopolitical Hypothesis and Its Alternatives

We cannot accept the suggestion from commentators that domestic oil-interests in the US have captured US foreign policy and used it to their own advantage by waging war in Iraq. Bush and his Texan oil associates want to drill at home, in Alaska, and at other domestic sites with high extraction cost. To attack Iraq and bring its oil wealth to market will

[47] This is reflected in efforts to locate stolen treasure in western museums and palaces. In 1857, troops from Britain and France entered Peking and deliberately burned the emperor's Summer Palace. The vandals brought back so much loot-vases, tapestries, porcelain, enamels, jades, wood carvings—as to set a fashion in Europe and America for Chinese art (see R.R. Palmer and Joel Colton. *A History of the Modern World*, New York: Alfred Knopf, 1969, 3rd edition, p. 653). Western troops looted during the Boxer Rebellion (see *South China Morning Post*, June 21, 2000, p. 7). For US participation in the Second Opium War and in the Boxer Rebellion Boot, and op. cit. (2002: Chapters 2 and 4).

[48] The new world view, accepted by the leadership in the mid-1980s was multilateral cooperation "in conformity with the objective needs of the development of modern productive force." Quoted in Wang Dinyong, "On tariffs and the GATT," *Social Science in China*, Vol. 15, No. 1 (1994), p. 13.

break the power of OPEC over prices. That will not boost the earnings of domestic producers. However, the US allies will be reminded, and may still remember from their own past, that brute countries have to pay an indemnity to their liberators-because they are brutes and were defeated.

Our geopolitical hypothesis about US power projection in West and CEA is that the US aims at securing energy supply to domestic society. This is the *first part* of our geopolitical hypothesis. How plausible is this part of the geopolitical hypothesis? US oil production from domestic sites peaked in 1970. An indebted US at the beginning of the twenty-first century may have driven itself into a corner. The US, continuing its mission to civilize the world, may in this way, search for a bold way out by unleashing preventive warfare in West and CEA. In 2000 the US took 20 million barrels of oil daily out of the global stock, which is about 50% of the combined daily use by the industrialized countries of the West and Mexico (International Energy Outlook 2002: 183). By waging war successfully, the US has transferred ownership over Iraq's oil resources and thus removed an obstacle to cross-border expansion to home-based non-state actors, including its businessmen in military uniform.

The US therefore aims at creating US loyalists in Iraq and to get them into power. This is the *second part* of the geopolitical hypothesis. State-supported, private faith-based organizations are reported to access Iraqis, hoping to make converts and thus create US-loyalists. NGO-activities are expected to shape host society into a form friendly to US interests and influence. The occupying force kills off armed opposition to it before Iraqis are called to vote. The British preceded the US in such an effort. The rulers of the British Empire, however, got stuck in the 1920s in warfare with the natives. Iraqis rejected the regime Britain had installed in Baghdad. The reality in the country today gives no indication that Iraq will not again be ruled by an extraverted regime. Extraverted regimes do not primarily rely on domestic support. Outside protectors keep such governments in power. The latter gets in return policies favorable to the protector:

> Iraq to cancel three oil deals with old regime. Thamir Ghadhan, US-appointed oil minister, said a contract with Russian ... (and) a Chinese company had been frozen. ... Phillip J. Carrol, the former Shell executive chosen by the Pentagon to advise the oil ministry, said there was some doubt whether existing foreign contracts 'gave the Iraqi people the full benefit of their oil wealth.

It is here that the US civilizing mission comes in, as explained by Ladeen 2002: 5

> First and foremost, we must bring down the terror regimes, beginning with the big Three: Iran, Iraq, and Syria. And then we have to come to grips with Saudi Arabia... Stability is an unworthy American mission, and a misleading

> concept to boot. We do not want stability in Iran, Iraq, Syria, Lebanon, and even Saudi Arabia; we want things to change. The real issue is not whether, but how to destabilize ... we must destroy [these regimes] to advance our mission.[49]

Britain supports that mission. Countries in Western Europe consumed daily 13.9 million barrels of oil, which is the same as daily consumption of Developing Asia (China, India, South Korea, and Southeast Asia). Europe's major continental elites cannot join the US war, and lack the guts to collectively launch an alternative world order concept more conducive to world stability.[50] Getting into proper shape the society, economy, and polity of Iraq will give the US the power to set conditions for third parties to access the energy sources of the country.[51] The opening up of society and state in Iraq and in West and CEA thus goes beyond US domestic oil needs. The end-objective is to prevent energy and transport integration among the industrial zones of Japan, Korea, China, Russia, and of the European Union.

This is the *third part* of the geopolitical hypothesis on the waging of war against Iraq. The leadership of France, Germany and Russia may see that they do have a non-war option. It consists of creating out of the dispersed industries in the EU, Russia, and East Asia, a single industrial system around the energy resources of Russia, West- and CEA. Integrating the Eurasian economies into a single industrial system by linking its components by road and rail and by connecting it to the energy sources of Russia, CEA and West Asia would turn Russia from an eastern peripheral power of Europe into the hinge between developed and military independent Western Europe and the developing East of the Eurasian landmass: "A minority of Russian policy experts say the widening Euro-American split could be the beginning of a permanent geopolitical shift that could see the end of the US-led NATO alliance and the growth

[49] Michael Ladeen. 2002. *The War Against the Terror Masters*. New York: St. Martin's Press, as quoted in *Le Monde Diplomatique*, April 2003, p. 5.

[50] Instead, European politicians have transferred to the EU level the authority to define the scope of market transactions, but maintain authority at the national level to decide on the policy of protection. The result is paralysis and right-wing populist movements throughout Europe.

[51] We do not investigate other plausible alternatives. One is that the war is an act intended to compensate for vulnerability revealed by the September 11th attacks. But why Iraq instead of, let us say the source country of the perpetrators, Saudi-Arabia? Acts of compensation are well known from 20th century European diplomacy. After the fall of Saigon, Kissinger recommended to Nixon that, to avoid the impression of weakness, the United States should carry out some "act" somewhere else in the world. Robert Jervis, "System Effects." *Complexity in Political and Social Life*. Princeton: Princeton University Press, 1997, p. 268.

of the EU into a military and political superpower in its own right" (*China Morning Post*, February 23, 2003).

Our expectation is that the US military moving into Iraq and beyond into West and CEA is the beginning of a struggle between Eurasian located industrialized countries and overseas America for establishing control over the industrial economies on the landmass of Eurasia and their energy provision. The US and Britain both want to prevent that unification from ever happening. Accordingly, oil is, in addition to an economic good, a strategic asset. Precisely due to the creation of a tripolar world economy, with roughly equal strength of each, the merging of the EU 11 trillion-dollar economy of the EU 25, with Russia's energy-military complex as a hinge to dynamic East Asia, the potential for an integrated Eurasian industrial complex is created. The politics of such a system has the potential for setting conditions for the US and its overseas ally in the North Sea to access the region. That would close the era in which the Anglo-Saxons determine the fate of polities and peoples on Eurasia and go on the rampage there with impunity. In this perspective, NATO-expansion in Europe serves US purposes. *First*, to create in the EU backyard faithful allies: "When the EU finally gets around to admitting candidate states, the latter can be expected to do their best to block the EU from becoming a strategic rival to the United States. They will be able to say no to EU ambitions that conflict with US interests" (the *International Herald Tribune*, August 3, 2001). *Secondly*, to use member states of NATO in the Black Sea region, in particular Rumania, as a take-off point for military power projection into the Caucasus and Central Asia, bypassing Turkey. Thus, a new leg is added to the US defense perimeter. Accordingly, the portion of the globe from which local actors can pose a threat to US security interests, as defined by the US, increases again as the US defense perimeter expands into former Russian CEA.

The Geopolitical Hypothesis and Northeast Asia

How plausible is the geo-political hypothesis in East Asia? Our first discussion focuses on strategic incentives and obstacles in Northeast Asia to such a development. This part will be followed by a review of Northeast Asian economic integration.

In the Russian Far East, Russia and China both face the US as adversary. China shares Russia's resistance against the missile shield. NATO expansion and military undertakings outside the Atlantic region charm neither the Chinese nor the Russians. Japan, however, seems to get along with the missile shield. But will Japan ever participate on the US side, assisted by Australia and maybe England, in a war in the strait of Taiwan against China? Will Japan get drawn into US preventive war against North Korea? That does not seem likely to ever happen. Japan normalized relations with the regime after it was declared by President Bush to be beyond the law. However, Japan fears unification. A unified Korea would probably expel US forces from the country. Unified Korea could become a nuclear armed independent actor in Northeast Asia, one with the rank of Germany. However, peaceful merger would be preferred in Japan over a US-led invasion, to be followed by war on the peninsula. Japan may find it more attractive to develop its own military capability than to participate in a US-led war (*International Herald Tribune*, April 28-29, 2001).

For some Americans, the depressing truth about Japan is that:

> Far from earnestly trying to Americanize itself, as American policy-makers have long fondly imagined, Japan continues to frustrate all American hopes of a convergence towards American values. The implications for American trade policy are unmistakable: the time for patience is gone.... Japan is at the core of the American trade disaster. Its contribution is not merely its huge surpluses—though these rank with China's as by far the world's largest. The nature of Japan's export activities has been particularly devastating for the United States because, in sharp contrast to the labor-intensive products that China exports, Japan has been concentrating its attack more and more in highly capital intensive industries that were once disproportionately important in maintaining America's edge in global competition. Moreover Japan is the undisputed ring-leader of East Asian mercantilism. Under Japan's influence—and often indeed its tutelage—other East Asian nations have one after another adopted remarkably similar approaches to trade. Not to put too fine a point on it, Japan taught the rest of East Asia how to cheat on trade; how to target American markets; how to winkle technology transfers out of the United States; how to circumvent America's dumping laws; how to mislead the American press about their true trade agenda; and how to stonewall every market-opening request from Washington. The significance of all this is hard to exaggerate. For, in the absence of East Asia's chronic trade surpluses, America's trade would today be in broad balance with the world. Even more important, America manufacturing would not be the hollowed out like it is today but rather would still enjoy the sort of lead it had in the 1950s and 1960s.[52]

This opinion, though reflecting innocence in matters relating to the international macroeconomics of savings and investments, is rather widespread among the US elite. It is not so likely, therefore, that the US will achieve its declared objective of turning Japan into America's England in the Far East. To clarify that statement, we highlight differences between Western Europe and Northeast Asia.

In Northeast Asia the *function* of NATO-expansion, that is to keep the region divided, is in East Asia enacted by divided Korea. The threat of preventive war against North Korea, implied in the doctrine of pre-emption, frustrates the efforts of South Korea[53] to bring both

---

[52] See (*http://www.unsustainable.org/about_un.asp*).

[53] The US resists the sunshine policy of the South Korean government. The US, by raising military threats against North Korea, is able to damage the economy of South

Korea's together and pushes Japan towards the US. South Korea drifts towards China economically. However, the current situation of no-war, no-unification, is probably more attractive for China than peaceful merger of both Koreas; but peaceful merger is no doubt preferred by the Chinese government over the US fighting a preventive war. For the North, Japan is an important trader. North Koreans in Japan are a source of hard currency for the regime no less important than trade. These events support the hypothesis that Japan sees as its interest to keep the Peninsula divided and thus to prevent the North from collapsing. In the face of this policy, the geopolitical hypothesis does not fare well. However, Japan would prefer peaceful merger of the Koreas to a US-led invasion to be followed by war on the peninsula. If the North is aware of the preference scheme of China and Japan, the US will find it difficult to impose its will by peaceful means. For the US, Japan is the most important ally, not South Korea. The South will not therefore get support from the US for a peaceful merger. South Koreans, however, tend to blame the US for the division of the country. Till 1994, the US general commanded the Korean Army, and thus the US government is blamed for human rights violations, in particular for the mass murder in Kwangju.

The conclusion from this reasoning is that both Japan and China prefer peaceful merger of the Peninsula over violent regime change imposed by the US, even thought both prefer a divided Korea over a unified Korea. The US moving to a preventive war doctrine and the North Koreans declaring themselves to have the status of a nuclear weapons state, threaten to push the first preferred option of China and Japan (keeping the continent divided) beyond the horizon. Both prefer peaceful merger to warfare. This creates a common interest between both northern powers. South Koreans share that preference. Peaceful merger requires economic cooperation between high-income Japan, rising China, and economically successful South Korea. The conclusion from this is that US policy of preventive war may push the northern powers to pursue the economic option of our geopolitical hypothesis. On both sides of the Eurasian landmass, Cold War alliances thus are adrift. A new pattern of alignments is in the making. Realignments try to adapt the post-war ocean-based defense perimeter to the post-Cold War reality of a new, postcolonial wave of competitive expansion.

Is there any indication that Northeast Asian countries are reorienting economic development policy away from a Pacific Ocean orientation

---

Korea. The Moody-Index of creditworthiness of South Korea responds significantly to American threats to escalate conflict with North Korea. A Taft-Katsura Memorandum in a dressing fit for modern times?

towards Northeast Asian integration? The cold-war economic structure in East Asia began to crumble when China in the 1978 normalized relations with the US. The process of realignment has not stopped there. After the normalization of relations between Indonesia and China in 1990, the Peoples Republic associated itself with ASEAN. China participates since 1996 in the ASEAN-Japan and ASEAN-Korea Summit Meetings. These meetings have evolved into the ASEAN+3 meetings. This regional gathering has put APEC to the sidelines. Russia participates in the ASEAN Regional Forum.

Figures for external trade for the Russian Far East are not easy to find.[54] However, it is plausible that transport costs (it takes a flight of 10 hours from Moscow), population dynamics, and the movement of people are indeed drawing the Russian Far East into the regional economy of the Pacific Rim. Six percent of the current population of 7.5 million of the Russian Far East are descendents of Chinese, Japanese, Vietnamese, and native peoples. Korean entrepreneurs in the Russian Far East are extending commercial and productive networks into Kazakhstan. The Russian Far East covers about 36 percent of Russia's territory and has great resource wealth.

There are good reasons for Japan to shift development policy towards Northeast Asia. Japan exports more to the world than the world exports to Japan. Part of the export surplus comes from Japan's negative trade balance with the East Asian region. By restricting imports, Japan compelled during the Cold War, its East Asian neighbors to export overseas. This trend is changing. The Cold War economic border in East Asia (as in Europe) is being brought down by investors, traders and governments. Whereas investors and traders from the west go east, Japanese, Koreans, Chinese go west and south. Will they ever meet, as our hypothesis from critical geopolitics requires?

The last 10 years have demonstrated that Japan cannot resume dynamic growth by exporting to the US and other high-income markets. The Japanese economy has grown too large to be fueled again by exporting to the US. Japan of the future will import more than its exports and pay for the difference from investment income. Where will the imports mostly come from? Japan has had since 1994 a negative trade balance with China and uses development aid as precursor to, or follow-up on, its overseas investors. Japan is also a very large aid provider to China and has become a substantial investor in that country. China's export trade is for a considerable part exports by affiliates of multinational companies. Last year, the sin-

[54] An informative work on the foreign and commercial policies regarding the Russian Far East is from Sue Davis (2003).

gle largest exporting company in China was from Taiwan. South Korea too is getting drawn into China's dynamic economy. South Korea is expanding economically into North Korea and beyond into the Russian Far East.

China earns today about 22 percent of its GDP by exporting. The US is untill 2003 its largest export market. Will the additional 350 + million people who live in the Far West of the country ever find employment in factories dependent on exports to high-income economies, the US and the EU in particular? To ask that question is to deny it. Accordingly, for Japan to resume growth and for China to maintain it, exporting to high-income economies is no option.

Newspapers report busy cross-border trade in Primorskii Krai. China buys large quantities of modern weapons, machine tools, aircraft and other transport equipment, as well as locally available raw materials, for China's industries from Russia. Much of cross-border trade is barter, and part of it is smuggling and thus is underreported or is not reported at all. China also absorbs lots of scrap metal originating in Russia's decayed Far Eastern industrial plant. Russian-Chinese trade sharply increased after the signing of the 2001 Friendship Treaty. Russia is turning into an important energy supplier to China, and Japan, with several pipelines under construction or being planned. The diplomacy of international economic relations of Japan, China, and Korea regarding the Russian Far East is compatible with the hypothesis of Northeast Asian integration.

Economic integration of the Russian Far East with the North Pacific is likely to reduce local pressures for cession and therefore serve Russia's national interest. China is not hostile to Russia' influence in the Pacific Northeast: the PRC sees it as a counterweight to US penetration and US military presence in Japan and Korea. The PRC did not profit from Russia's territorial weakness at the time of the disintegration of the Soviet Union. On the contrary, border disputes are reported to have been solved and borders largely de-militarized.

Since the mid-1990s relations between Russia and Japan have strongly improved and are now probably better than they were ever before. Japan is currently involved in oil and gas projects in Sakhalin, which sees its exports of fish, timber, and minerals almost all go to Japan. For Japan, the Russian Far East has the promise of cheaper trade routes via Russian ports and rail to European markets. Japan is involved in the upgrading of Far Eastern ports and rail.

Pressure from the international financial system is pushing to Northeast Asia towards regional financial cooperation.[55] Due to export dependence

[55] US resistance against an Asian IMF in the aftermath of the financial meltdown in East Asia in 1997-98 did not prevent the conclusion of 7 bilateral currency swap agreements

on the US, Japan and China share an interest in high dollar exchange rates relative to their own currencies. Japan alone has available for intervention on the currency market a reserve of about US $500 billion. China's dollar reserve is the second largest in the world. Large stocks of dollars held as reserve is an index of both dependence on exporting to the US and of the determination to maintain the price of that currency favorable to exporters.[56] To finance US consumption serves exports, and for Japan, investments in the US. The question is whether or not these powers are also willing to help finance the increase in US public debt caused by expenditures due to preventive warfare when US Treasury bonds have a yield as low as just over 1 percent and the dollar depreciates on the currency market. We know no other case from western history in which the greatest military power was at the same time the most indebted one to outsiders: about 40 percent of US debt is financed by savers abroad. It testifies to the remarkable strength of the US system that its allies and challengers help to finance US military expansion into their neighborhood, as China does, and that its enemies fuel the US war machine that conquered the country, as occupied Iraq does.

Nevertheless, the decline of industrial capitalism in the US has to be indexed by the severing of the investment-savings linkage in the private sector, which is the core of domestic capital accumulation, and the taxation-public debt connection, which is a pillar of representative democracy. In 2002, total savings by households, governments and enterprises declined to 1.6 percent of GDP. The increase in total US debt (= current account balance + net official foreign debt + debt accumulated in the past 40 year, including public and private sector debt) is up from US $10 trillion to US $30 trillion (Clairmont, Le Monde Diplomatique, April 2003: 2). The drop in savings will decrease long-term investment spending, slow down productivity growth and GDP growth. This trend towards lower domestic savings of the last 15 years is an early warning indicator of impending change in the power of the US to profit from the dollar being the reserve currency of the international financial system. That system may break when the largest holders of dollar reserves begin to sell. A bit of EU-US financial competition may cause severe disruption of the financial system of the world economy and thus affect the sustainability of the US debt-financed military expenditures as well as its capacity to absorb exports from Northeast Asian economies.

---

between July 28, 2001, and March 28, 2002, involving Japan, South Korea, the Peoples Republic of China, and three Southeast Asian countries. The ASEAN+3 framework, in which the US does not participate, has overshadowed APEC, which has been promoted by Australia and the US, but hardly got support from Japan.

[56] However, the Yuan is pegged to the dollar; the Yen is not.

Above, we discussed the notion of private interest politics for the US attack on Iraq and found it wanting. As an alternative, we proposed a geopolitical hypothesis. A third alternative sees humanitarian intervention as the cause of the war: Saddam's war crimes committed against its people are said to have hurt Anglo-Saxon humanitarian sensibility. Idealists do believe that Anglo-Saxons and their supporters in the present military effort have indeed moral virtues that others do not have. Realists, however, come to a different conclusion. Indeed, in 1986 the British Foreign Office published a document[57] on humanitarian intervention, arguing against the existence of such a right as well as against its creation. Accordingly, "state practice in the past two centuries, and especially since 1945, at best provides only a handful of genuine cases of humanitarian intervention, and on most assessments, none at all. The document argues against the creation of such a right 'on prudential grounds,' because of the scope it gives to politicians for 'abusing such a right'" (cited in Malanczuk 1993: 27).

The hypothesis of the US and Britain going to war with Iraq in order to bring to an end Saddam's violation of human rights, has, therefore, to be tested against its observable implications. For the hypothesis of hurt Anglo-Saxon humanitarian sensibilities bringing these powers to fight a just war to be true, one should not find Saddam as a US ally at the time when he used gas against Kurds; but he was precisely that.[58] US diplomatic personnel properly reported back home on that brutality, but were instructed to accuse Iran instead (International Herald Tribune, January 17, 2003). For the specific hypothesis to be true, the general hypothesis in which it is implicated should also be true. The proposition that Anglo-Saxons are particular sensitive to human rights violations is *prima facie* incompatible with the behavior of these powers in the recent past and present.[59] Britain has a well-documented (cf. Omissie, 1990; Omissie in *Guardian Weekly*, Vol. 144, No 5, 1991)[60] history of using gas against Kurds. Proud of

[57] Foreign and Commonwealth Office, Foreign Policy Document No 148; see also Rumage (1993).

[58] With consent of the Reagan government, US firms are reported to have supplied Iraq with microorganism for anthrax and insecticides, from which chemical weapons were produced.

[59] Saddam may take leaves out of the British book of recipes on how to deal with rebels, in particular the pages for the days between Easter Monday 1916 in Dublin and Bloody Sunday November 21, 1920. Instructive too is the fractioning of skulls, one after another, by British officers using clubs with iron-reinforced tips, of disarmed and peaceful defenders of salt deposits during the 1930 Indian Salt March.

[60] Blair thus finds his Ali Hassan Majid, or Chemical Ali, at home. Reflecting the remarkable transformation of social democracy under Blair's "New Labor," the British Prime Minister finds his verbal inspiration in his predecessors, repeating almost verbatim

its achievements, the RAF organized in 1927 an air show in Hendon, near London, demonstrating to the public it could wipe out a 'native village' from the air.[61] At the time of the writing of the Protocol of Geneva of 1925, England had invaded Mesopotamia. In June 1920 the indigenous protested against the ruler Britain had put on the Iraqi throne. Churchill's formula of—"dividing up local powers so that if we have some opponents, we also have at the same time some friends" (Glass 2002: 12)—did not have its intended effects. The results were predictable and summarized by T.E. Lawrence in the Sunday Times as follows: "We have killed ten thousand Arabs in this rising this summer. We cannot hope to maintain such an average: it is poor country, sparsely populated" (Glass 2002: 12). Churchill was convinced about the effectiveness of using mustard gas against Kurdish tribesmen in the oil-rich area of Iraq and therefore recommended time and again using more of it. Sir Arthur Harris received his training for the Second World War in Mesopotamia. Results on the ground of morale bombing were impressive indeed: "Arabs and Kurds now know what real bombing means in casualties and damage.... Within 45 minutes, a full sized village can be practically wiped out and a third of its inhabitants killed or injured ... blowing a lot of children to pieces, by delayed action fuses" (cited in the *Guardian Weekly*, February 6-12, 2003). In 1990 Sir Arthur saw his contribution to British justice rewarded with a statute in London. Continuity between Blair's contribution to the progress of civilization in Iraq and the efforts of his predecessor comes from pictures of British soldiers commemorating on an old war grave in Iraq the few deaths Britain suffered in the 1920s. Those at the receiving end, however, created the response that echoes in current news: "Europe's hordes with flame and fire, Desolate the world entire" (Muhammad Iqbal 1927, as quoted in Francis Robinson, "Islamic Responses to Centuries of Western Power," *Times Literary Supplement*, September 6, 2002: 15). Ideologues in the 1960s of the Iranian revolution a decade later, formulated the refrain: "Come friends, let us abandon Europe; let us cease this nauseating apish imitation of Europe. Let us leave behind this Europe that always speaks to humanity, but destroys humanity wherever it finds them" (Ali Shariati as quoted in Robinson 2002: 15).

Above, we commented upon British policy of starving Germans into compliance by a food blockade between November 11 and June 28, 1919. Within the next 20 years, Britain faced unanticipated consequences. The

---

words used in the *Proclamation of Baghdad*, March 19, 1917, that the British come as liberators. His liberal predecessor, Gladstone, bombed Alexandria into civilization. Blair uses cluster bombs and daisy cutters to same effect.

[61] For a picture (in black and white) of this event see Lindqvist (2000: 58).

economic sanctions imposed on Iraq by Security Council Resolution 661 did exempt in section 3 foodstuffs and medical supplies. However, that concession to humanity was *intended* to remain an empty gesture by denying Iraq the means to finance such imports until 1995. Labor M.P. George Galloway is quoted regarding the impact: "The peace we are keeping is starving the ordinary people of Iraq." Labor M.P. Dalyell declared, "the United Nations sanctions are causing deaths of more than 2000 people a week in Iraq through the lack of medicine, medical services, food, and diet supplements, bad water and the lack of equipment and parts needed for health care, good water and food processing.... UNICEF estimates that between 80,000 and 100,000 children under five will die in 1993 if sanctions remain."[62] During the Second World War, the British practiced morale bombing on Germany. It had learned that art in the 1920s in Iraq and practiced it with great effect as well on its "Hottentots" in Africa. Depriving a population of food aims at the same effect as morale bombing.

Our conclusion, therefore, is, first, that Anglo-Saxons have some difficulty in remembering past history of failures to achieve regime change by bombing a population or by starving it, second, that unintended effects[63] are distributed in time, and third, that initiating war against Iraq has nothing to do with any hurt feelings of humanitarian sensibility of those who so liberally appeal to the values of civilization and their self-created mission to spread its benefits.[64] Our conclusion is supported by finding cases of useful brutes that got support from western leaders, Suharto and Pol Pot being two recent examples of useful brutes under US protection. The CIA described methods used by later President Suharto to take over power as "one of the worst episodes of mass murder of the twentieth century, ranking with Germany's holocaust against Jews" (Prodes, posted in the *Guardian Weekly*, May 22-28, 2003: 34). Brutes getting punished, therefore, imply that these dictators lost their utility to serve the interests

[62] Geoff Simons. The Scourging of Iraq: Sanctions, Law and Natural Justice. London: Macmillan, 1996, p. xiv-xv.

[63] Did the US try to create loyalty to its occupation regime by exposing the population to widespread looting? The preference order of a population for any government over no government at all has been used by some occupiers as strategy to create loyalty. We have seen no evidence that the US has done so (see Federico Ferrara, "Why Regimes Create Disorder: Hobbes's Dilemma during a Rangoon Summer," *Journal of Conflict Resolution*, Vol. 47(3), 2003, pp. 302-326).

[64] At an earlier occasion, Clinton is cited for the statement that for Americans to provide humanitarian assistance, it does not matter where the victims lived, (International *Herald Tribune*, September 17, 1999). The same president has been reported in the same journal as having discouraged members of his administration to refer to the Rwanda massacre as "genocide" out of fear of getting involved.

of their protector, not because of their violations of human rights. That is why regime change is not limited to countries governed by brutes. The cases of Mussadiq and Allende, among others, come to mind.

The last hypothesis we wish to review is that US attacked Iraq to deprive that country from weapons of mass destruction. The preemptive war hypothesis has to be tested against its implications. These implications are that

(i) America supports inspection;
(ii) Inspectors locate weapons that the regime wants to retain;
(iii) The US moves in and destroys the weapons;
(iv) The US leaves the country.

The crucial point today is that the US did not support inspectors, its forces invaded the country, and no weapons of mass destruction have been found. Instead of leaving the country, America has set itself up as occupying power. The first thing the invading army did was securing Iraqi oil installations, leaving cities to looters. This sequence of events is rather different from what the hypothesis prescribes.

The hypothesis makes no direct reference to fear. The factor of fear may be important. Fear is auto-generated in case no external source of threat can be found as a cause of fear. Fear is real when such an external cause of threat is known to exist. The governments of Great Britain, the US, and the Kingdom of the Netherlands made definitive statements that the regime did have weapons of mass destruction, which they feared. These governments may have lied; they may have been in the grips of auto-generated fear. The impact of auto-generated fear on leaders of civilized countries that have used weapons of mass destruction on others is best illustrated for British leaders at the eve of the Second World War. Chamberlain was paralyzed into appeasement, among others, by the fear of a "knock-out blow" from the air in which London would be sprayed with gas by German planes. The origin of that fear is in British experience in Iraq. Wing Commander C. Edmonds at the Royal Service Institution discussed on December 12, 1923, the hypothesis that behavior of civilized man under air attack is the same as the response in primitive society: "The shocks and interruption, the inconveniences and indignity of it all, will tell in the end. The civilized nation will go through the same three phases, as did the semi-civilized tribe: alarm, indifference, weariness; followed ultimately by compliance with our will" (cited in Omissie 1990: 107).

"Shock and awe" thus seem to work on both in the same way. The difference is in the privilege of the civilized to conduct the experiment on brutes, who could not retaliate. However, by using those weapons on populations that cannot retaliate, the civilized may have created fear in their own minds that enemies who are capable of using such weapons against them will do so irrespective of the intention of the latter. The German Luftwaffe was not equipped for city bombing. The mission of the German dive-bombers was to serve the land army as mobile artillery in a war of the conquest in the East. Britain itself, on the other hand, equipped its air force for World War II for city bombing, the effectiveness of which had been demonstrated in Iraq. Chamberlain gave in to Nazi demands, not knowing that Germany's air force did not have such a mission. He is a victim of self-generated fear.

Blair and Bush repeated time and again that Iraq did have such weapons and was ready to use them at short notice. Yet, these governments decided to bring substantial numbers of their troops to Iraq. Did these leaders realize right from the beginning that Iraq did hot have these weapons? Were these leaders so impressed by the results of their own past use [65] of weapons of mass destruction that they may have decided to go to war out of self-generated fear that such weapons could ever be used against them? The fate of Iraq may serve in their minds as a generalized warning to third parties that only civilized powers have the natural right to procure, possess, and use such weapons.

Conclusion

We reviewed power projection by the US beyond its legal borders. First, by colonizing the North American continent, to be followed by maritime-colonial expansion and after World War II, by bringing the military border of the country overseas. Our focus is on power projection by the US in post-Cold War West- and CEA. We argued that the geographical extension underway in the US defense perimeter is part of a *multiple* transformation in the constitution of the multi-level world system created during the Cold War. In the history of the now high-income societies of the western world, war, constitutional change in the make-up of society and state, economic development by projecting state power and market forces beyond the borders of the continent, and international legal change are closely tied together (cf. Bobbitt 2002, Part II: 69-213). These changes are coordinated by selection among the diversity of domestically created state-society complexes of self-identifying groups that make up the international order. In the post-Cold War world, elites of high-income countries are

[65] We are not talking about events in a distant past. In 1990, the queen mother inaugurated a statue commemorating Sir Harris for his work. Germans, Iraqis, Berbers and Hottentots, who had played one time or another cricket from below with bombing RAF-pilots in the air, were not amused by the Royal happening. For activities of Western powers in Iraq between the 1920s till today, among others, are Cockburn and Cockburn (2003).

relieved from the domestic impact of the external Soviet threat. The US external policy is no longer constrained by Soviet extended deterrence or by the interests of allies. Accordingly, the elite in the US looks beyond borders for opportunities for enrichment of some of its members. [66]

The war against Iraq signals the end of the post-Cold War period. The September 11 attacks from those "down under" bring new social forces to impact world order. The terrorist attacks provided the US with a convenient excuse to expand its military presence in the area where over 60 percent of the world's fossil fuel stock of the planet is located. Britain's participation in that struggle provides the British Empire with the opportunity to secure its external role in exchange for keeping the continent divided. To safely operate outside Europe, its leaders are united in rejecting a concept of Europe equipped with an autonomous military capacity. Accordingly, the Blair government implements the foreign policy program as his post-war predecessor, Attlee. At that time, Britain was still too great in terms of colonial possessions for safeguarding the empire and simultaneously to protect the British Isles in the North Sea. Britain was therefore instrumental in getting US military power to Western Europe. Today, it is a British interest to nurture the US to be its watchdog on European integration.

In Western Europe, France and Germany seem prepared to extricate themselves from their cold war role of being objects of security. Former Chancellor Kohl, probably the best German Cold War friend of the US, is quoted as follows: "Those in Washington who dream about themselves being the new Rome and that the world will dance to their tune are out of touch with reality. The idea of a 'pax Americana' where we are all lined up in formation is out of the question. The multi-polar world has developed in recent years—look at China and Europe to name just a few regions." [67]

We reviewed several hypotheses on why the Anglo-Saxons waged war on Iraq, the first humanitarian intervention. Subsequently, we discussed the

[66] For domestic impacts in America, see Mishel et al. (2001: 55, table 1H). In the 1990s, incarceration in the US more than doubled from 3 per 1000 in 1990 to 7 per 1000 in 2000. In America, prisons satisfy the sense of Christian justice in the new stage of globalization; see Christian Parenti. *Lock Down America. Police and Prisons on the Age of Crisis*, London: Verso 1999, p. 237, on the mixture of economy and religion involved in incarating labor and its exploitation by private enterprise.

[67] Kohl as quoted in the *International Herald Tribune*, April 4, 2003. The strategy of the Christian Democratic government to cope with EU expansion consisted of the creation of a "core Europe." The Anglo-Saxons favor EU expansion with the objective of reducing the EU to a Free Trade Area. "Core Europe," therefore now implies conflict between CDU objectives and Anglo-Saxon policy of "divide and rule." For fragments of previous history, see Houweling and Siccama (1988; 1992: 197-215).

hypotheses of fear created by the possession and use of weapons of mass destruction. Third, we found the counter-hypothesis coming from public goods theory incompatible with the growing rift in the ranks of the West; that concept is itself is vanishing from diplomatic vocabulary. We brought in a new hypothesis from critical geo-politics, which brings together insights from realism, trade liberalism, and constructivism.

During the election campaign, challenger Kerry has said nothing incompatible with the geopolitical hypothesis. Voting in favor of US going to war, he subsequently criticized Bush for having deceived the country into war, without saying, however, that (i) he would have voted against the US going to war in case he had been properly informed about weapons of mass destruction, and thus (ii) declaring he would withdraw from the war in case he was elected. His position seems to be that (i) Bush deceived the nation by attacking Iraq; (ii) he accused the President of going to war against Iraq without knowing what his country was up to and (iii) promising the nation that in case he was elected, he would win the war in some smart way without revealing to the electorate how.

References

ACKERMANN, M.
2002 *Cool Comfort: America's Romance with Air Conditioning.* Washington: Smithsonian Institution Press.

VAN ALSTYNE, R.W.
1960 *The Rising American Empire.* Oxford: Basil Blackwell.

AMINEH, M.P.
1999 *Die Globale Kapitalistische Expansion und Iran: Eine Studie der iranischen politischen Ökonomie 1500-1980*, (Hamburg, Münster, London: Lit Verlag).

BAIROCH P. AND M. LEVY-LEBOY
1981 *Disparities in Economic Development Since the Industrial Revolution.* London: Macmillan.

BBC
2003 "Million' March Against Iraq War." February 16.
2003 "Global Protests Against Iraq War." January 19.

BEALE, H.K.
1956 *Theodore Roosevelt and the Rise of America to World Power.* Baltimore: The John Hopkins Press.

BERGESEN, A. AND R. FERNANDEZ
1999 "Who Has the Most Fortune 500 Firms?" Pp. 11-23 in Bornschier, V. and Chase-Dunn, C. (eds.), *The Future of Global Conflict.* Beverly Hills: Sage.

BERMANN, K.
1986 *Under the Big Stick: Nicaragua and the US since 1848.* Boston: South End Press.

BOBBITT, P.
2002 *The Shield of Achilles: War, Peace, and the Course of History.* New York: Knopf.

BOOT, M.
2002 *The Savage Wars of Peace: Small Wars and the Rise of American Power.* New York: Basic Books.

BORTHWICK, M.
1998 *Pacific Century: The Emergence of Modern Pacific Asia.* Boulder: Westview Press.

CARR, E.E.
1978 *The 20 Years' Crisis, 1919-1939: An Introduction to the Study of International Relations.* London: Macmillan.
CHACE, J.
1978 *"Tomorrow the World."* New York Review of Books. November 21.
CHOUCRI, N. AND R.C. NORTH
1975 *Nations in Conflict: National Growth and International Violence.* San Francisco: Freeman.
COCKBURN, A. AND P. COCKBURN
2003 *Saddam Hussein: An American Obsession.* London: Polity Press.
DAVIS, M.
2001 *Late Victorian Holocausts: El Nino Famines and the Making of the Third World.* London: Verso.
DOWER, J.
1986 *War Without Mercy: Race and Power in the Pacific War.* New York: Pantheon.
ECKES, A.E. JR.
1995 *Opening America's Market: U.S. Foreign Trade Policy Since 1776.* Chapel Hill: University of North Carolina Press.
ELLIS, E.R.
1995 *A Nation in Torment: The Great American Depression, 1929-1939.* New York-Tokyo: Kodansha Globe Book.
FERRARA, F.
2003 "Why Regimes Create Disorder: Hobbes's Dilemma During a Rangoon Summer." *Journal of Conflict Resolution* V, 47(3): 302-26.
GLASS, C.
2002 "Iraq Must Go: History of Regime Change in the Gulf." *London Review of Books.* October 3.
Guardian Weekly
1991 "British Bombing when the Natives were Restless" V, 144(5).
HASLAM, J.
2002 *No Virtue like Necessity: Realist Thought in International Relations since Machiavelli.* New Haven: Yale University Press.
HOCHSCHILD, A.
1999 *King Leopold's Ghost: A Story of Greed, Terror and Heroism in Colonial Africa.* New York: Houghton Mifflin.
HOGAN, M.J. (ED.)
2000 *Paths to Power: The Historiography of American Foreign Relations to 1941.* New York: Cambridge University Press.
HOUWELING, H.
1992 "Duitsland Tussen Expansie en Status Quo." (Germany Between Expansion and Status-Quo). Pp. 197-215 in Siccama, J.G. and Rood, J.Q.Th. (eds.), *Grenzen aan de Europese Integratie (Limits to European Integration).* Assen: Van Gorcum.
HOUWELING, H. AND J.G. SICCAMA
1985 "The Epidemiology of War: 1816-1980." *Journal of Conflict Resolution* 29(4): 641-63.
ISENBERG, A.C.
2000 *The Destruction of the Bison.* Cambridge: Cambridge University Press.
JERVIS, R.
1997 *System Effects: Complexity in Political and Social Life.* Princeton: Princeton University Press.
KAGAN, R.
2002 "Power and Weakness." *Policy Review* 113 (June).
KAUTSKY, J.
1982 *The Politics of Aristocratic Empires.* Chapel Hill: University of North Carolina Press.
KAUTSKY, K.
1983 "Der Weg zur Macht," in Goode, P. (ed.), *Karl Kautsky: Selected Political Writings.* New York: St Martin's Press.

KINDLEBERGER, C.
1973 *The World in Depression, 1929-1939.* Berkeley: University of California Press.
KISSINGER, H.
1994 *Diplomacy.* New York: Simon and Schuster.
LADEEN, M.
2002 *The War Against the Terror Masters.* New York: St. Martin's Press.
LEFFLER, M.
1984 "The American Conception of National Security, 1945-1948." *American Historical Review* 89: 350-51.
LINDQVIST, S.
2000 *A History of Bombing.* New York: The New Press.
1996 *Exterminate All the Brutes.* London: Granta.
LINKLATER, A.
2003 *Measuring America: How an Untamed Wilderness Shaped the United States and Fulfilled the Promise of Democracy.* New York: Walker & Co.
LUCE, H.R.
1941 "The American Century." *Life Magazine*, February 7.
MALANCZUK, P.
1993 *Humanitarian Intervention and the Legitimacy of the Use of Force.* The Hague: Martinus Nijhoff International.
MANNE, R.
2001 "In Denial: The Stolen Generation and the Right." The Australian Quartely Essay, Issue 1. Melbourne: Schwartz Paperback.
MCCLEARY, R. AND R.A. HAY
1980 *Applied Time Series Analysis for the Social Sciences.* Beverly Hills: Sage.
MCNEILL, W.H.
1982 *The Pursuit of Power: Technology, Armed Force, and Society since AD 1000.* Chicago: University of Chicago Press.
MISHEL, L. ET AL.
2001 *The State of Working America 2001/2002.* Ithaca: Cornell University Press.
OMISSIE, D.E.
1990 *Air Power and Colonial Control: The Royal Air Force, 1919-1939.* Manchester: Manchester University Press.
OSGOOD, R.E.
1953 *Ideals and Self-Interest in America's Foreign Relation: The Great Transformation of the Twentieth Century.* Chicago: University of Chicago Press.
PALMER, R.R. AND J. COLTON
1968 *A History of the Modern World.* New York: Alfred Knopf.
PARENTI, C.
1999 *Lock Down America: Police and Prisons on the Age of Crisis.* London: Verso.
PILGER, J.
2001 *The New Rulers of the World.* London: Verso.
POLANYI, K.
1944 *The Great Ttransformation: The Political and Economic Origins of Our Time.* Boston: Beacon Press.
POLLARD, R.A.
1985 *Economic Security and the Origins of the Cold War, 1945-1950.* New York: Columbia University Press.
ROBINSON, F.
2002 "Islamic Responses to Centuries of Western Power." *Times Literary Supplement*, September 6.
ROSENBERG, E.
2000 *Financial Missionaries to the World. The politics and Culture of Dollar Diplomacy, 1900-1930.* Cambridge: Harvard University Press.
RUIGROK, W. AND R. VAN TULDER
1993 "The Ideology of Interdependence: The Link Between Restructuring, Internationalization and International Trade." Ph.D. dissertation, Amsterdam.
RUMAGE, S.A.
1993 "Panama and the Myth of Humanitarian Intervention in U.S. Foreign Policy:

Neither Legal nor Moral, Neither just nor Right." *Arizona Journal of International and Comparative Law* 10(1).

SCHABERT, T.
2002 *Wie Weltgeschichte gemacht Wird: Frankreich und die Deutsche Einheit.* Stuttgart: Klett-Cotta.

SIMONS, G.
1996 *The Scouring of Iraq: Sanctions, Law and Natural Justice.* London: Macmillan.

SIMMONS, B.
1995 *Who Adjusts? Domestic Sources of Foreign Economic Policy During the Inter-War Years.* Princeton: Princeton University Press.

SPYKMAN, N.
1942 *American Strategy in World Politics: The United States and the Balance of Power.* New York: Harcourt Brace and Co.

STEEL COMMAGER, H.
1958 *Documents of American History.* New York: Appleton-Century Crofts.

STINNETT, R.
2001 *Day of Deceit: The Truth about the FDR and Pearl Harbor.* New York: Touchstone.

TUCK, R.
The Rights of War and Peace: Political Thought and International Order from Grotius to Kant. Oxford: Oxford University Press.

WOLFOWITZ, P.
2000 "Remembering the Future." *The National Interest*, Spring Issue.

ZIMMERMANN, W.
2002 *First Great Triumph: How Five Americans Made their Country A World Power.* New York: Farrar, Strauss and Giroux.

II. Caspian Energy: Oil and Gas Resources and the Global Market

MEHDI PARVIZI AMINEH AND HENK HOUWELING

ABSTRACT

This chapter develops several concepts of critical geopolitics and relates them to the energy resources of the Caspian Region. Energy resources beyond borders may be accessed by trade, respectively by conquest, domination and changing property rights. These are the survival strategies of human groups in the international system. The chapter differentiates between demand-induced scarcity, supply-induced scarcity, structural scarcity and the creation, respectively, transfer of property rights. Together, the behaviors referred to by these concepts create a field of social forces that cross state borders involving state and a variety of non-state actors. During World War II, the US began to separate the military borders of the country from its legal-territorial borders. By dominating the world's oceans, the Anglo-Saxon power presided over the capacity to induce scarcity by interdicting maritime supplies to allies and enemies alike. Today, overland transport increasingly connects economies and energy supplies on the Eurasian continent. The US has therefore to go on land in order to pre-empt the land-based powers from unifying their economies and energy supplies.

Introduction

The oil and gas reserves of the Caspian Region [1] are undeniably significant. According to the *Statistical Review of World Energy* (BP 2004), proven oil

[1] When we speak about the Caspian region, if not mentioned otherwise, we mean its five littoral states (Azerbaijan, Kazakhstan, Turkmenistan, Iran, and Russia).

reserves of the five Caspian littoral states (Azerbaijan, Iran, Kazakhstan, Russia, and Turkmenistan) are in total 216.4 billion barrels (billion bbl). The total gas reserves are estimated at 2819.2 trillion cubic feet (tcf). In terms of percentages, the five Caspian littoral states have about 18.8 percent of the world's total proven oil reserves and 45 percent of the world's total proven gas reserves.

The region's vast oil and gas resources have transformed it into a location in which the forces of interstate rivalry, enterprise competition, and responses by regional state and non-state actors intersect. All major industrialized powers and many of the multinational companies that have their home base in these countries meet in that part of the world. Contenders from late-industrializing countries try to get a foothold in the region. Local actors have to respond to social forces coming to the region and penetrating it from without. In such a complex matrix of social forces, competition and cooperation are ad hoc and multilevel. The region is not incorporated into the territorial sphere of security institutions of one of the major powers and its allies. This part of the world is not divided into *agreed upon*, and thus *stable, zones* of influence either. In the region under study here, extra regional state and non-state actors undertake to project their power and influence into the polities and societies of their hosts, interacting with local actors. Uncertainty and thus unpredictability are part of the rules of the game. "Multidimensional rivalry" is probably a suitable term for what is going on. Because everyone is involved and regime legitimacy is at stake, major power competition in the CEA region has the potential for aggravating instability of the world system as a whole.

Michael Klare (2001) captures the interstate dimension of competition for control over the region's energy sources. He argues that global politics of today evolves around the competition between state actors for getting access to natural resource wealth. This is what Klare calls the *Econocentric* approach to international security affairs. According to this view, the possession of a huge military arsenal and an extended alliance system is no longer necessary for state survival. Survival of state and domestic society instead depend on economic dynamism, on the cultivation of technological innovation, and getting access to raw material inputs required for both. We think that critical geopolitics is an improvement on Klare's *Econocentric* approach to security (see Introduction).

This article surveys the oil and gas reserves of the Caspian region in the matrix of competitive forces of the post-Cold War. It centers on the following three factors:

(i) The increasing global demand for oil and gas;
(ii) The scarcity of vital commodities such as oil and gas and;
(iii) The dispute over ownership rights of these resources.

Ownership Rights, Scarcity and Regime Change

Global demand and supply, scarcity and ownership rights over territory containing oil resources, respectively, over oil resources within a territory, refer to social forces at work at the transnational level. Sub-state and non-state actors create that level of interactions. State actors, however, participate in transnational interactions in a variety of ways. The reason is that inputs of energy into the power-wealth producing machinery of state-society complexes of high-income countries are traded in transnational networks that cross state borders. Wealth of society and power of these states originate in technological innovation and the incorporation of its fruits into *capital goods*, which are used in production of traded goods, and into *military capability*, which is used to conquer or control territory where finished goods, or input resources required for production, are located. Capital goods and weapons stocks become idle without fossil energy inputs to fuel them.

Controllers of territory can pick up what they want inside borders. Producers of wealth procure for local consumption or for trade. Territorial control and production/trade are strategies of human groups to survive. One strategy man shares with other animals. The other one is unique for humans. Unlike members of a flock of birds, humans do not forage in isolation; they survive in groups by developing a social division of labor in production and trade (cf. Jacobs 1994).

Both strategies, however, interact when goods picked up in territory are traded beyond borders, respectively when profit from production and trade is invested in military capacity for the conquest of territory that carries goods and resources. Conquest is followed by *diversion* of trade when traders of one group are ousted from the network to be replaced by traders from another group. These processes are currently underway in Iraq. The US has announced its intention for a repeat performance in Iran. After the fall of the Shah, the US engineered regime change *without* conquering the territory first. Instead, the US engineered internal regime change, which is the cost-effective way to divert trade, to be followed by the removal of Iranian state traders and the reinstatement of Anglo-Saxon-based companies. Regime change in Iran was followed by *trade creation* when the new regime bought large quantities of weapons from US-based private producers and arms dealers. This process of interaction between both survival strategies will be started in Iraq as soon as the US government has installed a government prepared to buy US weapons. A third case of interaction between survival strategies occurs when states collude, use, or threaten military force to create a barrier for third parties to enter the market. This happened to Iraq after the imposition of economic sanctions. For these reasons, markets are not institutionally separated in international relations from states and the distribution of military capability (see Chapter I on state-society complexes and on oil as a commodity and oil as a strategic asset).

As global consumption rises, per capita availability of oil and gas from a fixed stock will after some point in time begin to decrease. This effect is called *demand-induced scarcity*. Demand-induced scarcity is due to three factors:

(i) Population growth in consuming countries.

(ii) Rising per capita income in high-income countries, which are the major consumers and importers, and in late industrializing economies, particularly in South and East Asia where the bulk of world population lives. Demand-induced scarcity thus varies for groups at different levels of per capita income. Those who cannot afford market prices find themselves excluded without any actor deciding to exclude them. Due to the lopsided distribution of societies according to their level of per capita income, demand-induced scarcity will enter into the lives of high-income societies last. These are the countries that industrialized first on cheap energy. In the past, demand from these countries coincided with world demand. That is changing and will further change in the future.

(iii) Technological change. The history of technological change since the 1850s has rendered access to fossil energy more, not less, important for the production of wealth and power.

The process of sequential industrialization of human groups increases demand. This process may be compared with more panda bears that are on the move to the same bamboo field. The similarity is that both species have to survive and prosper for the foreseeable time *in one* resource field only. Since their emergence in the 1850s, industrialized societies have specialized in becoming dependent for their wealth and power on energy from fossil sources. Without energy, other resources cannot be mobilized or used. Technological innovation, governance, and households depend on it. Historically, wood and coal provided the resource base of industrialization. Today, most forests are gone. Coal came next in row. Oil and gas are replacing coal.

Supply-induced scarcity is caused by the dwindling of the stock. In reality, demand and supply-induced scarcity interact. Extraction cost, refinery and retail plus profit mark-ups determine offer price. The intersection of demand and supply determine consumer price. However, supply induced scarcity should be studied in its own right. One reason is that the dwindling of the stock is not translated by the price mechanism into gradual price increases. However, price volatility will increase as awareness spreads that stocks are dwindling. Supply-induced scarcity, or its anticipation, may be expected to provoke a process of competitive power projection by military-capable and import-dependent nations aiming at getting control over the stock, respectively, over territory in which stocks are located, by either internally engineered regime change or by conquest of territory. Domestic regime strength and military capability determine the capacity of target countries to ward-off unwanted penetration by outsiders.

This brings us to the third type of scarcity, called *structural scarcity*.[2] *Structural scarcity* is supply-induced by deliberate action of a major power, by non-state actors such as major oil companies, or by producer cartels such as OPEC. A major power that manages to get control over conditions of access by third parties to the stock has the option to induce scarcity for selected outsiders. In the US post-war planning, maritime control was deemed capable to interrupt food and oil supplies to post-war Japan.[3] In the current unipolar military order, the US has the option to induce scarcity for allies, competitors, and enemies alike by interdicting maritime transport of oil and gas. However, that option is available only after oil and gas have been brought to ship from the territory where it is extracted (see Chapter I on the geopolitical hypothesis). America, by creating an extension of the country's defense perimeter into the heartland of energy-supply, is equipping itself with the capacity to induce structural scarcity for contenders by diverting flows on land. This is the topic of pipeline diplomacy.

Have oil prices become more volatile over time? That is indeed the case. Historical price levels of a bbl of oil (expressed in 1999 US$) fluctuated in the time period between the 1880s and early 1920s in the rather narrow band between US$10 and US$20 per bbl. Between the 1920s and the late 1960s oil price declined to around US$10 per bbl. Power and wealth of countries that industrialized first is based on access to cheap oil. Since the early 1970s, oil price per bbl has been fluctuating wildly from the peak value of just over US$70 per bbl during the Iran Revolution, a bottom price of US$20 per bbl after the defeat of Iraq in Kuwait War, to passing the US$30 per bbl level in 2000.

Between 1985 and 2000 the share of OPEC in the total global annual oil production increased from about 17 percent to over 30 percent. In 2003, eleven OPEC members sat on 80.5 percent of the world's stock of oil. With the capital of OPEC in US and British hands and Iraqi territory containing the largest share of the world stock after Saudi Arabia, the occupying powers have gained leverage over energy security of third

[2] These distinctions are further developed and illustrated with several examples in Homer-Dixon and Blitt (1998).

[3] For archive based research on this point, see Cumings (1984: 18).

actors by influence over price and ownership rights of oil and gas sources in that country. Regime change, therefore, is not limited to the domain of politics. Iran found that out in the early 1950s.

The conclusion is that oil and gas are not just commodities traded on international markets. Control over territory and its resources are strategic assets. States as actors bring domestic-state society complexes to the international level (see Chapter I on the geo-political hypothesis). The world order is the level of interactions constituted by the process of selection among the diversity of state-society complexes. The Anglo-Saxons succeeded in a military take-over of Iraqi state territory and by this act transferred the property of the portion of the world stock located there. By controlling access to the portion of the oil stock, these actors also secured the resource niche in which state, enterprises, and households of domestic society subsist by creating trade and by diverting trade. The option to put in place barriers to entry gives these powers the capacity to induce structural scarcity for contenders.

Global Oil and Gas Demand

Over the next two decades, oil is expected to remain the main fuel for industries and households, accounting for about 40 percent of global energy consumption. It is expected that global oil demand will increase annually by about 1.9 percent from 77.1 MMbbl/d in 2001, to 120.3 MMbbl/d in 2025 (EIA 2004: 167). In the industrialized world it is expected that total oil demand will decline as gas use increases (see Table 2.1).

Global Oil Demand

Among industrialized countries, the largest increase in oil demand is expected in North America (US, Canada and Mexico). Petroleum product consumption in North America is projected to increase between 2001 and 2025 by 11.1 MMbbl/d at an average annual growth rate of 1.6 percent.

For Western Europe, oil is the largest energy source. However, its projected increase in demand is the lowest in the *International Energy Outlook* forecast (2004). It is expected that oil consumption in Western Europe will increase by about 0.5 percent per year or from 14.0 MMbbl/d to 15.7 MMbbl/d between 2001 and 2025. Its low growth in consumption is mainly due to increasing gas consumption (see Table 2.1). In industrialized Asia (Japan, South Korea, Taiwan, Australia, and New Zealand), oil demand is projected to increase in the same period by an average of 0.7 percent per year from 6.4 MMbbl/d to more than 7.5 MMbbl/d. Japan imports all the oil that it needs, accounting for 81 percent of the total oil demand in industrialized Asia. Since the disintegration of the Soviet Union, oil demand decreased steadily in

Table 2.1

Projected Global Oil and Natural Gas Consumption, 2001-2025

Region/Country	Oil			Natural Gas		
	2001 (MMb bl/d)	2025 (MMb bl/d)	Annual average growth 2001-2025 (in percent)	2001 (tcf)	2025 (tcf)	Annual average growth 2001-2025 (in percent)
North America	23.5	34.6	1.6	26.9	45.6	2.2
US	19.6	28.3	1.5	22.6	34.9	1.8
Western Europe	14.0	15.7	0.5	14.8	25.9	2.4
Industrialized Asia	6.4	7.5	0.7	3.9	5.6	1.5
Japan	5.4	5.8	0.3	2.8	3.6	1.0
Former Soviet Union and Eastern Europe	5.3	8.5	2.0	23.5	46.4	2.9
Developing Asia	14.8	31.6	3.2	7.5	21.6	4.5
China	5.0	12.8	4.0	1.0	6.1	7.9
India	2.1	5.3	3.9	0.8	3.4	6.1
South Korea	2.1	2.9	1.3	0.7	1.9	3.9
Central and South America	5.2	9.2	2.4	3.5	11.7	5.2
Middle East	5.4	9.1	2.2	7.9	13.9	2.4
Africa	2.6	4.7	2.5	2.3	5.3	3.6
World Total	**77.1**	**120.9**	**1.9**	**90.3**	**175.9**	**2.8**

Source: Based on EIA, *International Energy Outlook 2004* (April; 2004), tables A4 and A5, pp. 167, 186.

Eastern Europe and the former Soviet Union from 8.3 MMbbl/d to 3.7 MMbbl/d. Since 2000, however, economic prospects for the region are good and expected economic growth will lead to an increase in oil consumption with a projected increase at an annual average of 2.0 percent between 2001, and 2025, reaching 8.5 MMbbl/d in 2025. This expected rate is still well below the 9.0 MMbbl/d consumed in 1987.

The greatest increase in oil demand is expected for developing Asia. In 1985 China imported less than 800.000 tons of oil and oil products. In 2001 oil and oil product imports had increased to 5.0 MMbbl/d. China is the second largest oil consumer in the world behind the US. It is expected that in 2025 China's *aggregate* oil consumption may reach 46% of America's consumption level. Projections are that China's oil consumption will increase by 4.0 percent annually from 5.0 MMbbl/d in 2001, to 12.8 MMbbl/d in 2025.

Between 1962 and the present, India's per capita income growth lags behind growth in industrialized East Asia and industrializing China. In 2000, its population surpassed the 1 billion thresholds. Indian oil consumption is expected to grow by an annual average of 3.9 percent to almost 5.3 MMbbl/d in 2025. India imports about two-thirds of its crude oil requirements. For the rest of developing Asia, oil demand will increase at a slower rate for the projected period than during the 1990s. Central and South American, oil demand for the projected period will increase from 5.2 MMbbl/d to 9.2 MMbbl/d. However, the share of oil in the total energy demand in the region is declining due to substitutions of hydroelectric energy, natural gas, and coal. In the Middle East, oil demand will grow between 2001 and 2025 by an annual average of 2.2 percent from 5.4 MMbbl/d to 9.1 MMbbl/d. In Africa oil currently comprises 44 percent of total energy needs. It is projected that oil demand will increase from 2.6 MMbbl/d to 4.7 MMbbl/d between 2001 and 2025 (EIA 2004: 167, 186).

Global Gas Demand

Estimated global demand for natural gas will almost double from 90.3 tcf to 175.9 tcf between 2001 and 2025. It is projected that global natural gas consumption will rise by an annual average of 2.8 percent in the forecast period. In the developed countries gas demand will increase by an annual average of 1.5 percent. In North America, it is projected to increase by 2.2 percent per year, and in Western Europe by 2.4 percent. Western Europe, which holds less than five percent of the world's natural gas reserves, was responsible for 17 percent of the world's total gas consumption in 1999.

Between 2001-2025, industrialized Asia increases its demand for natural gas with an annual average of 1.5 percent. The increase is much slower than between 1970 to 1999, when gas demand in industrialized Asia increased by 11.2 percent per year. In the former Soviet Union and Eastern Europe, gas consumption will average a 2.9 percent annual increase in the forecasted period from 23.5 tcf to 46.4 tcf.

Developing Asia will account for an annual overage increase of 4.5 percent between 2001 and 2025 with China alone accounting for a 7.9 percent annual average increase. In Central and South America, the average annual growth rate in gas demand will be as high as 5.2 percent between 3.5 and 11.7. The Middle Eastern countries also seek to develop their domestic gas markets where consumption is expected to more than double in the projected period from 7.9 tcf to 13.9 tcf. Africa accounts for about 5 percent of the world's natural gas production but only consumes 2 percent of the world's demand. It is projected that African gas consumption will increase by an annual average of 3.6 percent from 2.3 tcf to 5.3 tcf between 2001 and 2025 (EIA 2004: 167, 186). Table 2.1 summarizes the trends underway.

Global Oil and Gas Reserves

The total global oil stock at the end of 2003 was estimated at 1,147.7 billion bbl proven oil reserves, 862.3 billion bbl in OPEC, and 285.4 billion bbl in non-OPEC countries (BP 2004). Fourteen countries account for 90 percent of the total global proven oil reserves: Saudi Arabia, Iraq, the United Arab Emirates (UAE), Kuwait, Iran, Venezuela, Russia, Mexico, the US, Libya, China, Nigeria, Norway, and the UK. Only five countries (Saudi Arabia, Iraq, the UAE, Kuwait and Iran) hold almost two-thirds of the global proven oil reserves.

Global natural gas reserves at the end of 2001 were estimated to be 6204.9 tcf. Almost 86 percent of global natural gas reserves are located in the Middle East and the former Soviet Union. The proven gas reserves for Azerbaijan, Iran, Kazakhstan, Russia and Turkmenistan are estimated at 2819.2 tcf, which is almost as much as the combined proven gas reserves of Europe, the US and the Middle East (see Table 2.2).

Iran and Russia alone account for about 41 percent of the global natural gas reserves (see table 2.3).[4] Due to its huge oil and gas reserves post-Soviet CEA has turned into one of the most important geopolitical areas in the world.

The Role of Caspian Oil and Gas in Global Oil and Gas Supply

The Caspian littoral states together hold one of the world's largest oil and gas reserves, which make them very significant in global markets. The estimates of proven oil and gas reserves in the Caspian region vary. For example, according to the US *Energy Information Administration* (EIA) the total proven oil reserves of the three new Caspian littoral states Azerbaijan, Kazakhstan and Turkmenistan are estimated at 7.2 billion bbl in 2002 and the total proven gas reserves at 170.4 tcf. The total proven oil reserves according to the *Statistical Review of World Energy* (BP 2004) is 16.5 billion bbl and the total proven gas reserves are 217.9 tcf. As has been estimated by energy consultant Wood Mackenzie, the five Caspian littoral countries (including only the Caspian off-shore sector of Russia and Iran) will have the potential to produce about 4 MMbbl/d by 2014 (McCutcheon, December 24, 2001). If the various oil projects boost production, then the Caspian region's oil exports might rise to 3 MMbbl/d in 2010 and an additional 2 MMbbl/d to 5 MMbbl/d in 2020

[4] For a detailed analysis of the role of the Caspian region in the global oil and gas market, see Amineh (2003: ch. 3).

Table 2.2

Proven Oil and Natural Gas Reserves in the Caspian Sea Region, Europe, US and the Middle East, 2003

Country	Proven Oil Reserves (billion bbl)	Proven Natural Gas Reserves (tcf)
Azerbaijan	7.0	48.4
Iran	130.7	942.2
Kazakhstan	9.0	67.1
Russia	69.1	1659.1
Turkmenistan	0.5	102.4
Total	**216.3**	**2819.2**
Europe	20.3	322
US	30.7	184.8
Middle East	685.6	2531.8
Total	**736.6**	**3038.6**

Sources: British Petroleum, *BP Statistical Review of World Energy 2004* (2004).

(EIA, February, 2002). At today's market prices, the potential oil reserves of the Caspian Sea zone have an estimated value of between US$2-US$4 trillion. The availability of the Caspian energy supplies on world markets will likewise enhance prospects for economic growth and political stability in the Caspian littoral countries (O'Connor, May 3, 1993).

Iran and Russia are the two main powers in terms of oil and gas reserves of the Caspian region and have the greatest energy reserves in the world. Iran is the world's second largest owner of proven natural gas reserves (estimated at 942.2 tcf) after Russia and ranking second in proven oil resources (11.4 percent, estimated at more than 136.7 billion bbl). In 2003, Iran produced 3.85 MMbbl/d. Russia's proven oil reserves are estimated at 69.1 billion bbl of oil (seventh largest in the world) and proven gas reserves at 1659.1 tcf (largest in the world). Russian oil production in 2003 was estimated at 8.54 MMbbl/d. Russia ranks second in oil production among the oil producing countries behind Saudi Arabia. Its gas production in 2003 was 578.6 billion cubic meters (BBcm) of gas (BP 2004). Russia is currently the world's largest gas producer.

Azerbaijan has been an important oil resource for more than a century. Azerbaijan's proven reserves of oil are estimated at 7 billion bbl and proven gas reserves at 48.4 tcf. After independence in 1991, Azerbaijan's oil production declined from 238,000 bbl/d to 180,000 bbl/d in 1997. Due to very substantial foreign investments in Azerbaijan's oil sector, this trend has been reversed. As shown in Table 2.4 output rose in 2003 to 313,000 bbl/d. It is expected that oil exports could exceed one MMbbl/d by 2010

Table 2.3

Top 20 Countries in estimated Oil and Natural Gas Reserves, 2003

	Regions and countries	Proven Oil Reserves		Proven Natural Gas Reserves[a]	
		Billion bbl (Share of world total in percent)	World Rank	Tcf (Share of world total in percent)	World Rank
	Caspian Sea countries				
1	Russia	69.1 (6.0)	7	1659.1 (26.7)	1
2	Kazakhstan	9.0 (0.8)	15	67.1 (1.1)	14
3	Azerbaijan	7.0 (0.6)	16	n/a	–
4	Turkmenistan	n/a	–	102.4 (1.6)	11
5	Iran	130.7 (11.4)	2	942.2 (15.2)	2
	Developed countries				
6	United States	30.7 (2.7)	10	184.8 (3.0)	6
7	Norway	10.1 (0.9)	14	–	–
8	Canada	16.0 (1.4)	12	58.7 (0.9)	18
9	The Netherlands	–	–	58.8 (0.9)	17
	Developing countries				
10	Saudi Arabia	262.7 (22.9)	1	235.7 (3.8)	4
11	Iraq	115.0 (10.7)	3	109.7 (1.8)	10
12	United Arab Emirates	97.8 (8.5)	4	213.9 (3.4)	5
13	Kuwait	96.5 (8.4)	5	55.0 (0.9)	19
14	Uzbekistan	–	–	65.3 (1.1)	15
15	Venezuela	78.0 (6.8)	6	146.5 (2.4)	9
16	Libya	36.0 (3.1)	8	46.4 (0.7)	20
17	Mexico	16.0 (1.4)	12	–	–
18	China	23.7 (2.1)	11	64.4 (1.0)	16
19	Nigeria	34.3 (3.0)	9	176.4 (2.8)	7
20	Algeria	11.3 (1.0)	13	159.7 (2.6)	8
21	Brazil	10.6 (0.9)	14	–	–
22	Angola	8.9 (0.8)	15	–	–
23	Oman	5.6 (0.5)	17	–	–
24	Qatar	–	–	909.6 (14.7)	3
25	Malaysia	–	–	84.9 (1.4)	13
26	Indonesia	–	–	90.3 (1.5)	12
	World Total	**1,147.7 (100)**		**6204.9 (100)**	

Source: British Petroleum, *BP Statistical Review of World Energy* (2004).

and two MMbbl/d in 20 years. Azerbaijan's natural gas production was 4.8 BBcm in 2003 (Table 2.5). This production is rather low, due to the country's lack of a suitable infrastructure to deliver natural gas to markets. Given the necessary infrastructure, it can be expected that Azerbaijan's natural gas production could increase to as much as 600 billioncubic feet (bcf) by 2010.

Table 2.4

Caspian Sea Region Oil Production and Exports

Country	Production			Net Exports		
	Production, 1993 Mbbl/d+	Production, 2003 Mbbl/d+	Possible Production, 2010 Mbbl/d# [high]	Net Exports, 1990 Mbbl/d†	Net Exports, 2001 Mbbl/d†	Possible Net Exports, 2010 Mbbl/d†
Azerbaijan	209	313	1,140	77	175.2	1,000
Kazakhstan	490	1,106	2,400	109	631	1,700
Iran	3,712	3,852	0*	0*	0*	0*
Russia	7,173	8,543	150*	0**	7**	300**
Turkmenistan	92	210	964	69	107	150
Total	**11,676**	**14,024**	**4654**	**255**	**920.2**	**3150**

Notes:
*Only the regions near the Caspian Sea are included.
+Based on *BP Statistical Review of World Energy 2004* (June 2004).
#Based on EIA. *Caspian Sea Region: Key Oil and Gas Statistics* (August 2003).
†Based on EIA, *Caspian Sea Region: Reserves and Pipelines* (July 2002).
Sources: British Petroleum, *BP Statistical Review of World Energy 2004* (June 2004); EIA, *Caspian Sea Region: Key Oil and Gas Statistics* (August 2003); EIA, *Caspian Sea Region: Reserves and Pipelines* (July 2002).

Table 2.5

Caspian Sea Region Natural Gas Production and Exports

Country	Production			Net Exports		
	Production, 1993 (BBcm), per year+	Production 2003 (BBcm), per year+	Possible Production, 2010 (bcf)#	Net Exports, 1990 (bcf)†	Net Exports, 2000 (bcf)†	Possible Net Exports, 2010 (bcf)†
Azerbaijan	6.3	4.8	600	–272	00.0	500
Kazakhstan	6.2	12.9	1,700	–257	–176	350
Iran	27.1	79.0	n/a**	n/a**	n/a**	n/a**
Russia	576.5	578.6	N/A**	N/A**	N/A**	N/A**
Turkmenistan	60.9	55.1	4,200	2,539	1,381	3,300
Total	**677.0**	**730.4**	**6100**	**2010**	**1205**	**4150**

Notes:
**Only the regions near the Caspian are included.
+Based on *BP Statistical Review of World Energy 2004* (June 2004).
†Based on EIA, *Caspian Sea Region: Reserves and Pipelines* (July 2002).
#Based on EIA. *Caspian Sea Region: Key Oil and Gas Statistics* (August 2003).
Sources: *BP Statistical Review of World Energy 2004* (June 2004); EIA, *Caspian Sea Region: Reserves and Pipelines* (July 2002); EIA, *Caspian Sea Region: Key Oil and Gas Statistics* (August 2003).

Kazakhstan has much larger reserves than were estimated during the Soviet period. Kazakhstan is considered, after Russia, to be the richest of the former Soviet republics in oil resources, with proven oil reserves of 9 billion bbl and also an enormous natural gas reserve, estimated at 67.1 tcf. Kazakhstan's oil production dropped to 415,000 bbl/d during the first few years following the collapse of the Soviet Union. Foreign investments in Kazakhstan's oil sector have helped the country boost its oil production to 1.1 MMbbl/d in 2003. Production is expected to reach 1.2 MMbbl/d in 2005, 2.4 MMbbl/d by 2010, and as much as 2.5 MMbbl/d by 2015. Kazakhstan exported 631,000 bbl/d of oil in 2001. The country's remoteness from world markets, along with its lack of export pipelines, has hindered faster growth of exports. In 2001, most of Kazakh oil exports were shipped mainly via the Atyrau-Samara pipeline through Russia, with additional supplies being shipped by rail and by barge across the Caspian Sea.

Kazakhstan's gas industry is significantly under-developed and hampered by a lack of infrastructure. In August 1999 the Kazakh government passed a law requiring TNOCs to include natural gas utilization projects in their development plans. As a result, Kazakhstan increased its natural gas production to 314.3 bcf in 2000 the highest level in the past decade and to 12.9 BBcm of natural gas in 2003. If the domestic natural gas demand remains stable gas production is expected to reach 1,700 bcf in 2010.

Turkmenistan has one of the world's major natural gas reserves and also significant oil reserves. According to recent investigation, it is claimed that the country's possible gas reserves may be as high as 102.4 tcf and its proven oil reserves 0.5 billion bbl. After independence, oil production decreased to 81,000 bbl/d in 1995 and then almost doubled to 156,400 bbl/d in 1999. In 2003, Turkmenistan produced 210,000 bbl/d. The expected oil production for 2010 is 964,000 bbl/d. In Turkmenistan, the production of natural gas fell sharply in the first decade after in dependence. In 2003 the country produced 55.1 BBcm. The recent trend is positive-mainly due to a major gas export deal with Russia and the resumption of supplies to Ukraine — and the country is expected to produce up to 4,200 bcf in 2010 (BP 2004).

It is expected that without the Caspian exports, oil exports from the Persian Gulf to Europe would increase by 0.5 MMbbl/d in 2010. If the Caspian region fully participates in the market exports, oil from the Persian Gulf to Europe will have decreased to 1.5 MMbbl/d by 2010 (Emerson 2000: 178, 184).

Are Proven Reserves Able to Satisfy Demand?

Above we distinguished supply-induced shortage from demand-induced scarcity. Supply-induced scarcity arrives from the moment the world stock begins to decrease. Experts differ on the timing of arrival of supply-induced scarcity, though not by very wide margins. In the mid-1990s analysts began to take a serious look at the projection methods based on fitting data to growth curves whose underlying generating mechanisms are well studied. Hubbert used these methods in the mid-1950s to anticipate US domestic oil output peak. Most experts hold the view that the peak in world production will arrive suddenly and somewhere in the end of this decade, early next decade. Demand for gas is expected to peak at the end of the century. Oil prices expressed in terms of the quantity of oil required to produce and transport to consumers a barrel of oil multiplied between 1950 and mid 1980's from 3 litres to 20 litres.

According to estimates of the *Energy Information Administration,* world oil supply in 2020 will exceed the 2000 level by 41 MMbbl/d. Production increases are expected for both OPEC and non-OPEC countries (EIA 2002, 31). The rise in non-OPEC oil supply over the last two decades has resulted in a substantial decline of OPEC's market share, once at a historic high of 52 percent in 1973. However, it is projected that by 2020 only about one-third of the total oil production increase will come from non-OPEC areas. OPEC oil production is expected to reach 57.2 MMbbl/d in 2020, with oil supply growing at an annual average rate of 3.3 percent. Its capacity utilization will increase immensely after 2000, reaching 95 percent in 2015.

Momentarily, the OPEC sees itself in a dilemma, especially in regard with the uncertainty concerning the future of Iraq within the organization. Iraq could be the world's second largest supplier of crude after Saudi Arabia. It has 115 billion bbl of crude oil in reserve and OPEC worries that the world market might demand more oil from Iraq. OPEC fears that a rise in Iraq oil supply could drown markets, forcing prices to slump. OPEC would like to see prices balanced between US$22 to US$28 per bbl (AGOC, May 15, 2003).

In 2000, the industrialized countries imported 15.8 MMbbl/d from the OPEC countries, 9.9 MMbbl/d of which came from the Persian Gulf region. OPEC members exported 70 percent of their oil exports to industrialized countries, of which almost two-thirds came from the Persian Gulf region. It is expected that OPEC's exports to industrialized countries in 2020 will be about 6.2 MMbbl/d higher than in 2000. More than half of this increase will come from the Persian Gulf countries. However, despite this growth, the share of OPEC total petroleum exports to industrialized countries in 2020 will be 14 percent below the share in 2000. Persian

Gulf exports to industrialized countries will fall to about 40 percent. At the same time OPEC oil exports to developing countries will increase by more than 17.0 MMbbl/d between 2000 and 2020, half of which will go to developing Asia. China alone is expected to import about 7.2 MMbbl/d from OPEC by 2020, most of which will come from the Persian Gulf region (EIA 2002: 31).

Non-OPEC oil supply is expected to increase steadily from 46.0 MMbbl/d in 2000 to 61.1 MMbbl/d in 2020 (Emerson 2000: 175-76). For the period 1998-2010, the three new Caspian littoral states Azerbaijan, Kazakhstan, Turkmenistan alone will account for 18 percent of the total increase in non-OPEC production. The North Sea will account for 4 percent of the total increase, Latin America for 9 percent, and Africa for 14 percent (Emerson 2000: 174).

As the Middle East is politically unstable, alternative oil resources will be important for reducing dependence on this region. However, the shift in oil production from the Persian Gulf to other areas does not guarantee that the new sources will be more secure. Colombia and Nigeria have recently experienced considerable internal violence, and Venezuela is undergoing a difficult political transition (Klare 2001: 46). China has similar experiences. To protect its oil supplies, China has tightened its hold on the Xinjiang Uighur Autonomous Region (XUAR). Because of the high concentration of an ethnic minority population, the Chinese leadership views the Xinjiang region as particularly susceptible to foreign anti-Chinese influences. It fears that the radical Islamic and separatist forces operating in CEA could stir up separatist aspirations of minority groups in China. Xinjiang is important to China for various reasons. Instability in Xinjiang could undermine China's control of the region and thus threaten the integrity of the country as a whole. The region has vast open spaces and a relatively small population that makes it perfect for nuclear testing and large-scale conventional military exercises of the People's Liberation Army. Xinjiang is a significant domestic source of oil and gas.[5] Bordering Mongolia, Russia, Kazakhstan, Kyrgyzstan, Tajikistan, Pakistan, and India makes the region an important springboard for China to strengthen its influence on other countries (Amineh 2003: ch. 5; see also Amineh 1999).

The central question is how conflicts and social struggle around the Caspian oil and gas resources will be articulated and who will gain the upper hand. Will multilevel conflicts and cooperation in the region be controlled by the US as the dominant military actor in cooperation with

[5] With estimated oil reserves of 20.9 billion tonnes and natural gas deposits of 10.3 trillion cubic meters, the Xinjiang region could develop into China's second largest oil producing region (Chan, J., January 3, 2001).

local governments? How will Russia and China respond to its policies? Will transnational battles by non-state actors force the major powers involved in the region to cooperate and collectively impose a local peace order to their common advantage? One thing is sure, an increase in oil production and export will be essential for economic prosperity and political stability in the Caspian region.

References

ALEXANDER'S GAS AND OIL CONNECTION
2003 "Uncertainty of Iraq's Future Adds to OPEC's Dilemma."
AMINEH, M.P.
2003 *Globalization, Geopolitics, and Energy Security in Central Eurasia and the Caspian Region.* Den Haag: Clingendael International Energy Programme.
1999 *Towards the Control of Oil Resources in the Caspian Region.* New York: St. Martin's Press.
BRITISH PETROLEUM
2004 BP Statistical Review of World Energy (*http://www.bp.com*).
CHAN, J.
2001 "China Pushes into Central Asia for Oil and Gas." *World Socialist Web*, January 3.
O'CONNOR, R. ET AL.
1993 "Future Oil and Gas Potential in the Southern Caspian Basin." *Oil and Gas Journal*, May 3.
CUMINGS, B.
1984 "The Origins and Development of the Northeast Asian Political Economy: Industrial Sectors, Product Cycles and Political Consequences." *International Organization*, vol. 38(1).
MCCUTCHEON, H. AND R. OSBORN
"Risks Temper Caspian Rewards Potential." *Oil & Gas Journal*, December 24.
HOMER-DIXON, T., AND BLITT (EDS.)
1998 *Ecoviolence: Links Among Environment, Population and Security.* Boston: Rowan & Littlefield.
EIA
2002 *International Energy Outlook 2002,* on-line version. Retrieved August 20, 2002 (*http://www.eia.doe.gov/oiaf/ieo/*).
2003 Caspian Sea Region: Key Oil and Gas Statistics (*http://www.eia.doe.gov*).
2002 Caspian Sea Region: Reserves and Pipelines (*http://www.eia.doe.gov*).
EMERSON, S.A.
2000 "The Relevance of Caspian Oil for the World Market," in *Caspian Energy Resources: Implications for the Arab Gulf.* Abu Dhabi: ECSSR.
JACOBS, J.
1994 *Systems of Survival: A Dialogue on the Moral Foundations of Commerce and Politics.* New York: Vintage Books.
KLARE, M.T.
2001 *Resource Wars—The New Landscape of Global Conflict.* Markham, Ontario: Metropolitan Books.

PART TWO

LOCAL DYNAMICS

III. Nation-State Building in Central Asia: A lost Case?

PINAR AKÇALI

ABSTRACT

With the collapse of the Soviet Union in 1991, the five Central Asian republics of Kazakhstan, Kyrgyzstan, Tajikistan, Turkmenistan, and Uzbekistan have entered a period of nation-state building, which had been started to a large extend by the political elites of the former Soviet Union. These republics were not prepared for independence that came to the region suddenly. The former communist leaders of the Soviet era became the new national elites to take their countries by publicly declared goals and policies through the path of independent nationhood and independent statehood. However, it remains unclear whether this top-to-bottom approach will prove successful in the long run. This article discusses nation-state building in the region by first looking at problems of external sovereignty. Second, domestic state building policies and structures, more specifically, the newly formulated official discourse on nation-building and the political-legal framework to develop that discourse, are analyzed. Then, the limitations of this process with specific emphasis on supranational identities (basically religious identity of Islam), subnational identities (local and/or tribal identities), and ethnic minorities (with a specific on the Russians in these five countries) are examined. It is concluded that the process of nation-state building in Central Asia is not complete yet and that each republic has unique problems that may challenge this process. For the time being, there exist certain frictions between the goals of the official discourse and nonofficial levels of identity that may hinder the success of the nation building process in the region.

General Introduction

When the Soviet Union collapsed in 1991, the five Central Asian republics of Kazakhstan, Kyrgyzstan, Uzbekistan, Tajikistan, and Turkmenistan were not prepared for independence, which came to the region suddenly, even as a surprise event, thus without any national struggle or demand. The local administrations were not equipped with necessary fiscal, military, political, or economic framework to deal with this newly gained status of independence. There was now the need of transformation from being one of the Soviet Socialist Republics of the former Soviet Union to the legal status of independent statehood, without having in place the institutions of a nation-state. Furthermore, establishing secular and democratic regimes were considered to be the *sine qua non* of these new nation-states.

At the beginning of this era, ex-communist Central Asian political and bureaucratic elites had to transform themselves into national elites protecting the interests of the newly independent states. All of the post-Soviet era leaders of Central Asia, with the exception of Askar Akaev of Kyrgyzstan, were the First Secretaries of the Communist Parties in their own republics. [1] They were now the presidents of the newly independent countries and the former Communist Parties were turned into republican parties.

The successful model for the post-Soviet transition was the nation-states of Western Europe, which, historically speaking made nation and state spatially congruent (Smith 1998: 17). So a process of nation state-building from the top (that is the state apparatus) would be started in the five Central Asian republics. However, today, it remains unclear whether this top-to-bottom approach will prove successful in the region. When a distinction is made between nation-state building by the state apparatus and nation-state building from below, it is possible to observe certain frictions between government initiative at centralizing the power of administration and responding society. This article attempts to analyze these frictions in the case of post-Soviet Central Asia. In the first part, nation-state building in the region will be analyzed by its external and domestic dimensions. Here, the emphasis will be on the problems of external sovereignty and domestic state building policies and structures. In the second part, limitations of the process of nation state building in Central Asia will be analyzed; the issue will be examined from the perspectives of supranational identities, subnational identities, and ethnic minorities. The conclusion of the study will highlight frictions between the official discourse and the nonofficial levels of Central Asian identity in the post-Soviet era.

---

[1] Even Askar Akaev is considered by some scholars to be a member of the Soviet era elite (Gleason 2002).

Problems of External Sovereignty

When Central Asian countries became independent, there were several issues to be solved regarding external sovereignty. First of all, it has been pointed out that these countries suddenly found themselves in a "confusion of preference and influence" (Bacık 1999: 62). The debate, coming out of this uncertainty, evolved to a large extent around the existence of several developmental models that have been discussed regarding Central Asia. These models would range from the Turkish secular political model versus Iranian theocratic model on the one hand, and the Chinese model of gradual economic reform versus Russia's shock therapy approach on the other. Some suggested East Asian economic models adopted in Japan and South Korea. There was even the alternative of the "Kuwaiti model" for Turkmenistan that would include a strong, antidemocratic tradition, exchanging political submission with material and financial wealth for the people. [2]

Second, Central Asian countries had important problems of border demarcation, uncontrolled border crossings, and territorial claims coming from the neighboring countries, not only with some of the regional powers such as Russia and China, but also among themselves. For example, those Central Asian countries, which share borderlines with China, attempted to solve their problems with several accounts, agreements, and declarations. China got commitments from Central Asian republics about not supporting the Muslim separatists in the Xinjiang Uygur Autonomous Region. In exchange, the Chinese government promised to respect the current borders of Central Asian republics. [3]

Another important issue is related to the border with Afghanistan, which is still protected by Russian troops. Despite such protection, there is a big problem of drugs and weapons trafficking in the area. Nevertheless, if Russia withdraws completely from the area, its departure is expected to have disastrous effects for the fragile status quo in the region (Rumer 2000: 77). Finally, the border problems between Central Asian states themselves are not completely solved. Turkmen-Uzbek, Uzbek-Kazakh, Uzbek-Tajik, Uzbek-Kyrgyz and Kyrgyz-Tajik borders are still disputed,

---

[2] For a detailed discussion on these models, see Rafis Abazov (1998), "Central Asian Republics' Search for a Model of Development," The Slavic Research Center, from the web page: *http://src-h.slav.hokudai.ac.jp/publicn/CentralAsia/rafis/rafis/html*

[3] For more information on these accounts, agreements, and declarations, the following websites can be visited: *http://www.fmprc.gov.cn/eng/4372.html*, *www.stimson.org/cbm/china/crplus.htm*; *http://english.peopledaily.com.cn/200109/14/eng20010914_80232.html*; *http://english.pravda.ru/main/2002/05/22/29121.html*; *http://russia.shaps.hawaii.edu/fp/russia/joint-statement980703.html*—11k.

and these countries have attempted to solve these issues by mutual talks, memorandums, and decisions. In other words, "popular usage has not legitimized many of these borders" (Gleason 2001a: 1090). According to a diplomat, who lives in Kyrgyzstan and asked not to be identified, the borders among the Central Asian countries have still to achieve their final status due to these boundary problems (Balbay 2001: 130).

Another important problem concerns the relationships with regional and nonregional actors. For the time being, Russia and China seem to be the most important regional powers for Central Asian republics. Being Russia's near abroad (*blizhnee zarubezh'e*), the Central Asian states are still quite dependent on Russia in economic, political, military, and cultural terms. According to one observer, "Not one of the countries seeking to carve out of a sphere of influence in Central Asia is ready for Russia to leave" (Rumer 2000: 10). China, on the other hand, being perceived by its neighbors as the "giant, nontransparent, dogmatic neighbor with great power ambitions" has interest in the region (Rumer 2000: 11). To these two most powerful countries, one may add the United States, the most important geopolitical rival of these two regional powers. According to the President of Kazakhstan, Nursultan Nazarbaev, the United States will "always be interested in everything that is going on not only inside of these states but also beyond them."[4] After the September 11 attacks, the US interest and actual presence in the region increased dramatically, and a new pattern of relations in economic and military terms has started to develop. Kazakhstan, Tajikistan, and Uzbekistan have offered their territories for the basing of US troops, whereas Kyrgyzstan and Turkmenistan have tendered airspace rights (Starobin 2001: 58-61).

All these problems created a serious doubt about the extent to which these countries would be able to establish independent nation states, which enjoy external sovereignty. However, it is possible to suggest that, despite all these problems, Central Asian countries made serious attempts to be involved in the international area as sovereign states on their own. For example, on the occasion of the eighth anniversary of Turkmenistan's independence, President Saparmurad Niyazov Turkmenbashi characterized the five Central Asian countries as "reliable and adequate partners" in the international arena. According to him, the global community "should realize that Central Asian countries are no longer pawns in someone else's

[4] From an interview made with Kevin Baerson and Barbara Ferguson of the *Washington Times* (*http://www.internationalspecialreports.com/cicentralasia/99/kazakhstan*).

game. They are carrying out, and will be doing so in the future, their own policies in international affairs."[5]

In the post-Soviet era, the Central Asian countries did join several major international organizations, including the United Nations, the World Bank, and the International Monetary Fund. Furthermore, since they would have different foreign policy priorities, these countries did not follow any particular foreign policy line, but instead, adopted what Kazakh President Nazarbaev described as "multi-vectored" attitude. This concept referred to the prescription to create stable relations and partnerships both with the closest neighbors and with more developed countries, which are of great importance in world politics.[6] It is also possible to observe the different directions in the foreign policies of the five Central Asian countries. For Kazakhstan and Kyrgyzstan, relations with Russia are still of primary importance, but they have also entered into contact with other powers as well. For Tajikistan, there was an inevitable dependence on Russia, due to the civil war. After the war, this dependence can be expected to decrease. As was indicated above, Turkmenistan follows the policy of neutrality and avoids regional cooperation attempts of any type. Finally, Uzbekistan tries hard to get out of Russia's sphere of influence and to be accepted as a regional leader. These different orientations have resulted in the emergence of two blocs within the Commonwealth of Independent States (CIS). The first bloc was pro-Moscow and included Belarus, Armenia, Kazakhstan, Kyrgyzstan, and Tajikistan. The second bloc was closer to the United States and NATO and included Georgia, Ukraine, Uzbekistan, Azerbaijan, and Moldova (GUUAM). Therefore, the two main states of Central Asia, Kazakhstan and Uzbekistan, now belong to different alliances (Rumer 2000: 11).

The Central Asian countries are also involved in regional cooperation attempts such as the Central Asian Union and the Shanghai Cooperation Organization. However, even these attempts themselves were realized as "independent nation-states" on the part of these countries. According to a prominent regional policymaker, the late Umirserik Kasenov of Kazakhstan, even such regional cooperation was actually for the end purpose of reinforcing the sovereignty of the newly emerged states (Gleason 2001: 1078). It is also important to point out that, Turkmenistan, right from the beginning of its independence, pursued a policy of neutrality and stayed away from any type of alliance. However, Tajikistan could not start

[5] See Human Rights Watch Annual Report 2002 at *http://www.hrw.org/wr2k1/europe/turkmenistan.html*

[6] From an interview made with Kevin Baerson and Barbara Ferguson of the Washington Times (*http://www.internationalspecialreports.com/cicentralasia/99/kazakhstan*).

any meaningful initiative for regional cooperation for a long time, basically because of the civil war between 1992-1997.

Domestic State Building Policies and Structures

After 1991, the Central Asian republics attempted not only to achieve external sovereignty as separate and internationally recognized nation-states, but they also adopted certain domestic policies and established certain structures to facilitate the process of nation-state building. In other words, these republics had to adopt a policy of "de-Sovietization," and to that end, the nationalizing elites have been removing the previous symbols and political representatives belonging to the Soviet era and replacing them with new "national" symbols, institutions and practices. For the political elites, distancing themselves from the previous regime and adopting national codes was inevitable (Smith 1998: 14). There were basically two simultaneously adopted mechanisms: one was the *official discourse* developed on the theme of being new "nations"; the other was the *political-legal framework* to form the base of such discourse. Both of those mechanisms were expected to mutually strengthen each other and facilitate the transition process.

Official Discourse on Nation

One rather unique and interesting thing about Central Asian nation-state building process is the fact that the carving out of new polities actually started during the Soviet era. Before the establishment of the Soviet Union, the Central Asian people did not exist as "nations" but rather as loose ethnic groups under clan leadership. With the Soviet policy of national delimitation (*natsional'noe razmezhevanie*) between 1924-1936, the five Central Asian republics were created with separate boundaries. This was the first time the concept of territorial-based nationality (*natsionalnost*), as introduced by Stalin, was implemented (Roy 2000: 64). This stage, however, would be only temporary because the final goal was to create the new Soviet person (*novy Sovetski chelovek*) for whom national attachments would not be meaningful. Accomplishing this goal would result in the emergence of the Soviet people (*Sovietski narod*), a common identity for all the citizens of the country, including the Central Asians (Abazov 1998).

During the Gorbachev years, the Central Asian elites started to feel the necessity of establishing their own "national" identities, and they slowly started to leave aside the ideal of *Sovietski narod*, which was basically "a dream." In reality, the Soviet Union was a political system in which many different civilizations did exist together with deep divisions among themselves (Nazarbaev 1997: 33).

The official discourse on nation-state building, therefore, had to respond to this reality. The new state builders focused basically on three issues: 1) history, 2) new alphabets and new linguistic policies, and 3) national glory and pride. *History* refers to the rediscovery, that is, the creation of the "national" past. During the Soviet era, the official historiography put severe censorship and control over works of Central Asian history. The censor only allowed Marxist interpretations of the region's past, banning the name "Turkestan" and many of the accounts of the Jadids. Similarly, it was officially frowned upon, if not totally forbidden, to write about Central Asian historic figures such as Tamerlaine, Shaybani Khan, Amir Temur, and Chinggis Khan (Allworth 1998: 73-74). In some cases, certain national and historical figures of Central Asia were declared "enemy of the people" (Bacık 1999: 94).

In the post-Soviet era, there is a great deal of effort by the Central Asian historians to rewrite their history devoid of the old version of Soviet ideological interpretations. There is a big emphasis on the past events, former historic figures and Central Asia's role in the general history of the world. One Kazakh scholar, for example, would suggest that the Turks "made history but they could not write it" (Abdülvahap 2002). Therefore, it was now the time to start writing "national" histories.

One other basic issue that is being constantly emphasized has to do with the adaptation of *new alphabets and new linguistic policies*. The Central Asian republics want to reduce the impact of Russian language and have discussed adopting Latin or Arabic script instead of the Cyrillic one. Turkmenistan and Uzbekistan switched to the Latin alphabet in 1993, but for the other countries, this is not achieved yet. Kazakhstan and Kyrgyzstan have considerable Russian minorities, and the policy of linguistic Russification was strongest in these two republics. Thus, for the time being, the issue is simply postponed. As for Tajikistan, this country could not deal with such policies adequately because of the pressures of civil war and the restoration period afterwards.

One important implication of new linguistic policies had to do with the institutionalization and promotion of the titular language in state bureaucracy and politics. There has been a serious debate as to the ascription of official status to the titular language. Central Asian republics have passed several laws to that end. Even in Kazakhstan and Kyrgyzstan, titular languages were accepted as state languages. In fact, in Kazakhstan the first law, which accepted Kazakh as the "state language," goes back to 1989 (Büyükakıncı 2002: 363). However, after independence, Kazakhstan and Kyrgyzstan gave Russian official status by accepting it as the language of "interethnic communication" (Gürsoy-Naskali 2002: 57). In today's Central Asia, there is a big emphasis on renaming the cities, public squares, streets, and official buildings by dropping their Russian names and replacing them with national names. This linguistic policy is a direct outcome of the official discourse on nation-state building process.

Finally, the Central Asian leaders put great effort in creating an atmosphere of *national glory and pride*. The most obvious example of this can be observed in the national flags of these countries. New state flags contain national symbols. For example, the Kazakh flag has a national ornamentation on the hoist side, and an eagle under the big sun symbolizing Kazakh nomadic traditions. The Kyrgyz flag, which depicts a yellow sun with forty rays and the *tiunduk* (the top part of the Kyrgyz yurt), represents the forty Kyrgyz tribes united under one common history (Caspiani 2000: 236). The Turkmen flag has five *guls* (national design of Turkmen carpets) as symbols of unity among the five major Turkmen tribes (Bohr 1998: 145).

The Central Asian leaders have also placed great importance on declaring several national days to be celebrated, such as those dates on which their countries declared their sovereignty and independence. There are many holidays and festivals. In this sense, Turkmenistan provides a rather unique and extreme case. There are more than fifteen new national holidays, including Turkmen Racehorse Day, Turkmen Carpet Day, and even Turkmen Melon Day (Bohr 1998: 145). Turkmenistan also puts emphasis on its status as a "neutral" nation state. The twelfth of December is celebrated as the "Neutrality Day," as on that day in 1995, the General Assembly of the United Nations approved a resolution recognizing Turkmenistan's neutrality (Ochs 1997: 350).

Central Asian leaders also glorify their newly created national and historical events, heroes, and experiences. For example, for the Kyrgyz leadership the celebration of the age-old epic trilogy of Manas was a big event. Similarly, the Uzbeks glorify Amir Temur and the Kazakhs glorify the Kazakh Khanate of the fifteenth century. What matters here is the fact that the Central Asian leaders seem to be emphasizing national identity a lot *more* than the other identities. This is more so for the leaders of Uzbekistan and Turkmenistan. The Uzbek leadership tries to restore national tradition by setting certain limits as to who can be included in the Uzbek nation and who cannot. In this sense, the unique example of the so-called "Mahalla Fund" can be given. *Mahalla*, the traditional Uzbek neighborhood community, is now given an official recognition, having several new political, economic, and financial responsibilities and powers. The Fund is a new state body established at the national level headed by the Uzbek President Islam Karimov himself. As for the Turkmen President Saparmurad Niyazov Turkmenbashi, being "Turkmen" is more important than being a member of a particular tribe. In his book *Address to the Peoples of Turkmenistan* published in 1994, he says: "To have our state united in the future we must completely eradicate the epidemic habit of talking about tribal relations. No matter what tribes we come from, we remain ... sons of the one big family of Turkmenistan" (quoted in Ochs 1997: 317). It has further been pointed out that the personality cult created around the President himself can be seen "as a peculiar nation-building enterprise, a device to address concerns about Turkmenistan's unity when people still tend to see other as Tekke or Yomud, as opposed to Turkmen" (Ibid: 330).

Political-Legal Framework

Even though nation building in Central Asia was first started in the 1920s by the Soviet regime, in the post-Soviet era it acquired a new dimension that had little resemblance to its earlier version. After independence, the Central Asian countries, in addition to nation building, also started state building. This was unprecedented in Central Asian history, since there did not exist the practice of territorial rule in the highly mobile steppe societies. The Soviet state was the first one to impose a system of territorial governance. Central Asia was carved up in administrative units of the Soviet Union. States in the real sense of the term did not exist in the region prior to 1991. In the post-Soviet era, nation building and state building go hand in hand, as there is now an attempt to build an independent state that derives its legitimacy and support *from* the nation.

Part of this project is the adoption of national constitutions, one after the other. Turkmenistan led the way in 1992, followed by Uzbekistan, which adopted a new constitution in December of the same year. Kazakhstan accepted its constitution in January 1993 and Kyrgyzstan in May 1993. Even in Tajikistan, where civil war brought the country into near collapse, a new constitution was adopted in November 1994.

These constitutions would explicitly state in their preambles that the countries were based on the principle of being nation states. For example, in the preamble of the Turkmen Constitution, it is pointed out that the people of Turkmenistan adopted the constitution "possessing the goal of protecting the national values and interests and securing the sovereignty of the Turkmen people." Similarly, the preamble of the Kyrgyz Constitution states that the constitution is adopted for "endeavoring to ensure the national rebirth of the Kyrgyz" and that it confirms the devotion to the rights and freedoms of the people and the idea of *national statehood.*

These national constitutions have also established the legal foundation for new legislative, executive, and judiciary bodies. However, in all Central Asian republics, the presidents are the most powerful political actors. It is appropriate, therefore, to consider these countries as presidential republics. Kazakhstan's *Ulu Kenges* (Parliament), Kyrgyzstan's *Zhogorku Kenesh* (Parliament), Turkmenistan's *Halk Maslahaty* (The People's Council),

Tajikistan's *Majlisi Oli* (Supreme Legislature), and Uzbekistan's *Oli Majlis* (Supreme Legislature) are all basically less powerful than the presidents of their countries, since they do not have the de facto power to withhold consent to decision making in the executive branch of government.[7]

Limitations on the Process of Nation-State Building

Even though the Central Asian political leaders have attempted to develop an official discourse on nation-state building and adopted a new political legal framework to that end, there are certain limitations placed on this official goal coming from the people themselves. These limitations are related to *religious fundamentalism, local and/or tribal identities,* and *the minority groups.*

Religious fundamentalism

Many scholars who analyzed Islam in Central Asia have agreed on the fact that Islam has considerable influence in historical, cultural, and traditional terms in the region. Overall, religion is believed to have an impact on the life-cycle rituals of marriages, births, circumcisions, deaths, and funerals. In Central Asia, families are more traditional in the sense that they are larger, bringing two to three generations together. Females usually marry off young (average age being 16-17), and divorce rates are low. In many cases, there is strict morality in matters concerning gender relations (Rywkin 1990: 89-90).

In the new discourse of state building, Islam is acknowledged as an integral part of the cultural heritage. But it is rejected as a guiding principle of public or political life. Nevertheless, there are important differences in Islamic practices and understandings not only among different countries but also within the same country between different regions in Central Asia. In general, it is possible to claim that traditionally nomadic people—the Kazakhs, the Kyrgyz, and the Turkmen—have less religiously defined cultures throughout history. The sedentary populations of the region, basically the Uzbeks and the Tajiks, on the other hand, were always more religious, having established Islamic centers such as madrasas and mosques.

Depending on the regions, religious practices may also vary within one country as well. The typical examples of this can be seen in Kazakhstan and Kyrgyzstan. In Kazakhstan, generally speaking, in the northern parts

[7] However, it must be clearly stated that the degree to which presidential rule is applied shows important variations in Central Asian republics. In Turkmenistan and Uzbekistan, Turkmenbashi and Karimov are much more authoritarian than Nazarbaev's Kazakhstan and Akayev's Kyrgyzstan. In Tajikistan, opposition parties are in the coalition government.

of the country people are less religious, whereas in the south they identify themselves more with Islam. The southern Kazakhs have been more sedentary since the eighteenth century and they have been more influenced by the Uzbek and the Arabic tradition of madrasa (Büyükakıncı 2002: 359). There is a similar division in Kyrgyzstan. The northern parts have been more cosmopolitan and less religious, whereas the southern parts have been more under the influence of the settled people. As such, they have been less cosmopolitan and more religious.

Since independence, countries with more religious traditions, that is, Uzbekistan and Tajikistan, started to move in the direction of more radical versions of Islam. In Tajikistan, as early as 1990, two leaders of the underground Islamic Revival Movement, Davlat Usmon and Said Ibrahim Gado, formally requested the permission of the authorities to hold a regional founding congress of the all-Union Islamic Renaissance Party (IRP) in the capital city of Dushanbe. Even though this request was rejected, the Tajik IRP was formed, and in a short time it became one of the most influential opposition groups in Tajikistan. During the anti-government demonstrations in the spring of 1992, the Tajik IRP played the leading role, as most of the demonstrators were the party's supporters. After the dissolution of the Soviet Union, the Tajik IRP would emerge as the most powerful opposition party, being one of the contenders in the Tajik Civil War. The party would also play a major role in the negotiations process between the opposition groups and the authorities, and become one of the major partners of the newly established government after the signing of the peace accord in 1997.[8]

Uzbekistan is another interesting case in terms of Islamic radicalism, where, immediately following independence, the Wahhabi movement would emerge. Wahhabism is an Islamic fundamentalist sect established in Saudi Arabia in the eighteenth century by Muhammad ibn Abd al-Wahhab (1703-1792). It is the puritanical offshoot of the Sunni branch of Islam, and it became the religion of the Saudi royal family (Critchlow 1991: 179). The Wahhabis believe in the establishment of a Muslim community similar to that which existed at the time of the Prophet. During the Soviet-Afghan war and after Gorbachev's coming to power, the Wahhabi activism increased in Central Asia. However, it was basically after the collapse of the Soviet Union that the Wahhabis emerged as a determined and well-organized fundamentalist group. Once highly secretive, since late 1980s the Wahhabis have been involved in open confrontation with the authorities across Uzbekistan with the aim of defeating Karimov's government and

[8] For an analysis of the Tajik IRP, see Haghayeghi (1995), Caspiani (2000), and Rashid (1994).

spreading an Islamic revolution in Central Asia. Wahhabism stands in rigorous opposition to Sufism. As the Wahhabis do not take into consideration any doctrines other than those expressed by the generation of Prophet Muhammad, they consider the Qur'an and the hadith (the sayings of the Prophet) as the only authoritative sources by which the Islamic community may conduct its affairs (Haghayeghi 1995: 190). The Sufi tradition in Central Asia is regarded as "nothing but a Zionist and Turkish conspiracy to undermine Islam" (Rashid 1994: 104).

Alongside the Wahhabis there is the radical Islamic group called the Hizb-ut Tahrir. This is a secretive organization that aims to unite all Muslims by creating a caliphate ruled by the Islamic law of Shariah. The organization emerged in the Middle East in 1991, and after the collapse of the Soviet Union, it spread to former Soviet Central Asia and Azerbaijan. [9] The group seeks to establish an Islamic state across Central Asia. Its leaders tend to consider September 11 as "God's punishment to Americans." [10]

However, the most powerful group to emerge in Uzbekistan is the Islamic Movement of Uzbekistan (IMU), established in 1998 by Takhir Yuldash and Jumabai Namangani (sometimes also called as Jumabai Khojaev). These two Uzbek leaders had earlier joined the Tajik Islamic opposition forces, but upon their initial defeat, had escaped to Afghanistan. In 1998, Yuldash and Jumabai returned to Uzbekistan and established the IMU. The organization first became known when it organized an assassination attempt on Islam Karimov in Tashkent in February 1999. Karimov was almost killed by the explosion of six bombs. The IMU claimed responsibility for the attacks (Gleason 2002: 7). The IMU members also took hostage four Japanese scientists in the Batken region of Kyrgyzstan in the same year. The leaders of the movement were claimed to be in close relation with Osama bin Laden; Jumanbai Khojaev is believed to be bin Ladin's deputy (Starobin 2001). The IMU had 3,000 guerillas basically located in Tajikistan and Afghanistan prior to September 11 attacks. It aims to overthrow Karimov administration and establish strict Islamic law in Uzbekistan (Ibid).

For the time being, such radical groups are, to a large extend, marginal in Central Asia. However, in time they may increase their power and influence basically resulting from the economic problems and/or repression of opposition. In that case, they may be a direct threat to the process of nation-state building in Central Asia, since such fundamentalist groups view "nation" as an irrelevant entity of political and legal loyalty. What really

[9] AP Worldstream (24 July 2002).

[10] *The Christian Science Monitor* (19 July 2002: 7).

matters, in other words, is loyalty to the Muslim international community and its leaders, not the artificial division of nation.

Another rather interesting development that is taking place in Central Asia is related to the activities of the missionaries. New Christian groups, some originating in the United States, provide financial incentives to the Muslims on the condition that they convert to Christianity. These evangelizing groups such as the *Word of Faith* and *Mission of Mercy* became especially active after the September 11 attacks. Their emergence has created another division and tension among Central Asians: Muslim versus Christian Central Asians. Even though the overwhelming majority of Central Asians belongs to the Sunni branch of Islam, in the long run these missionary groups may be expected to increase their impact, further limiting the process of nation-state building.

Local and/or Tribal Loyalties

In addition to the potential threat of religious fundamentalism, one of the most important limitations of the nation-state building process in Central Asia is the importance of local and/or tribal identities in the region. Such identities have played important political and economic roles in Central Asian societies, both before and after independence. During Soviet times, such affiliations were effective in the party and government hierarchy even though carefully controlled by Moscow. Regional Communist Party leaders would attempt to reinforce their power by bringing their fellow tribesmen or people from their own region to important positions. After independence, such ties did not lose their importance. In fact, they may be one of the greatest threats in the long run for the process of nation-state building in Central Asia.

One of the most striking examples of this threat could be observed during the civil war in Tajikistan, which broke out in the spring of 1992 and which ended in 1997. In Tajikistan, the main cleavage is territorial, even though there are religious cleavages in the Gorno Badakhshan region, where a Shi'ite sect, called the Ismailis, resides. The Ismailis, sometimes also called Pamiris because they live around the Pamir Mountains, consider themselves a separate group not only religiously but nationally as well (Prazauskas 1998: 54). The Tajik regions are diverse in ethnic, topographic, linguistic, cultural, economic and religious terms. These differences, known as *mahalgaroi* (regionalism), have been both the cause and the consequence of historical, geographic, political, and economic rivalries. Economic benefits and political power were distributed unequally throughout the history of modern Tajikistan. The northern regions of the country, basically Khojend (formerly Leninabad), had most of the republic's investment. This region was economically richer and more industrialized. Khojendi clans also had political power. They had a patron-client relation with the Kulyabis in the south and the two allied themselves with Moscow. In the north, there is also a sizeable Uzbek population. The south, on the other hand, remained basically agricultural and impoverished. The region known as Kyrgan Tyube (where the Garm Valley is located) is one of the most conservative areas of the country. Finally, Gorno Badakhshan is the most isolated region in Tajikistan, with few roads and other facilities of interrepublican means of transformation and communication.

When the civil war broke out in Tajikistan in spring 1992, the conflicts and rivalries between these regions were clearly seen by everyone interested. Even though the war was basically portrayed by the western media as a conflict between the communists and the Islamic opposition, regions and clans representing those regions were the most active participants in the conflict. Khojend and the Kulyab regions supported the old Communist Party elite, whereas the opposition enjoyed the loyalty of the people from the Kyrgan Tyube and Gorno Badakhshan regions. The war appeared to be fought on ideological grounds; however, "the fault lines were ... regional in nature" (Gleason 2001b: 127).

Another example of regionalism comes from Uzbekistan. In this country, until 1937, there was a relative balance between the different regions of Bukhara, Tashkent, and the Fergana Valley. However, in 1937 the Fergana faction started to take over, and it increased its power when it established an alliance with Tashkent. The Tashkent-Fergana axis maintained power until 1959, when Sharaf Rashidov from Samarkand would come to power. The Samarkand faction would once again be influential in the election of Islam Karimov as the First Secretary of the Communist Party of Uzbekistan in 1989 (Roy 2000: 18).

A similar situation can be observed in Turkmenistan, where the most important social units are the tribes of the Tekke and the Yomud, followed by the Ersary, Salyr, Karyk, and the Choudur (Prazauskas 1998: 54). During Soviet times, tribal affiliation played an important role in recruiting people for political and administrative positions. The First Secretary of the Turkmen Communist Party would put his tribesmen into prominent posts (Khazanov 1994: 148). In Turkmenistan, the Tekke tribe from Ashkhabad was the most influential.

Kyrgyzstan is another interesting case in which clan and tribal memberships always had great importance. During the Soviet era, first the southern Kipchak tribe and then the northern Sary-Bagysh tribe dominated political life. After independence, President Akaev, a northerner, would come to power and form his whole team from the northern regions of Talas and Chu (Roy 2000: 115). In fact, the division between the northern and southern regions of Kyrgyzstan is one of the biggest threats to the nation-state building process in the country. The north is politically and economically more powerful and developed, and less conservative and less religious. Most Slavs reside in this part of the country (Anderson 1999: xii). People in the south, on the other hand, are impoverished, less developed in economic terms and more excluded from decision-making. Traditionally speaking, they have been more sedentary and agricultural and more prone to conservative and traditional Islamic influences.

In the post-Soviet era, the regional and tribal divisions between the south and the north of Kyrgyzstan emerged as the most fundamental aspect of Kyrgyz's split identity. Politics, economics, and social conditions are shaped to a large extend by this division (Achylova 1995: 326-327).

Kazakhstan also has its own regional and tribal divisions. Since Russian influence in Kazakhstan was much stronger than in the other Central Asian republics, tribal opposition is less powerful and allocation of power positions is more of an ethnic (*Russian versus Kazakh*) nature than regional or tribal (Roy 2000: 114). Nevertheless, the Kazakhs too were divided into three tribal confederations among themselves, called the *Ulu Zhuz* (Great Horde), the *Orta Zhuz* (Middle Horde) and the *Kishi Zhuz* (Little Horde). It is generally accepted that people from the Great Horde would enjoy more political power than the others. Kazakh President, himself a member of the Great Horde, usually encouraged, even in the late Gorbachev years, his fellow tribesmen to come forward so that he could favor them over the others in recruitment for the state offices. These loyalties may be expected to become even more important in the post-Soviet era (Olcott 1997: 225). As such, the loyalty to the hordes may be a factor hindering the establishment of nation-state in Kazakhstan.

The Minority Groups

The process of nation-state building is further complicated in Central Asia by the issue of minorities. Central Asia is a region in which all of the republics have sizable minority groups. Two serious ethnic clashes prior to the disintegration of the Soviet Union—the conflict between the Meskhetian Turks and the Uzbeks in the Fergana Valley in June 1989 and the conflict between the Kyrgyz and the Uzbeks in the city of Osh in June 1990-ended in the death of many people. These two clashes caused serious concerns about ethnic stability in Central Asia in the early years of the post-Soviet era.

As a response to the societal divisions mentioned above, Central Asian leaders attached great importance to ethnic unity and harmony in their nation building efforts. The case of Kazakhstan is one of the most interesting examples in terms of a government dealing with minority groups. All of the administrative divisions of the country have significant minority populations, which sometimes are actually dominated by non-titular groups. Of the 17 million people living in the country, 46 percent are Kazakh, 35 percent are Russian, 5 percent are Ukrainian, 3 percent are Volga Tatar, and the rest belong to different other ethnic and national minorities. The existence of so many different groups in the country makes Kazakhstan a "poly-ethnic" society.[11] President Nazarbaev, therefore, has been very cautious in his ethnic policies and attitude. He has, on a number of occasions, emphasized that ethnic harmony is vital for Kazakhstan's political, economic, and social development as an independent state. Nazarbaev also was very harsh on extreme nationalist organizations such as Alash. Kazakh Constitution does not allow the formation or operation of social associations promoting to change the constitutional system and the territorial integrity of the state, as well as exacerbating social, racial, national religious, class, or tribal animosities (Article 5.3).

Kazakhstan has been portrayed as the only republic in which an ethnic conflict can emerge between the Russians and the Kazakhs (Roy 2000: 177). The so-called "Alma-Ata events" of 1986 emerged from controversies over a Russian being appointed as the Kazakh Communist Party First Secretary. These controversies are usually accepted as being the first ethnic clash in the Soviet Union, which started a chain reaction throughout the country, contributing to the final collapse of the Soviet regime.

After 1991, President Nazarbaev was also very much concerned about the stability between the Kazakhs and the Russians. He would openly declare that "whoever tries to stir up discord and harmony between the Kazakhs and the Russians will be the common enemy of the two nationalities" (Liu 1998: 81). Even though Nazarbaev did not grant dual citizenship to the Russians of Kazakhstan, Russian is given an official status as "the language of interethnic communication" (Kangas 1995: 282).

Kyrgyzstan also has a sizeable Russian minority, though not as large as in Kazakhstan. Out of the 4.4 million people living in Kyrgyzstan, about 53 percent are Kyrgyz, 18 percent are Russian and 13 percent are Uzbek, and the rest are of various other ethnic groups. Similar to Nazarbaev, Kyrgyz President Askar Akaev, too, has stressed the unity of all nationalities as the primary condition of nation building. According to him, without ethnic harmony, Kyrgyzstan could not make any progress (Liu 1998: 81). Similar to the situation in Kazakhstan, President Akaev did not give dual citizenship to the Russians but made Russian a special language for ethnic communication.

[11] One floor of the Central State Museum of the Republic of Kazakhstan in Almaty is actually dedicated to the ethnic groups in Kazakhstan, who are portrayed to be living side by side in peace and harmony.

Uzbekistan also has its own ethnic minorities. Out of the 23.5 million people living in Uzbekistan, 80 percent are Uzbek, 5.5 percent are Russian, about 10 percent are Tajik, 3 percent are Kazakh, and 2.5 percent are Karakalpak. According to the preamble of the Uzbek Constitution, "the people of Uzbekistan are the citizens of the Republic of Uzbekistan, regardless of their nationality."[12] President Karimov, similar to the other Central Asian leaders, has frequently emphasized ethnic harmony and stability, calling for reconciliation if disagreements emerge between ethnic groups. However, in Uzbekistan, Uzbek is the official language and the predominance of Uzbek is on the rise in social, political, and public life of the country (Kangas 1995: 282).

The situation in Tajikistan is similar to the situation in Uzbekistan. Here, out of the 6 million people living in the country, 65 percent are Tajik, 25 percent are Uzbek, 5 percent are Pamiri, and 2 percent are Russian (down from 10 percent in 1990). The Tajik leadership has been quick in making Tajik the official language and switching the alphabet from Cyrillic to Arabic shortly after independence. However, the civil war prevented the active application of these measures. Even though the preamble of the Tajik Constitution stresses the "unwavering freedoms and rights of people" and "respect for equal rights and friendship of all nations and peoples," the real-life situation is actually more complicated than that.

Turkmenistan seems to be the most stable country in terms of ethnic relations. The population of the country is 4.3 million out of which 77 percent are Turkmen and 9.2 percent are Russian. President Turkmenbashi, too, has emphasized the importance of ethnic harmony and unity in nation building process, despite the fact that the country did not show any signs of extreme nationalism (Freitag-Wirminghaus 1998: 157). Turkmen, however, has been accepted as the state language.

In conclusion, ethnic relations in Central Asian republics are more stable than one would expect, given internal divisions and the rather short period to settle these. However, events that may threaten the status quo and endanger the nation state building process are not difficult to find. Even though ethnic harmony was seen as the most fundamental condition for political stability (Liu 1998: 73), the share of non-indigenous groups decreased considerably in political, administrative and cultural spheres. It is possible to talk about a general process of "indigenization" of post-Soviet Central Asian societies and governments, creating a new mechanism

[12] The Uzbek Constitution has a specific section on the Autonomous Republic of Karakalpakistan, which is considered to be a sovereign republic within Uzbekistan. According to Article 74 of the Constitution, Karakalpakistan possesses the right to withdraw from Uzbekistan on the basis of a general referendum of the people of Karakalpakistan.

of "ethnocracies" (Prazauskas 1998: 58). In political processes and in the executive and legislative policymaking, the minority groups are not sufficiently represented, and they do not have any substantial influence.[13] Non-indigenous groups are steadily being excluded from public, civil, and social services and being replaced by titular nationals. In other words, there is a new process of nationalization "by stealth" (Bohr 1998: 142).

Finally, it should be pointed out that, even though all non-titular minorities are influenced by such policies, 10 million Russians living in the region may be worst off. Having no longer the privileges that they had, that is, being protected by Moscow, they may be feeling more discriminated than other ethnic groups. They do not know the titular language, which is now the appointed language of position and status, just as Russian once was in the recent past. So they may have little choice but to learn it. They may even feel that other minorities such as ethnic Uzbeks in Tajikistan or ethnic Tajiks in Uzbekistan are more readily accepted than the Russians, leading to further frustration and exclusion (Smith 1998: 17).

Conclusion

The process of nation-state building in Central Asia, which was started by the ex-communist party elites in the post-Soviet era, is a long and painful one that is not complete yet. The leaders of the five republics have adapted themselves to new conditions, and the existing bureaucratic institutions ensured smooth political transition. The elites became defenders of national independence, justifying their post-Soviet existence by national sovereignty and utilizing national identity in order to legitimize their rule. However, as was analyzed above, there are certain limitations and unique problems in each republic that may challenge the process of nation-state building.

All these limitations and unique problems, however, may further hinder this process in the region if *economic development* is not achieved. Though all Central Asian countries (with the notable exception of Tajikistan because of the 1992-1997 civil war) have introduced "national" economic programs, none has achieved great success yet. With independence, they have lost the subsidies from the union budget. The Central Asian leaders supported the idea of preserving the Soviet Union until the last moment, and even after the disintegration, they sought ways of preserving their economic relations with Russia. When the disintegration of the Soviet Union became inevitable, the Central Asian politicians, led by the Kazakh President Nursultan Nazarbaev, initiated economic, political, and military

[13] See: *http://www.unesco.org/most/kyrgyz2.htm*

reintegration within the CIS. After independence, reform packages had mixed results. In addition to the problem of economic dependence on Russia, the problem of failure in regional economic cooperation further decreased the changes for successful transformation.

In the last couple of years, the Central Asian republics have reoriented their economics toward exports. Economic growth of the region became entirely driven by demand for hydrocarbons and cotton. The economies in the region were essentially defenseless in the face of economic shocks from the outside. They also depended increasingly upon the influx of foreign capital. So they do not have any substantial domestic potential to propel their economic development. Therefore, there is no real change for economic union in the short run (Rumer 2000: 11).[14]

State building is the costly process of erecting government administrative offices and staffing these throughout the country. Without a tax-structure to tap rising incomes, investment in infrastructure will be limited. The high level of dependence on revenues coming from exports creates outward oriented governments. All these developments may have a negative impact on the nation-state building process in Central Asia. Under these circumstances, official discourse may lose whatever credibility it has. As was put forward by a cab driver in Bishkek, the celebration of Manas did not put bread on the table, even though the scholars and politicians speak of the role of the Manas epic in Kyrgyzstan in developing fraternity, independence, and national pride in the country (Anderson 1999: 61).

References

ABAZOV, R.
1998 "Central Asian Republics' Search for a Model of Development." The Slavic Research Center from the web page: *http://src-h.slav.hokudai.ac.jp/publicn/CentralAsia/rafis/rafis/html.*

ABDÜLVAHAP, K.
2002 Paper presented at the "Second Meeting of the World Kazakhs." Almaty.

ACHYLOVA, R.
1995 "Political Culture and Foreign Policy in Kyrgyzstan." Pp. 318-336 in Tismeneanu, V. (ed.), *Political Culture and Civil Society in Russia and the New States of Eurasia.* New York: M.E. Sharpe Inc.

ALLWORTH, E.
1998 "History and Group Identity in Central Asia." Pp. 67-90 in Smith, G. et al., *Nation-Building in Post-Soviet Borderlands: The Politics of National Identity.* Cambridge: Cambridge University Press.

ANDERSON, J.
1999 *Kyrgyzstan's Island of Democracy?* The Netherlands Harwood Academic Publishers.

[14] According to the United Nations Development Program Resident Coordinator in Almaty, Fikret Akçura, however, in the long run the Central Asian countries need to form an economic union in order to develop and survive in global economy (personal interview, 29 October 2002, Almaty).

Bacık, G.
1999 "Türk Cumhuriyetlerinin Kimlik Sorunu," in Öke, M.K. (ed.), *Geçiş Sürecinde Orta Asya Türk Cumhuriyetleri*. İstanbul: Alfa Basım Yayım Dağıtım.
Balbay, M.
2001 *Ortadaki Asya Ülkeleri*. İstanbul: Cumhuriyet Kıtap Klubü.
Bohr, A.
1998 "The Central Asian States as Nationalizing Regimes." Pp. 139-164 in Smith, G. et al., *Nation-Building in Post-Soviet Borderlands: The Politics of National Identity*. Cambridge: Cambridge University Press.
Büyükakinci, E.
2002 "Kazakistan'da siyasal bütünleşme ve ulus devlet olma süreci." Pp. 346-369 in Gürsoy-Naskali, E. and Şahin, E. (eds.), *Bağımsızlıklarının 10. Yılında Türk Cumhuriyetleri*. Haarlem, Holland (SOTA).
Caspiani, G.
2000 *The Handbook of Central Asia: A Comprehensive Survey of the Republics*. London: Tauris Publishers.
Critchlow, J.
1991 *Nationalism in Uzbekistan: A Soviet Republic's Road to Sovereignty*. Boulder: Westview Press.
Freitag-Wirminghaus, R.
1998 "Turkmenistan's Place in Central Asia and the World." Pp. 157-176 in Atabaki, T. and O'Kane, J. (eds.), Amsterdam The International Institute for Asian Studies.
Gleason, G.
2002 "The Politics of Counterinsurgency in Central Asia." *Problems of Communism* 49: 1-11.
2001 "Inter-State Cooperation in Central Asia from the CIS to the Shanghai Forum." *Europe-Asia Studies* 53: 1077-1095.
2001 "Power Sharing in Tajikistan: Political Compromise and Regional Instability." *Conflict, Security and Development* 1: 125-134.
Gürsoy-Naskali, E.
2002 "Bağımsız Türk Cumhuriyetlerinde Dil Politikaları." Pp. 55-61 in Gürsoy-Naskali, E. and Şahin, E. (eds.), *Bağımsızlıklarının 10. Yılında Türk Cumhuriyetleri*. Haarlem, Holland: SOTA.
Haghayeghi, M.
1995 *Islam and Politics in Central Asia*. New York: St. Martin's Press.
Hiro, D.
1995 *Between Marx and Muhamad: The Changing Face of Central Asia*. London: Harper Collins Publishers.
Kangas, R.D.
1995 "State building and civil society in Central Asia." Pp. 271-291 in Tismeneanu, V. (ed.), *Political Culture and Civil Society in Russia and the New States of Eurasia*. New York: M.E. Sharpe Inc.
Khazanov, A.M.
1994 "Underdevelopment and Ethnic Relations in Central Asia." Pp. 144-163 in Manz, B. (ed.), *Central Asia in Historical Perspective*. USA: Westview Press.
Liu, G.
1998 "Ethnic Harmony and Conflicts in Central Asia: Origins and Policies." Pp. 73-92 in Zhang, Y. and Azizian, R. (eds.), *Ethnic Challenges Beyond Borders*. Great Britain: Macmillan Press Ltd.
Nazarbaev, N.
1997 *Yüzyılların Kavşağında*. Ankara: Bilgi Yayınları.
Ochs, M.
1997 "Turkmenistan: The Quest for Stability." Pp. 312-359 in Dawisha, K. and Parrot, B. (eds.), *Conflict, Cleavage and Change in Central Asia and the Caucasus*. Cambridge: Cambridge University Press.
Ollcott, M.B.
1997 "Democratization and the Growth of Political Participation in Kazakhstan." Pp. 201-241 in Dawisha, K. and Parrot, B. (eds.), *Conflict, Cleavage and Change in Central Asia and the Caucasus*. Cambridge: Cambridge University Press.

PRAZAUSKAS, A.
1998 "Ethnopolitical Issues and the Emergence of Nation-States in Central Asia." Pp. 50-69 in Zhang, Y. and Azizian, R. (eds.), *Ethnic Challenges Beyond Borders*. Great Britain: Macmillan Press Ltd.
RASHID, A.
1994 *The Resurgence of Central Asia: Islam or Nationalism?* Karachi: Oxford University Press.
ROY, O.
2000 *The New Central Asia: The Creation of Nations*. London: I.B. Tauris.
RUMER, B.
2000 "In Search of Stability." *Harvard International Review* 22: 44-48.
RUMER, E.
2000 "Fear and Loathing in the 'Stans'." *Christian Science Monitor* 92: 11.
RYWKIN, M.
1990 *Moscow's Muslim Challenge* (revised edition). Armonk, NY: M.E. Sharpe.
SMITH, G.
1998 "Post-colonialism and Borderland Identities." Pp. 1-20 in Smith, G. et al., *Nation-Building in Post-Soviet Borderlands: The Politics of National Identity*. Cambridge: Cambridge University Press.
STAROBIN, P.
2001 "The 'Stans' Seize the Day." *Business Week* 3753: 58-61.
WRIGHT, R.
1992 "Report from Turkestan." Pp. 53-75. *The New Yorker*.

IV. Political Processes in Post-Soviet Central Asia

Shirin Akiner

Abstract

The Central Asian republics inherited a high degree of civic order. The challenge that confronts these republics today is whether or not, as independent states, they will be able to maintain order and stability; and if so, by what means will they achieve this. It is unlikely that in the near future these newly independent states will be able to mobilize sufficient resources to sustain the level of development achieved under Soviet rule. Standards in health care and education are showing signs of severe erosion; the general social and technical infrastructure, especially in rural areas, is deteriorating owing to inadequate maintenance and investment. If this regression continues, it is likely to have potentially disastrous political consequences. Pauperization, frustrated hopes, bitter competition for scarce resources, as well as the increasing gap between aspirations and the ability to satisfy them, could provide a fertile breeding ground for intercommunal violence, as has already occurred in Tajikistan. In some areas, particularly in the south, this is finding expression in a militant form of Islam. When the Soviet Union collapsed in 1991, there were many who cherished the same hopes for the Central Asian republics that had previously been held for the newly decolonized world of the 1950s, namely, that it would be possible to launch these states on a smooth process of cultural change, economic growth, and stable democracy. The reality here, as in other parts of the developing world, has already proved to be far more complex. To consolidate genuine political and economic reform will require a fundamental shift in social and cultural attitudes.

Introduction

The Soviet Union formally ceased to exist at the end of December 1991. However, like other Soviet republics, the five Central Asian republics—Kazakhstan, Kyrgyzstan, Tajikistan, Turkmenistan, and Uzbekistan—had already proclaimed, in 1990, their sovereignty and had instituted the office of president to head the republican administration.[1] In the second half of 1991, they went further and, one by one, proclaimed their independence. They were prompted to take this action by the attempted *coup d'état* in Moscow in August of that year, aimed at removing President Gorbachev from office and reversing the trend towards greater liberalization. The coup failed, but it severely undermined the authority of the central government, thereby encouraging the republican leaders to augment their own status. The declarations of independence in Central Asia were unilateral and were made in an almost impromptu manner, with no prior debate or preparation. In effect, they were symbolic gestures, lacking constitutional substance. Thus, when genuine independence dawned in January 1992, it was unexpected. It signaled a cataclysmic upheaval, the consequences of which could not be fully envisaged.

The level of development in Central Asia had always lagged behind that of other Soviet republics, and standards of living were lower. Consequently, this region had been more dependent on all-Union structures and on subsidies from the central government. Thus, the abrupt termination of budgetary transfers from the central government (one of the principal sources of funding for welfare services) and the dislocation of interrepublican trade and transport links caused greater distress here than elsewhere. The most immediate challenge that confronted these new states was the construction of viable national economies, based on free-market principles. This was not a theoretical issue, but one that directly affected people's lives. Yet there were no blueprints to follow, no simple models that could be applied. There was also little time to reflect on the best policies to adopt,

[1] In March 1990, Mikhail Gorbachev, formerly General Secretary of the Communist Party of the USSR, became President of the USSR; shortly after, the First Party Secretaries of the Communist Parties of the Central Asian republics were also transformed, by means of an internal administrative procedure, into presidents; later that same year, Uzbekistan (June), Turkmenistan and Tajikistan (August), Kazakhstan (September) and Kyrgyzstan (December) made declarations of sovereignty. The main constitutional implication was that republican laws would henceforth take precedence over federal laws. However, in the referendum on the future of the Soviet Union, held on 17 March 1991, the Central Asian republics returned a vote of over 90 percent (in Turkmenistan 98 percent) in favor of preserving the Union. There were some complaints of intimidation and ballot-rigging, but on the whole, this result was accepted as a reasonably accurate reflection of public opinion at the time.

as the pace of change was largely dictated by external factors. Predictably, progress was chaotic and results variable.

Nevertheless, the new states initially were remarkably successful in coping with these upheavals. Economic reforms, including privatization, enterprise restructuring, and commodity price liberalization, were set in motion (EBRD 1994). Likewise, there was some attempt to build new institutions (Akiner 2000). Issues of regional cooperation and integration into the international community were also addressed (Akiner 2001b). There was a high degree of social cohesion, which, to some extent, alleviated the hardships caused by rising levels of unemployment and rapid pauperization. In Tajikistan, factional rivalries plunged the country into civil war within a few months of independence (Akiner 2001a), but elsewhere in the region, stability was maintained. The new states possessed many assets, not least highly educated populations and substantial resource bases. This led many to believe that it would be relatively easy for them to make a painless transition to democracy and economic prosperity.

The reality proved to be very much more complex. After the euphoria of independence had subsided, it became clear that there were deep-seated systemic problems that would hinder genuine political and economic reform. Economic issues are discussed at length in other papers in this volume and, hence, will not be explored here. However, it is important to note that the agenda for economic reform was largely determined and controlled by the ruling elites. This provided them with new opportunities for the exercise of patronage, since they were able to reward supporters with privileged access to lucrative business transactions. This in turn served to consolidate their hold on power and thereby became a covert element in the political process. Thus, from the outset, hopes of creating more open, better regulated, societies were subverted by insider deals that not only strengthened the black economy (which had existed during the Soviet era), but also fostered parallel power structures, invisible and unaccountable, that operated behind the façade of formal institutions (Akiner 1998).

Local factors, such as culture and the size and diversity of population, influenced the pace and style of post-Soviet adaptation. In the first few years, this influence resulted in rapid and striking differentiation, with each state following its own path. Gradually, however, it was the similarities, particularly in the political sphere, that became more prominent. In part, these were retrospective reflections of the common Soviet legacy, in part a resurfacing of pre-Soviet traditions. Thus, rather than the sharp break with the past that some had anticipated, authoritarian modes of governance were perpetuated and even reinvigorated.

Leadership

In all the Central Asian states, the presidents have dominated the post-independence political scene. Four came to power before the collapse of the Soviet Union; only in Tajikistan, in the wake of a civil war, was there a change of leadership. The Kazakh, Turkmen, and Uzbek presidents were former Communist Party chiefs; the Kyrgyz president, though a senior academic, was likewise drawn from the Soviet establishment. These ruling elites, far from being discredited on account of their links to the previous regime, acquired even greater legitimacy as symbols of continuity in a time of flux and uncertainty. They did not have the revolutionary credentials of Asian and African nationalist leaders who had led liberation struggles against colonial regimes, but even so, they succeeded in assuming the mantle of founding fathers of the new states. By the same token, they were regarded as guarantors of unity, independence, and stability.

The embryonic political movements that emerged in the last years of the Soviet era were unable to consolidate the support that they had enjoyed when they first appeared. This was partly because in some states—notably Uzbekistan and Turkmenistan—after independence they were subjected to even greater harassment, official and unofficial, than had been the case under Soviet rule (US DoJ 1993, 1994). Yet there were other reasons for their lack of success, including organizational weaknesses and an inability to adapt to the new environment. More importantly, perhaps, they were distrusted because they were seen by many as divisive, and thus potentially dangerous, at a time when the society was beset by internal and external threats to stability (Akiner 1997b). Had there been a longer, more orderly period of transition, it might have been possible to dispel such fears. As it was, independence did not prove to be a stimulus for political liberalization but instead, stifled the first cautious experiments in pluralism.

With no effective domestic opposition, the heads of the new states began to operate in an increasingly autocratic manner. Their power was at first modified by newly created democratic (at least in intention) institutions and also by the influence, direct and indirect, of international organizations. However, the constitutional checks and balances, including limitations on presidential tenure of office that were introduced in the early years of independence were steadily eroded by referendums, decrees, dubious legal rulings, and flawed electoral proceedings. At the same time, national security came to be regarded as synonymous with regime security, which in turn was understood as maintaining the incumbent presidents in power. Thus, criticism of the leadership, no matter how legitimate, was interpreted as an attack on the state. Articles were introduced into the criminal codes of all these countries making it an offence, punishable by fines or even lengthy prison sentences, to impugn "the honor and dignity" of the president

(McCormack 1999). This placed severe curbs on the freedom of the media. It also effectively muzzled any would-be political rivals.

During the past decade, striking sociocultural parallels have emerged between the style of government of the contemporary Central Asian leaders and that of the rulers of the pre-Soviet khanates. Today's presidents, like the khans of the past, exercise personal control over all public affairs. In theory, there is a division of institutional functions and powers. In practice, senior figures in the state apparatus are selected by the presidents, generally from among their own networks of dependents and clients; these in turn surround themselves with loyal subordinates of their own. This pattern is repeated at every level of the administrative hierarchy, thereby creating chains of interlocking personal loyalties that can readily circumvent proper procedures. Such a system inevitably fosters jealousies and rivalries between individuals as well as government departments. Partly as an antidote to these negative tendencies and partly in order to strengthen their own hold on power, the leaders frequently reshuffle ministers and other senior personnel. The high turn-over rate ensures that they alone represent continuity and authority and remain the ultimate dispensers of patronage (Akiner 1998: 21). However, this constant movement has exacerbated the sense of personal insecurity and created an atmosphere in which individuals at every level of the administration seek to "milk" their office for private gain. Most independent assessments now place the Central Asian states among the most corrupt countries in the world.[2]

A major priority for the Central Asian leaders has been to preserve stability. This preoccupation must be placed in the context of the situation in the early 1990s, when there were fears, both in the region and abroad, that the strains caused by the sudden collapse of the Soviet Union might fatally weaken the social fabric, opening the way to anarchy and violent conflict. The ongoing civil war in Afghanistan and the outbreak of conflict in Tajikistan gave added potency to these anxieties. The new states were ill-equipped to defend themselves against such dangers. The army units that they had inherited from the Soviet era were weak and lacked cohesion. Only the internal security forces, such as the police and the Ministry of Interior troops, were relatively strong, with organizational structures still intact and experienced cadres in place. Increasingly, the authorities came to rely upon these units to fight crime and terrorism. However, as during the Soviet period, there was a blurring of the distinction between the

[2] In 2002, out of 102 countries assessed according to the TI Corruption Perceptions Index, Kazakhstan was rated 88th, close to such countries as Cameroon, Nigeria, and Bangladesh. Uzbekistan was rated 65th; the other Central Asian states were not listed, owing to insufficient data (TI 2002: Table 1).

need to combat genuine and specific threats to law and order, and the more general aim of controlling society by the exercise of fear. With or without official sanction, the security forces frequently indulged in brutal and corrupt behavior (ICG 2002). The courts routinely turned a blind eye to such excesses. The result was the widespread violation of human rights, including arbitrary arrests, denial of legal representation to detainees, and the use of physical and psychological torture. Gradually, following persistent international pressure, the Central Asian governments began to address these issues, and in most states steps were taken to eradicate at least some of the worst abuses. However, after the terrorist attacks on the USA in September 2001, the random use of state violence increased sharply, particularly in Uzbekistan, prompted by fears that militant Islamism, linked to international terrorism, was becoming entrenched in the region (AI 2001). These fears were increased by a spate of suicide bombings in Uzbekistan 2004, first in March-April and then again in July. The second set of attacks targeted the Israeli and US embassies in Tashkent, as well as the Uzbek Prosecutor-General's Office. Militant Islamists were blamed for these incidents. Several individuals were arested and put on trial.

The characteristics outlined above are to be found in all the Central Asian states, but the manner and degree in which they are manifested vary (Ishiyama 2002). The most extreme example of monocratic rule is to be found in Turkmenistan, where President Saparmurad Niyazov (b. 1940) is the chief, and usually sole, decision maker in economic, political, social, and cultural matters. Ministers and other senior figures are appointed and dismissed with bewildering speed, rarely surviving more than a few months in office. By popular acclaim, he was awarded the title *Turkmenbashi* (Leader of the Turkmen People). In 1994, his mandate was extended by nationwide referendum (according to official reports, with the support of 99.9 percent of the electorate). However, before the specified term had expired, he was made president for life, though he later indicated that he might step down before his seventieth birthday (FH 2002: 389-99). The focus of a colossal personality cult, his image is reproduced in portraits, statues, and gigantic monuments throughout the country; even such personal items as wristwatches and bottles of aftershave fragrances are embellished with his picture. Media output in Turkmenistan is almost entirely devoted to reporting his activities and pronouncements. His living family members play little role in public life, but his late mother has been elevated to a status resembling that of a tutelary spirit.

In the other four states, personality cults are somewhat less flamboyantly displayed. Nevertheless, images of the incumbent presidents dominate public places, and presidential power permeates public life. Uzbek President Islam Karimov (b. 1938) presides over a regime that is regarded by many as the most brutal and repressive in the region. Allegations abound of mass arrests and the systematic use of torture on prisoners suspected of involvement in banned religious organizations (HRW 2003). There is no indication that he will step down from office in the near future. When elections were held in January 2000, some 90 percent of the electorate (including—by his own admission—the only challenger) voted for Karimov. In January 2002, his term of office was extended by referendum from five to seven years; however, even senior government officials did not know when the seven-year period was to commence (i.e., from the date of the previous election, the referendum, or the next election).

In the first years of independence, Kazakhstan and Kyrgyzstan seemed set to follow a very different trajectory. The Kazakh president appeared eager to embrace reform. In Kyrgyzstan, too, the leadership professed firm commitment to a path of democratic development. Yet abuses of power soon surfaced in both countries. The mandates of Kazakh President Nursultan Nazarbayev (b. 1940) and Kyrgyz President Askar Akayev (b. 1944) were extended several times by dubious manipulation of the respective electoral laws. In mid-2000, the Kazakh parliament granted Nazarbayev extraordinary powers and privileges for life (i.e., whether or not he steps down from office). Initially it was anticipated that he would not seek re-election, but he later made clear his intention to stand again in the 2006 presidential race. The Kyrgyz president likewise indicated that he would be a contestant in the next round or elections. In both countries a handful of independent leaders have emerged, mostly from within the ranks of the post-Soviet political elite. However, few have been able to sustain active opposition for long. Some have been co-opted back into the ranks of the presidential entourage, while others have been removed from the political arena following criminal investigations that have resulted in serious charges being laid against them; the outcome has been that they have been sentenced to long terms in prison.[3] In both countries, nepotism is rife. This has enabled relatives of the presidents to acquire

---

[3] For example, Akezhan Kazhegeldin, a former Prime Minister of Kazakhstan (1994-97) and founder of an opposition party, would have been the only serious rival to Nursultan Nazarbayev in the presidential elections of January 1999; however, he was debarred from standing by means of a minor legal technicality. He left Kazakhstan shortly after to live in self-imposed exile in Europe. Criminal charges were later brought against him, and in September 2001 he was sentenced *in absentia* to ten years in prison; in addition, a substantial fine was imposed and his property was confiscated. Similarly, in Kyrgyzstan, Feliks Kulov, a leading political figure (previous posts included that of Vice-President and Minister of National Security) was arrested in September 2000, soon after announcing his intention to run in the forthcoming presidential elections; in January 2001 he was sentenced to seven years in prison; this term was later increased to ten years.

extensive business interests, amounting to virtual monopolies in some areas. There is a widespread perception that they are heavily involved in corrupt business transactions (TI 2001: 109-23). Censorship in these countries is less obvious than in Turkmenistan and Uzbekistan, but there are still very strong limitations, direct and indirect, on press freedom (FH 2002: 215-17, 231-32). In Kazakhstan, members of the president's family control almost all media outlets, print and electronic; consequently, there is little opportunity to air independent views, particularly if they are critical of the incumbent regime (HRW 1999: Kazakhstan). Thus, although the political systems in Kazakhstan and Kyrgyzstan appear to be more open than in Uzbekistan or Turkmenistan, in reality, there is scarcely more scope for contesting the current order here than in the states where repression is much more overt (Anderson 1999; Olcott 2002).

The only country in which a new leader has come to power during the past decade is Tajikistan. During the civil war (see below) the post of president was suspended. When it was reinstated in November 1994, Imomali Rahmonov (b. 1952) was elected by a fairly slender margin for a five-year term. A relatively young and inexperienced figure, he was seen at the time as the stooge of powerful regional warlords; by the end of the decade, however, he had strengthened his own power base and was able to act more independently. He then began to conform to the pattern of strong autocratic rule that was prevalent elsewhere in the region. He was reelected in 1999, winning some 97 percent of the vote. Prior to the election, amendments to the constitution had increased presidential powers and extended the term of office to seven years (Akiner 2001a: 58).

Constitutional Change

The Central Asian states inherited an array of Soviet civic institutions, but after independence these were no longer acceptable from either a functional or an ideological standpoint. Ostensibly, one of the goals of the post-Soviet constitutions was to create new parliamentary structures that would ensure a proper division of executive, legislative, and judicial powers. In fact, they created overweening executive presidencies. The new institutions of state management lacked autonomy. Consequently, the primary function of the deputies was to confirm presidential decisions, with little or no possibility of exercising any form of critical scrutiny. This was most marked in Turkmenistan and Uzbekistan, but even in Kazakhstan, Tajikistan, and (to a somewhat lesser extent) Kyrgyzstan, the executive branch succeeded in marginalizing the parliaments, depriving them of any real authority. The judiciary in all five states was likewise brought firmly under presidential control, with virtually no scope to hand down independent judgments.

Turkmenistan was the first Central Asian state to embrace institutional change. It adopted a new constitution in May 1992. A fifty-seat legislative body was created, also two innovative institutions that symbolically drew on pre-Soviet traditions. One of these was the People's Council. Comprising senior ministers, judges, and representatives from each region of the country, sitting under the chairmanship of the president, this body has among its formal powers the right to declare war and to amend the constitution, thus in effect, representing a fourth branch of state power. The other new institution, the Council of Elders, has among its powers the exclusive right to nominate presidential candidates. The members of this body are respected members of society, personally selected by the president to provide "wise guidance" in affairs of state (Akiner 2000: 100).

Uzbekistan assumed a more conservative stance on institutional change. The post-Soviet constitution, adopted in December 1992, provided for the creation of a 250-member legislative body, the Supreme Council. Yet this new entity did not differ greatly from the Soviet model. Currently, it is not a standing body, though it meets on a regular basis several times a year (some sessions are public, others are closed). In theory, it is supposed to initiate and pass legislation (FH 2002: 422-25). However, in practice its chief function is to approve laws that have been drafted by the executive. In January 2002 the decision to introduce a bicameral system was approved by referendum. Elections to the new body were held in December 2004.

In Kazakhstan and Kyrgyzstan, constitutional reforms were introduced in several stages. Both countries eventually bicameral parliamentary systems. Yet, far from strengthening democratic procedures, the introduction of two chambers substantially increased the power of the respective presidents, enabling them to stack the second chamber with their own supporters (some directly nominated, some elected). In Kazakhstan, the first post-Soviet constitution was adopted in January 1993. The Soviet-era legislature continued to function until the end of that year. It was dissolved as a prelude to institutional reform, but independent commentators and opposition activists suspected that the real reason for abolishing this body was that it was not amenable to presidential control. A new parliament was elected in 1994, but disbanded a year later. The president governed alone, by decree, until a new constitution (reportedly drawn up under his personal supervision) was approved by referendum in August 1995; this introduced a bicameral parliament, but also greatly increased the powers of the executive (Olcott 1997: 226-37). In Kyrgyzstan, the post-Soviet constitution was adopted in 1993. It was amended in 1994 and a bicameral parliament was created in January 1995 (Anderson 1999: 49-55). Additional changes to the constitution were adopted in 1996 and 1998, most of which substantially weakened the legislature. In February 2003, the draft of a new constitution was approved by referendum; among other amendments, a unicameral parliament was introduced. The authorities justified this move on the grounds that there was a need to redress the balance between the executive, legislative, and judicial powers. However, opposition leaders were convinced that the real aim was to strengthen yet further the position of the president. In Tajikistan, institutional change was disrupted by the civil war. The post-Soviet constitution was adopted in November 1994; important amendments were introduced in September 1999, including the creation of a bicameral parliament.

A semblance of democratic participation in government has been created in all the Central Asian states by the holding of nationwide referendums to approve constitutional amendments (e.g., extending the term of office of the incumbent president). Likewise, elections, presidential and parliamentary, are held at regular intervals. In general, however, the population evinces little enthusiasm for these proceedings, regarding (with some justification) the outcome of such polls as a foregone conclusion. In the referendums, presidential proposals have received almost unanimous endorsement. In presidential and parliamentary elections, there have been innumerable violations of electoral laws, including instances of bribery, intimidation, proxy voting, stuffing of ballot boxes and falsification of the count. The media consistently exhibit blatant bias in favor of the incumbent leader. Whenever international agencies (e.g., the OSCE) have sent observers to monitor elections in the Central Asian states, they have been highly critical of the proceedings; on several occasions they have refused to attend, in order to avoid, by their presence, lending spurious legitimacy to these flawed exercises in popular participation.

Civil Society and Political Parties

Civil society is at a rudimentary stage of development in the Central Asian states. The most vigorous informal associations are based on traditional structures and draw on family and neighborhood ties. Western-style Non-Governmental Organizations (NGOs) are a new phenomenon. This form of activity first appeared in Central Asia in the early 1990s. There has been a proliferation of such organizations in recent years, particularly in Kazakhstan and Kyrgyzstan. They are involved in such areas as child support, women's rights, civil liberties, environmental protection and business training (Ruffin et al. 1999: 235-321). These developments may look impressive on paper, but closer scrutiny reveals a somewhat different picture. Many NGOs are crypto-government organizations, their aims and programs designed to underpin official policy directives. Another large group of NGOs consists of bodies funded by overseas sponsors and, in part at least, run by foreign staff. Many of these are linked to bilateral assistance programmes. Some do provide valuable training and technical support, but the activities of others have little relevance to community needs. Frequently, such groups are poorly integrated into local society and are regarded with suspicion by the general public. There are a number of indigenous NGOs, but these tend to be small, poorly managed and chronically underfunded; most are incapable of implementing their stated objectives and suspend activities soon after registration.

Political parties have also attracted little support from the public at large. The majority of the population in the Central Asian states has little respect for politicians and little interest in the political process. Where independent parties have emerged, they tend to revolve round an individual who has a strong personality and/or sufficient wealth to establish a power base. Membership is almost always small, drawn from a very narrow social and geographic base. Programs and party platforms are generally vague, consisting of little more than idealistic platitudes. The only parties of any size are pro-presidential parties. These have official backing and closely reflect government policies.

The greatest progress towards developing multiparty systems has been made in Kazakhstan and Kyrgyzstan. In the Kazakh parliamentary elections of October 2000, ten parties participated. According to official reports, the pro-presidential bloc gained a substantial majority, but independent assessments cast doubt on these results. The share of votes won by the Communist Party was almost certainly underestimated. The only other opposition party of any size, the Republican People's Party of Kazakhstan, withdrew its candidates on the eve of the election owing to harassment from government officials (Akiner 2002a: 281-83). By early 2002 there were over twenty political organizations in Kazakhstan. However, later that year new regulations introduced stringent registration criteria for political parties. By mid-2004 twelve parties had been (re-)registered and were thus eligible to participate in the parliamentary elections later that year. As expected the pro-presidential parties von a decisive victory.

In Kyrgyzstan, there was an extraordinary blossoming of independent political parties in the early 1990s. Thereafter, however, the government adopted an increasingly repressive attitude towards such activities. At the same time, there was a massive rise in corruption in public life, and this tainted the reputation of politicians across the spectrum. Factors such as these stunted the development of a multiparty system. Elections to the Supreme Council were first held in February 1995 and contested by over 1,000 candidates from twelve parties. In the next round of elections, five years later, a similar number of parties participated, though the number of candidates was greatly reduced. There were many irregularities, such as last-minute changes to the electoral law, which prevented some parties from fielding nominees (Akiner 2002a: 289-91). Nevertheless, despite these setbacks, this is the one state in Central Asia where there is a genuine parliamentary opposition, albeit weak and often ineffectual.

In Turkmenistan and Uzbekistan, there has been very little political liberalization. Turkmenistan remains a one-party state. The Democratic Party of Turkmenistan was founded at the end of 1991, on the basis of the Communist Party. It has since dominated all political activity in the country. In December 1993, the authorities announced that a Peasant Party would shortly be granted official registration, but this did not happen. There are some Turkmen opposition organizations abroad; the best known is *Agzybirlik* (Unity of Voice), but it is difficult to judge how much support it actually has in the country (Akiner 2002a: 472-73). In Uzbekistan, the People's Democratic Party was likewise founded on the basis of the Communist Party. It has remained the dominant political force in the country, though there has been an attempt to create a semblance of pluralism, albeit under strict state control. By the end of the 1990s there were five registered parties, all pro-president and pro-government in orientation (Akiner 2002a: 518-19). The main opposition parties, *Birlik* (Unity) and *Erk* (Freedom), founded at the end of the Soviet era, were banned in 1992. Their leaders subsequently sought refuge abroad; they have tried to maintain links with supporters in Uzbekistan, but it is impossible to judge how effective they have been in creating underground organizations within the country.

In Tajikistan, after the outbreak of civil war, all opposition parties were banned. Only the pro-presidential People's Party of Tajikistan (later renamed the People's Democratic Party of Tajikistan) remained active. In 1997, a few of the former opposition parties were allowed to re-register. New parties and movements also appeared. By the time that parliamentary elections were held in February 2000, there were some ten active political organizations (though not all succeeded in obtaining official registration). Six parties were allowed to nominate candidates, but only three passed the threshold 5 percent of the vote required to win proportional representation. The People's Democratic Party of Tajikistan, as anticipated, won a huge majority (Akiner 2002a: 456-58).

Civil War in Tajikistan

The only major conflict in Central Asia in the 1990s was the five-year civil war in Tajikistan. Social and economic conditions here had long been deteriorating. By the 1980s, there was high unemployment, especially among the young. Weapons and drugs from Afghanistan were readily available, and violent crime and other forms of antisocial behavior were prevalent.

Corruption had reached such proportions that it was having a serious impact on matters that directly affected the well being of the population, such as the provision of housing and transport. Nevertheless, the intellectual and cultural environment in Tajikistan was buoyant, and this was the only Central Asian state that showed, at the end of the Soviet era, genuine signs of democratization. The presidential elections held in November 1991 offered a real choice of candidates and, by comparison with electoral proceedings elsewhere in the region, the ballot was free and fair. The winner was Rahmon Nabiyev, a previous First Secretary of the Communist Party of Tajikistan. However, he had been out of office for some ten years by this time and was unfamiliar with the current political scene. The collapse of the Soviet Union left him bereft of institutional support. His political maneuverings were clumsy and he failed to establish a working relationship with the various opposition groupings (Akiner 2001a: 35-40). Within months, the country was sliding inexorably towards civil war.

Initially, the conflict was between three factions: old-style communists, aspiring democrats, and Islamic activists. Soon, however, regional, local, and even personal rivalries were grafted on to the struggle, melding with ideological issues. In September 1992, Nabiyev was forced to resign and the post of president was suspended. Two groupings emerged as the prime combatants: the government and the United Tajik Opposition (supposedly an umbrella organization for all the anti-government parties, but in reality, overwhelmingly dominated by the Islamic Rebirth Party). Violent conflicts broke out in several places. Most of the fighting was confined to the Tajik-Afghan border region, but there was severe economic and social disruption throughout the republic. The only area that escaped relatively unscathed was the northern province of Leninabad (later renamed Soghd). Elsewhere, the economy was devastated and thousands of refugees were left homeless and destitute. Within a year, nearly one sixth of the population (over 778,000 people) had fled Tajikistan: some 145,000 sought refuge in the Russian Federation, another 100,000 in Afghanistan (many of these were opposition fighters with their families), and the remainder in neighboring Central Asian republics and elsewhere in the CIS. By the time hostilities ceased, 35,000 houses had been destroyed. The death toll was estimated at 60,000, with many more individuals missing, lost without trace. Some 55,000 children were orphaned and thousands of women widowed; 26,000 families were left without their primary breadwinners (Akiner 2001a: 43-44).

Several foreign actors played a part in the Tajik civil war. Yet their involvement was never sufficient to give a decisive victory to any one faction. Uzbekistan, which shares a long border with Tajikistan, and Iran, which shares cultural and linguistic roots with the Tajiks, had an interest in the outcome of the conflict, but Russia was the most prominent external power. This was inevitable, not only for historic reasons, including the presence of a large Russian civilian population in the country, but also for the fact that only professional troops in Tajikistan in 1992 were under Slav command.[4] Official Russian policy regarding these men was one of neutrality, though it is alleged that they helped the pro-government forces with arms and vehicles. In October 1993, a joint CIS peacekeeping force—predominantly Russian, with contributions from Kazakhstan, Kyrgyzstan, and Uzbekistan—was deployed in Tajikistan (Akiner 2001a: 45-47).

The first round of intra-Tajik talks on national reconciliation was held in Moscow in April 1994, under the auspices of the United Nations. The UN Mission of Observers in Tajikistan was established in December 1994 to monitor the cease-fire that had been negotiated a few months earlier. Some degree of law and order had been restored in most parts of the country by this time. The post of president was revived and elections were held in November 1994. Imomali Rahmonov, a southerner from Kulyab province, won by a relatively (in Central Asian terms) narrow 25 percent majority. However, the government was still weak, and field commanders were increasingly prone to act on their own initiative. In 1995, the rise to power of the Taleban in neighboring Afghanistan posed a new threat, raising the possibility that this militant Islamist regime might invade Tajikistan. Pressures such as these finally convinced President Rahmonov of the need for compromise with the opposition. In December 1996, the warring factions in Moscow signed a preliminary protocol on the formation of a Commission for National Reconciliation. In June 1997, the Peace Agreement and Protocol of Mutual Understanding were concluded, which formally brought to an end the civil war. Many difficulties and issues of contention remained, but at the time of writing, some seven years later the peace was still holding. Moreover, the situation was stable enough for some degree of economic and political reform to be initiated (Akiner 2001a: 51-62).

Post-Soviet Nation-Building

The Central Asian states, within their present boundaries, were formed in the early years of Soviet rule.[5] The territorial division was accompanied by

[4] In September 1991, three elements of the Soviet armed forces were stationed in Tajikistan: the 201st Motor Rifle Division (with bases in Dushanbe, Qurghonteppa and Kulyab), a regiment of the Air Defense Forces, and the KGB Border Guards (along the Afghan and Chinese frontiers). After a brief period under CIS control, they were taken under Russian command.

[5] In 1924, as a result of the National Delimitation of Soviet Central Asia, five territorial-administrative entities were created. Of these, Uzbekistan and Turkmenistan immediately

a vigorous campaign to reshape the cultural and intellectual environment of the region. The aim was to eradicate the "harmful" and "obsolete" traditions of the past and in their place to inculcate "progressive" Soviet norms (Akiner 1998: 11-19). When the Soviet Union ceased to exist, the validity of this legacy was called into question. Thus, along with the problems and complexities of political and economic readjustment, the reshaping of the nation—its heritage, its very identity—was also an essential component of the process of consolidating independence.

One manifestation of this nation-building project was the rejection of Soviet-era state symbols in favor of images that reflected national aspirations. These included new flags, emblems, and anthems. Other symbols of statehood, such as postage stamps and banknotes,[6] likewise proclaimed the reality of independence. There was extensive renaming of cities and streets; where possible, pre-Soviet titles and institutions were revived. New ceremonies and rituals were created, and public holidays were introduced to mark events of historic national significance. Artists and scholars helped to provide intellectual validation for the new identities. Revisionist histories were produced that re-interpreted the past in a manner more in keeping with nationalist sentiments; particular emphasis was accorded to episodes that demonstrated resistance to Tsarist or Soviet rule. National pride was also expressed and elaborated in new architectural ensembles, as well as in the monumental paintings and sculptures that were commissioned to decorate public spaces.

Another aspect of nation-building was the importance accorded to language usage. During the Soviet era, Russian was widely used by people of all ethnic origins, particularly in urban areas. Towards the end of the 1980s, there was a reaction against this practice. The languages of the respective titular peoples were given the formal status of "state language" (e.g., Uzbek in Uzbekistan); efforts were made to promote the use of these languages in the public sphere. This trend continued after independence, with increased legal and administrative backing (Landau and Kellner-Heinkele 2001). An important symbolic step was the decision, adopted by four countries (excluding Tajikistan), to abolish the Cyrillic script,

---

acquired the status of full Union republics; Tajikistan became a Union republic in October 1929, Kazakhstan and Kyrgyzstan in December 1936. These republics, formed on the basis of ethno-linguistic criteria, were without historical precedent in the region. However, without any movements of population, the great majority of the main Turkic groups were encompassed within their own eponymous unit; the Tajiks fared less well, as some traditional areas of Tajik settlement were allocated to Uzbekistan (Akiner 2001a: 13-15).

[6] National currencies were introduced in Kyrgyzstan (*som*) in May 1993, in Kazakhstan (*tenge*) and Turkmenistan (*manat*) in November 1993, in Uzbekistan (*som*) in January 1994, and in Tajikistan (*somoni*) in October 2000.

introduced during the Soviet period, in favor of the Latin script.[7] By 2002, the shift to the new script had already been accomplished in Turkmenistan; in Uzbekistan, progress was slower, but consistent. In Kyrgyzstan and Kazakhstan, however, the Cyrillic script was still in common use.

The developments described above helped to consolidate new national identities, but they also encouraged the rise of ethnic nationalism. All the Central Asian states are multiethnic.[8] After independence, the energetic promotion of the language and culture of the titular peoples made the minorities feel insecure. They felt that they were being treated like second-class citizens and soon came to believe that they had no future in the region; this triggered a major exodus. Many of the emigrants were highly qualified professionals, and their sudden departure caused severe problems for the new states. Some effort was then made to reassure the minorities that their civil rights would be respected. This slowed the outflow somewhat, but did not succeed in halting it (Akiner 1997a: 16). The result has been that over the past decade there has been a gradual change in the ethnic composition of the Central Asian states, resulting in greater national homogeneity.

Role of Islam

One of the most difficult nation-building issues was the fashioning of new ideologies. Under Soviet rule, Marxism-Leninism had been an integral feature of society, shaping values and aspirations, as well as legitimizing the regime. After independence, this ideology was discredited in the eyes of many and was soon abandoned by the incumbent political elites. This led to a loss of social orientation and consequently, intellectuals and government officials alike came to regard the fashioning of "national ideologies" as a priority. In the absence of an explicit political doctrine, history and culture were mined to provide material for new national visions.

Islam, for centuries the dominant religion in Central Asia, had been suppressed during the Soviet period. After independence, however, in each of the republics, it was co-opted as one of the chief planks of the national

[7] In the pre-Soviet era and during the first decade of Soviet rule, these languages were written in the Arabic script; the Latin script was used c. 1930-40, but then replaced by the Cyrillic script.

[8] The percentage of the titular ethnic group in these states is approximately 50 percent in Kazakhstan and Kyrgyzstan, 70 percent in Tajikistan, 77 percent in Turkmenistan, and 80 percent in Uzbekistan. In each of these states there are numerous ethnic minorities (in Kazakhstan alone, there are over 100 different ethnic groups), but many such communities are quite small (Akiner 1997a: 19-37).

ideology. It soon acquired a status resembling that of an established state religion. Muslim dignitaries were regularly invited to participate in public ceremonies. Many new mosques and madrasa (Islamic college) were opened, and restrictions on making the hajj (prescribed pilgrimage to Mecca) were relaxed. The political elites embraced Islam as an alternative source of national legitimization (Akiner 2002b: 79). This trend was most marked in Uzbekistan and Kyrgyzstan, where the respective presidents took their oath of office on both the Qur'an and the Constitution to underline the dual importance of religion and law in society. Over the next few years, each of the Central Asian presidents performed the pilgrimage to Mecca.[9]

The separation of mosque and state is enshrined in the post-Soviet constitutions of all the Central Asian states. Yet the official promotion of Islam has led to a blurring of the boundaries between the secular and religious spheres. In Kazakhstan, the president at first avoided aligning himself too closely with the Islamic establishment, but gradually, he too began to transgress the limits of strict impartiality. In Turkmenistan, the president did not hesitate to appropriate religious honorifics and was referred to in the press by the titles traditionally reserved for the Caliph, and even the Prophet (Akiner 2002b: 83).

Historically, the overwhelming majority of Central Asian Muslims have professed orthodox Sunni Islam, following the Hanafi school of jurisprudence. This is the form of Islam that currently receives the support of the political leadership throughout the region. During the Soviet period, knowledge of Islam was reduced to a very low level; the great majority of Central Asians had little, if any, understanding of the tenets of the faith and the ritual obligations. However, since independence, thousands of mosques and hundreds of part-time and full-time Muslim schools and colleges have been opened.[10] Many of the younger generation receive Muslim instruction and attend the mosque regularly. Thus, there is now a greater awareness of Islam and a more active commitment to the faith. This is increasingly reflected in aspects of personal behavior, for example, the observance of dietary prohibitions. Some girls have adopted a Muslim dress code, wearing the *hejab* (headscarf) and long, loose outer garments.

Yet there is a profound ambivalence towards the faith. On the one hand, there is a general acknowledgement that Islam is not only an integral part of the national culture, but represents 'good' values that are beneficial

[9] They have undertaken the *umrah* (lesser pilgrimage), which involves fewer rituals than the hajj and can be made at almost any time of the year.

[10] In Kyrgyzstan, for example, there were only 34 mosques open for worship in 1987, but about 1000 in 1994; in Uzbekistan, in the same period, the number rose from 87 to 3,000. There were similar increases in places of worship in the other Central Asian states (Akiner 2002b).

for the development of the state; on the other hand, there is acute fear of militant Islam. The political leadership, especially in Uzbekistan, tries to counteract the latter by promoting an "acceptable" version of Islam, while repressing alternative interpretations (Akiner 2002b: 81). This is not regarded as a matter either for public debate or for personal choice. Rather, it is seen as an issue that directly impacts on national security and consequently, must be rigorously controlled and monitored.

Anxiety over the rise of Islamic militancy did not emerge in a vacuum. When the Soviet Union collapsed, powerful Islamists in Afghanistan (and reportedly, in Pakistan, too) were eager to foment insurrection in the Central Asian states. During the Tajik civil war, the main opposition faction was avowedly Islamist and received military support from Muslim groups abroad. In Uzbekistan, meanwhile, clandestine groups soon began challenging the authority and legitimacy of the official Muslim establishment. This resulted in civil disturbances and several armed clashes. The geographical centre of the Islamist activity has long been the Ferghana Valley, a fertile, densely populated region where Kyrgyzstan, Tajikistan, and Uzbekistan converge. This is an area that has a history of disaffection and rebellion, stretching back to pre-Soviet as well as Soviet times.

Currently, the main Islamist groups in Central Asia are *Hezb-i Tahrir* (Liberation Party) and the Islamic Movement of Uzbekistan (Akiner 2002b: 86-89). Hezb-i Tahrir (HT) is an international organization, founded in Jerusalem in 1953.[11] It is proscribed in many countries (including in the Middle East), but acts openly in Western Europe and parts of the CIS. The Islamic Movement of Uzbekistan (IMU) is a home-grown group that developed out of Soviet-era Islamic revivalist movements. Given the dearth of reliable information, it is impossible to know the degree to which HT and the IMU are linked. It is also difficult to estimate their relative size; anecdotal evidence suggests that HT is the larger (possibly numbering some 50,000 adherents), with a broader geographical spread. Both groups aim to establish an Islamic state based on the model of the early Caliphate. The IMU espouses military confrontation; HT, at least in its publications, advocates the use of peaceful means to achieve its goals, though there are hints that it would sanction the use of force if necessary.

Since the late 1990s, the government of Uzbekistan has been accusing both groups of acts of terrorism, also of plotting to overthrow the government and the constitutional order of the country. Human rights organizations are dubious as to the validity of the evidence that has been produced against them (HRW 2001). However, other Central

[11] The founder was a Palestinian, Sheikh Taki ad-din Nabhani (1909-78), a judge in the Sharia court, Haifa; he later moved to Nablus.

Asian governments have come to the conclusion that these movements do represent a danger to regional security. Consequently, they have adopted increasingly repressive measures towards them. Some Western governments have concurred with the view that the Islamist threat is serious. In September 2000, the US State Department placed the IMU on the list of international terrorist organizations to which US citizens are forbidden to give assistance, and whose members are denied entry to the USA. With the launching of the Western-led "War on Terror" in late 2001, the Uzbek government and, to a somewhat lesser extent, the other Central Asian governments, intensified their campaign against suspected Islamic radicals. This has resulted in a curtailment of civil liberties for all and an increase in reports of gross violations of human rights (AI 2001; HRW 2003). However, it does not appear to have seriously deterred the spread of Islamist activity in the region.

International Relations and Regional Cooperation

Before independence, the Central Asia republics were largely isolated from the external world. The international borders with neighboring countries were tightly controlled, and there were almost no direct communication or transport links. All foreign relations were handled by Moscow. Few Central Asians traveled outside the Soviet bloc; equally, not many foreigners visited Central Asia. Thus, after independence, a process of mutual exploration and fact-finding was required before international relationships could be established.

The first step towards international integration was the accession of the five states to the United Nations on March 2, 1992. Thereafter, they joined several UN-affiliated special agencies, funds, and programs (e.g., UNDP, UNESCO, WHO, IMF, and World Bank). Kyrgyzstan is to date the only Central Asian state that has been accepted as a member of the World Trade Organization, though Uzbekistan and Kazakhstan are current applicants, and Turkmenistan and Tajikistan have observer status. The new states also joined several non-UN international governmental organizations, including the Organization for Islamic Conference;[12] the North Atlantic Cooperation Council; the NATO Partnership for Peace programme; and the Organization for Security and Cooperation in Europe. Turkmenistan and Uzbekistan are members of the Non-Aligned Movement.[13] All five states have joined the Asian Development Bank, the European Bank for Reconstruction, and the Islamic Development Bank.

[12] Kyrgyzstan, Tajikistan, and Turkmenistan joined in 1992, Kazakhstan in 1995 and Uzbekistan in 1996.

[13] Uzbekistan joined in 1992, Turkmenistan in 1995.

In the early 1990s, foreign policy in the Central Asian states was mainly reactive, a response to external overtures. However, two sets of national priorities gradually emerged. One was to secure capital investment and technical assistance from abroad. As discussed in other papers in this volume, some progress has been made in this direction, particularly as regards the development of the oil and gas sectors. The other priority was the diversification of transport routes, to facilitate access to world markets. Symbolically, this represented the recreation of the ancient "Silk Roads" and, hence, the reintegration of Central Asia into the global economy. Road, rail, and air links are already in place with China in the east, and with Iran in the southwest. Thus, the main segments of a trans-Asian transport corridor, stretching from the Yellow Sea to the Gulf, now exist and could become functional when political and economic conditions are ripe. There is also an ambitious EU-funded plan to create a transport corridor linking Europe to the Caucasus and Central Asia (TRACECA). At present, the main emphasis of the project is to improve facilities in the southern Caucasus. In the longer term, however, it is hoped that a network will be created that will encompass the Central Asian states.

After the collapse of the Soviet Union, the Central Asian leaders firmly stated their intention to avoid becoming part of any exclusive ideological or political bloc. Nevertheless, there was much speculation, particularly in the West, as to whether Iran or Turkey would emerge as the dominant influence, and by extension, whether the Central Asian states would adopt the "Iranian model" of Islamic monocracy, or the "Turkish model" of secular democracy. In fact, neither country gained ascendancy. Iran, a Shi'a Muslim state, has had very limited influence in religious matters, since Central Asia is overwhelmingly Sunni. In the economic sphere, it does not have the resources to provide the investment that the new states require. By contrast, Turkey seemed to be in a more advantageous position. This was partly because of a strong sense of kinship, based on ethnic and linguistic affinities, between the Turks of Turkey and the Turkic peoples of Central Asia. However, there were also other considerations that inclined the Central Asians to favor Turkey. Chiefly, there was an expectation that Turkey would be able to provide unlimited economic assistance. This was prompted by what, in reality, was an overestimation of Turkey's level of development. Moreover, Western governments, eager to counter any possible Iranian influence, encouraged and supported Turkey's efforts to play a leading role in the region; indeed, Turkey was often described as the bridge between the Central Asian states and the international community. Yet this was not a sustainable policy (Bal 2000). When the Central Asians gained a better understanding of world affairs, they were soon disabused of their illusions regarding Turkey's capabilities. Meanwhile, there was resentment at the lack of respect shown by some Turks towards Central Asian languages and culture; this provoked accusations of "cultural imperialism:" Gradually, Turkey's image in the region lost some of its early luster. Diplomatic relations remained cordial, apart from bouts of friction with Uzbekistan, and Turkish business interests continued to be well represented in the region, but no pan-Turkic "special relationship" emerged.[14]

Other countries have shown varying degrees of commitment to developing links with the Central Asian states. There was an initial burst of interest from the Arab world, but thereafter the engagement was relatively slight. The one Middle Eastern country that has had significant success in the region is Israel; virtually unnoticed by the outside world, it has established close diplomatic, economic, and security ties with all the Central Asian states (Boucek 1999). Among Asian countries, South Korea was a pioneer investor, with a particular interest in construction and manufacturing enterprises; however, domestic crises subsequently caused some retrenchment. Japan made a slower, more phased entry into the Central Asian field, concentrating initially on humanitarian projects and only later engaging in large-scale economic ventures. Relations with other Asian countries have, for the most part, been spasmodic, though some, such as India, have been working towards the creation of an institutional framework to facilitate the deepening of bilateral relations (Stobdan 1999; Gopal 2003).

The EU presence in Central Asia has largely focused on technical and humanitarian assistance, though some member states have established substantial economic interests.[15] US involvement in the region during the 1990s was predominantly directed towards the oil industry in Kazakhstan; elsewhere in Central Asia, activities were mainly linked to aid programs. This situation changed dramatically in autumn 2001. With the launching of the campaign to destroy the bases of the Taleban and al-Qaeda

[14] In 1992, Turkey initiated regular meetings between the heads of the CIS Turkic states (Azerbaijan, Kazakhstan, Kyrgyzstan, Turkmenistan, and Uzbekistan). However, the Central Asian leaders evinced little enthusiasm for integration, and rejected plans for such projects as the creation of a Turkic Common Market and a Turkic Development and Investment Bank. In his address to the seventh Turkic Summit (Istanbul, April 2001), Turkish President Sezer spoke of the role of these meetings in promoting bilateral and multilateral cooperation between member states. Yet in the past decade there has been little structural evolution. Moreover, the Turkic Summits do not appear to have developed mechanisms for resolving, or defusing, tensions between member states. The strengthening of ethno-linguistic ties has also not proceeded as rapidly as anticipated (Akiner 2001b).

[15] EU Partnership and Cooperation Agreements (PCAs) with Kazakhstan, Kyrgyzstan, and Uzbekistan came into force in July 1999; an EU PCA was signed with Turkmenistan in 1998, but has not yet been fully ratified. Tajikistan signed an EU Partnership and Cooperation Agreement in October 2004.

in Afghanistan, the Central Asian states acquired strategic importance for the Western alliance. Two US air bases were established, one in Uzbekistan and one in Kyrgyzstan. In 2002, plans were announced for the refurbishment and expansion of these facilities. At the time of writing it seemed unlikely that the USA was contemplating an early reduction of its military presence in the region.

Regional cooperation was pursued concurrently with the development of broader international relations. All the Central Asian states were signatories of the Minsk Accord of December 30, 1991, whereby the Commonwealth of Independent States (CIS) was created. It soon became evident, however, that this entity was not of itself a suitable vehicle for enhancing regional cooperation. Consequently, during the 1990s the Central Asian states explored different approaches to regional partnerships (Akiner 2001b). These included developing bilateral cooperation with Russia and with other CIS members. Turkmenistan adopted a stance of neutrality and remained outside most of the new groupings.

By 2002, a number of regional bodies had been created, though only four had been institutionalized. Two of these, the Central Asian Cooperation Organization (Kazakhstan, Kyrgyzstan, Tajikistan, and Uzbekistan)[16] and the Eurasian Economic Community (Kazakhstan, Kyrgyzstan, Tajikistan, Belarus, and the Russian Federation)[17] were created by CIS member states. Both developed out of efforts, initiated in the early 1990s, to strengthen economic cooperation. However, despite several phases of restructuring, internal problems of planning and coordination continued to hamper progress. The Central Asian Cooperation Organization in particular was plagued by dissension over goals and policies, as well as by hegemonic rivalry between Uzbekistan and Kazakhstan. In 2004 the nature of this organization was unexpectedly changed by the decision to invite the Russian Federation to become a member. This signaled a desire on the part of the Central Asian states to strengthen their links with Moscow and equally demonstrated Moscow's intention to remain engaged in the region.

[16] In 1993, the Central Asian Union was established between Uzbekistan and Kazakhstan, later joined by Kyrgyzstan. In 1998, after the accession of Tajikistan, the organization was transformed into the Central Asian Economic Community. In March 2002, this became the Central Asian Cooperation Organization.

[17] In early 1995, a customs union was concluded between Kazakhstan, Kyrgyzstan, Russia, and Belarus. This was the basis for the 1996 quadripartite agreement on "The Regulation of Economic and Humanitarian Integration"; Tajikistan acceded to this treaty in 1998. In April 2001 this five-member group was formally transformed into the Eurasian Economic Community (Akiner 2001b).

The membership of two other organizations extends beyond the CIS borders. The Economic Cooperation Organization, formed on the basis of previous alliances between Iran, Pakistan, and Turkey,[18] includes all five Central Asian states, likewise Azerbaijan and Afghanistan. An intergovernmental organization, it aims to promote economic, technical, and cultural cooperation among member states. The Shanghai Cooperation Organization (SCO) also has an economic dimension, but it is more directly concerned with political and security issues. It developed out of initiatives to resolve border issues between China and its CIS neighbors. The first step was the signing of the Treaty on Deepening Military Trust in Border Regions by the heads of state of the Russian Federation, Kazakhstan, Kyrgyzstan, Tajikistan, and China, in Shanghai in April 1996. In June 2001, the group was formally established as a regional organization; Uzbekistan, though not a border state, was also admitted at this time. The declared aims of the SCO included the creation of a new international order, based on the principle of multipolarity. A separate Shanghai Convention on Combating Terrorism, Separatism, and Extremism, also signed by the SCO heads of state during the June 2001 summit meeting, underlined the importance accorded to regional security. In June 2004, the SCO Regional Antiterrorism Centre was inaugurated in Tashkent.

These organizations are potentially of crucial importance for Central Asia, since many of the most urgent problems that face these new states are of a transnational nature, such as water management, environmental degradation, terrorism, trafficking in drugs, arms and humans, and illegal migration. The stabilization and reconstruction of post-Taleban Afghanistan is also a matter that directly concerns the neighboring states. Issues such as these can only be addressed on the basis of close cooperation (IISS 2002). However, the regional organizations mentioned above are still at an embryonic stage of development, and it is not yet clear how effective they are likely to be. At present, they comprise overlapping sets of members and are pursuing overlapping agendas. To date, there has not been a significant clash of interests, since activities have focused mainly on internal structural and organizational issues. In the future, however, choices will certainly have to be made if these bodies are to survive and be compatible. This will require strategic planning on the part of member states in order to prioritize their objectives and clarify their commitments.

[18] The first was the Baghdad Pact of 1955, which, with some change of membership, became the Central Treaty Organization (CENTO) in 1958. In 1985, Iran tried to revitalize the alliance, now renamed the Economic Cooperation Organization. However, there was little activity until February 1992, when Iran hosted the first summit meeting of heads of member states in Tehran.

There is also a need to find a means of engaging with non-member states, such as Turkmenistan, that wish to retain political distance from a given grouping, but, by virtue of their geography, are physically involved in many common issues. Thus, although some progress has been made towards establishing the modalities of regional cooperation, the real challenges of joint action lie ahead.

Conclusions

The prognosis for the Central Asian states today is more equivocal than it was ten years ago. The difficulties of negotiating the complex process of social, political, and economic readjustment, and yet preventing the onset of chronic instability and civil disorder, are now more apparent. The harmful aspects of the Soviet legacy are proving to be more tenacious than had been anticipated. At the same time, many of the positive features of Soviet rule are rapidly being undermined, particularly as regards achievements in education and healthcare. These internal processes are further complicated by external factors. Instability in the global context, and particularly in the broader region nearby and neighboring states, is having a deleterious effect on Central Asia.

The dominant trends, as recorded in numerous regional reports and surveys, convey an impression of deteriorating conditions. Yet it is important to balance this assessment by noting that, nevertheless, there have been some encouraging achievements. In Kazakhstan, Kyrgyzstan, and Uzbekistan, progress has been made on a number of social, economic, and legal issues, albeit slowly, erratically, and often on a case-by-case basis. In Tajikistan, the fact that the peace agreement concluded in 1997, whatever its shortcomings (and admittedly, there were many), has remained in force to date is a remarkable accomplishment. It has permitted a significant degree of recovery and normalization to take place in a relatively short period. In Turkmenistan, despite the high level of political repression, until 2002, in some respects, the private individual here had more freedom than citizens in other Central Asian states, particularly as regards traveling aboard or engaging in private business. However, this changed in early 2003, as crippling new restrictions were introduced to control such activities. There was a new wave of repression, marked by summary arrests and confiscation of property. The right to hold dual Russian-Turkmen citizenship was suddenly revoked in April, triggering a panic-stricken exodus not only of ethnic Russians, but also of representatives of other ethnic groups, including Turkmen, who had acquired Russian passports as a form of insurance, believing that this would allow them to leave the country voluntarily, if and when they should so wish. The loss of dual citizenship provoked fears of being trapped in an increasingly unstable environment, subject to an arbitrary, unpredictable regime.[19]

Looking ahead, the leaders of the Central Asian states face many daunting challenges. Some of these matters (e.g., environmental and economic issues) are discussed elsewhere in this volume. Regarding the questions that have been raised in this paper, there are four main areas of concern. One is the need to protect and strengthen civil liberties, while at the same time combating the very real dangers of terrorism and organized crime. This is not an easy balance to achieve; measures introduced in "mature democracies" of the West in the wake of the events of September 2001 have demonstrated how readily cherished freedoms can be sacrificed when national security is perceived to be under threat. The Central Asian states are in a far more precarious position, not only because independence is a very new experience and statecraft is still at an early stage, but also because the abrupt change of system that followed the collapse of the Soviet Union has placed these societies under immense strain. Yet this is not an issue that can be avoided. If these states are to develop in an orderly fashion, then a better balance must be achieved between security and freedom than exists at present.

A second area of concern is that of political succession. This may not at present seem to be a matter of urgency. Compared to other world figures, the Central Asian leaders are not unusually elderly. Moreover, the Turkmen president has a mandate for life. The terms of the other presidents are due to expire in a few years, but past experience has shown that they could well choose to remain in office for several more years. However, eventually there must be a transfer of power. It is highly unlikely that "open and fair elections" will be the mechanism by which this is achieved. It is possible that the current incumbents will try to promote a protégé as successor; this might ensure a relatively smooth transition, though the problems of authoritarian governance would be perpetuated. More probably, the succession will be decided by internecine struggles between powerful interest groups in business (legitimate and criminal) and the security forces. There are already signs that "palace coups" are being hatched. So far they have been easily crushed and have scarcely been commented on outside the immediate entourage of the ruling elites (unless the latter have exploited them for propaganda advantage, as happened in November 2002 following the attempted assassination of the Turkmen leader). There may yet be time to devise a more rational strategy for

[19] The protocol abolishing dual citizenship of Turkmenistan and the Russian Federation was signed on 10 April 2003, during President Niyazov's state visit to Moscow (RFE/RL Newsline, 18 April 2003).

managing the succession, but to date there is little sign of this in any of the Central Asian states.

A third area of concern is the deterioration in relations between the Central Asian states themselves. As described above, efforts to create regional institutions are still at a rudimentary stage, and meanwhile, issues of interstate conflict are increasing. In addition to problems such as the management of transnational water resources and the regulation of frontiers, there are now growing fears of cross-border political interference. In Tajikistan, there has long been concern that Uzbekistan has been fomenting unrest, and even supporting rebel attempts to overthrow the government. Similar accusations were made against Uzbekistan by the Turkmen authorities after the assassination attempt mentioned above in late 2002. The Uzbek government has consistently denied all such charges, yet the suspicion remains in some Central Asian circles that Uzbekistan is seeking to enhance its own security by attempting to install client regimes in neighboring states.

Finally, there is the issue of great power rivalry in Central Asia. During the 1990s, this was scarcely perceptible. Post-September 2001, however, has given such rivalry a new immediacy. The major players are China, Russia, and the USA. There has been no coordinated Central Asian response to this situation. Rather, each of the Central Asian states has sought to develop an independent relationship with one or more of the great powers. At the same time, these powers are competing against each other to draw the Central Asian states into their exclusive orbit of influence. This is having two consequences. First, it heightens the possibility of military confrontation in the region. Russia acquiesced in the establishment of US bases in Uzbekistan and Kyrgyzstan, but in October 2003 expanded its own presence in the region by inaugurating a military base in Kyrgyzstan, near the US location. In late 2004, Russian troops were withdrawn from the Tajik-Afghan border, but overall deployment in Tajikistan was upgraded and a permanent base established. China has to date remained silent. Yet, already concerned by the eastward expansion of NATO, it has viewed the US penetration of the region with dismay. Its present capabilities are very limited, but it has provided military aid to the Central Asian states (notably Tajikistan) and is seeking to enhance cooperation between the defense forces of members of the Shanghai Cooperation Organization. Should relations between these three world powers deteriorate, possibly in conjunction with developments elsewhere, then Central Asia would become a frontline zone of confrontation. The second consequence of great power rivalry in the region is the impact that this is having on domestic politics. Foreign governments seek the support of sympathetic leaders to promote their interests and block those of rival states; equally, local actors, whether in power or hoping to come to power, court foreign sponsors in the hope of crushing their opponents. Thus, alliances of convenience are being formed on the basis of different sets of ambitions. This can but heighten the tendency to intrigue and covert manipulation of internal tensions, which in turn risks triggering prolonged conflict and instability in the region.

Today, the situation in each of the Central Asian states is more fragile than at the time of the disintegration of the Soviet Union. Internal and external pressures have combined to produce conditions of extreme vulnerability and fluidity. There are too many variables for it to be possible to predict with any degree of certainty the likely course of events, even as regards the near future. In these circumstances, it would be simplistic to assume that the task of consolidating nationhood and statehood could be anything other than arduous. It will require consummate skill, fortitude and no small measure of sheer good luck to ensure a peaceful and prosperous outcome.

References

AKINER, S.

2002 "Islam in Post-Soviet Central Asia: Contested Territory." Pp. 73-101 in Strasser, A. et al. (eds.), *Zentralasien und Islam/Central Asia and Islam.* Hamburg: Deutsches Orient-Institut.

2002 "Country Entries" in Day, A. (ed.), *Political Parties of the World* (5th edition). London: John Harper Publishing.

2001 "Regional Cooperation in Central Asia." Pp. 187-208 in Weichhardt, R. (ed.), *Economic Developments and Reforms in Cooperation Partner Countries: The Interrelationship Between Regional Economic Cooperation, Security and Stability.* Brussels: NATO.

2001 Tajikistan: Disintegration or Reconciliation? *Royal Institute of International Affairs.* London.

2000 "Emerging Political Order in the New Caspian States: Azerbaijan, Kazakhstan and Turkmenistan." Pp. 90-128 in Bertsch, Gary K. et al. (eds.), *Crossroads and Conflict: Security and Foreign Policy in the Caucasus and Central Asia.* New York: Routledge Inc.

1998 "Social and Political Reorganization in Central Asia: Transition from Pre-Colonial to Post-Colonial Society." Pp. 1-34 in Atabaki, T. and O'Kane, J. (eds.), *Post-Soviet Central Asia.* London: Tauris Academic Studies.

1997 "Potential Threats to Stability and Social Cohesion in Central Asia." Pp. 47-60 in Weichhardt, R. (ed.), *Economic Developments in Partner Countries: Social and Human Dimensions.* Brussels: NATO.

1997 *Minorities in a Time of Change: Prospects for Conflict, Stability and Development in Central Asia.* London: Minority Rights Group.

AMNESTY INTERNATIONAL (AI)

2001 "Central Asia: No excuse for Escalating Human Rights Violations" in AI-index EUR. Retrieved November 21, 2002 (*http://www.web.amnesty.org/ai.nsf/Recent/EUR040022001*).

ANDERSON, J.

1999 *Kyrgyzstan: Central Asia's Island of Democracy?* Amsterdam: Harwood Academic Publishers.

BAL, I.

2000 *Turkey's Relations with the West and the Turkic Republics: The Rise and Fall of the 'Turkish Model.'* Aldershot, UK (Ashgate).

BOUCEK, C.
1999 "An Impact Greater than its Size: Israeli Foreign Policy in Central Asia." Masters Dissertation, School of Oriental and African Studies, University of London, UK.
EBRD
1994 *Transition Report*. London: European Bank for Reconstruction and Development.
FREEDOM HOUSE (FH)
2002 *Nations in Transit*. Retrieved November 21, 2002 (*http://www.freedomhouse.org/research/nattransit.htm*).
GOPAL, S.
2003 "India and Central Asia: the Preceding Decade." Unpublished paper.
HUMAN RIGHTS WATCH (HRW)
2003 "Europe and Central Asia Overview" in *HRW World Report 2003*. Retrieved February 13, 2003 (*http://www.hrw.org/wr2k3/europe.html*).
2001 "Memorandum to the US Government Regarding Religious Persecution in Uzbekistan" in *HRW World Report 2001: Europe and Central Asia*. Retrieved November 21, 2002 (*http://www.hrw.org/backgrounder/eca/uzbek-aug/persecution.htm*).
1999 Kazakhstan: "Violations of the Right to Freedom of Expression" in *Kazakhstan: Freedom of the Media and Political Freedoms in the Prelude to the 1999 Elections*. Retrieved November 21, 2002 (*http://www.hrw.org/reports/1999/kazakhstan/Kaz1099b-04.htm*).
INTERNATIONAL CRISIS GROUP (ICG)
2002 "Central Asia: The Politics of Police Reform." *Asia Report* no. 42. Osh/Brussels (ICG). Retrieved November 21, 2002 (*www.crisisweb.org*).
INTERNATIONAL INSTITUTE FOR STRATEGIC STUDIES (IISS)
2002 *Central Asia and the Post-Conflict Stabilization of Afghanistan*. London: International Institute for Strategic Studies.
ISHIYAMA, J.
2002 "Neopatrimonialism and the Prospects for Democratization in the Central Asian Republics." Pp. 42-58 in Cummings, S. (ed.), *Power and Change in Central Asia*. London: Routledge.
LANDAU, J.M. AND B. KELLNER-HEINKELE
2001 *Politics of Language in the Ex-Soviet Muslim States*. London: Hurst and Co.
MCCORMACK, G. (ED.)
1999 *Country Entries in Media in the CIS: A Study of the Political, Legislative and Socio-Economic Framework* (2nd edition). Düsseldorf: European Institute for Media.
OLCOTT, M.B.
2002 *Kazakhstan: Unfulfilled Promise*. Washington DC: Carnegie Endowment for International Peace.
1997 "The Growth of Political Participation in Kazakhstan." Pp. 201-241 in Dawisha, K. and Parrott, B. (eds.), *Conflict Cleavage and Change in Central Asia and the Caucasus*. Cambridge: Cambridge University Press.
RUFFIN, M. ET AL. (EDS.)
1999 *Civil Society in Central Asia*. Seattle: University of Washington Press.
STOBDAN, P.
1999 *Building a Common Future: Indian and Uzbek Perspectives on Security and Economic Issues*. New Delhi: Institute for Defense Studies and Analyses.
TRANSPARENCY INTERNATIONAL (TI)
2001 *Corruption Perceptions Index*. Retrieved November 21, 2002 (*http://www.transparency.org/pressreleases_archive/2002/2002.08.28.cpi.en.html*).
2002 *Global Corruption Report*. Retrieved November 21, 2002 (*http://www.transparency.org*).
US DOJ (DEPARTMENT OF JUSTICE) IMMIGRATION AND NATURALIZATION SERVICE
1994 *Uzbekistan: Political Conditions in the Post-Soviet Era*. Retrieved November 21, 2002 (*http://heiwww.unige.ch/humanrts/ins/inservie.htm*).
1993 *Turkmenistan: Political Conditions in the Post-Soviet Era*. Retrieved November 21, 2002 (*http://heiwww.unige.ch/humanrts/ins/inservie.htm*).

V. The Economic and Social Impact of Systemic Transition in Central Asia and Azerbaijan

MICHAEL KASER

ABSTRACT

The economies of Azerbaijan, Kazakhstan, Kyrgyzstan, Tajikistan, Turkmenistan, and Uzbekistan differ from the other states that quit the Soviet Union in 1991 by their inheritance of poor productivity growth and high demographic pressure for job creation. Moreover, since their incorporation into the Russian Empire during the nineteenth century, their production has been geared to primary goods—cotton and hydrocarbons—that in the 1930s Stalin's policy towards autarky was directed to Soviet domestic consumption. The six countries hence gained independence, but with high export dependency on markets that all suffered severe demand recessions. The corresponding production decline in the six states was modified during the 1990s by diversifying the direction of trade and was not as deep as indicated by the official GDP data by reason of the substantial growth of unmeasured production. That 'shadow economy' goes untaxed and all six states show government revenue inadequate for the social expenditure required to maintain the stock of human capital inherited from Soviet planning priorities and to reverse the widening of income differentials, as well as for capital formation to employ the expanding labor force. Some improvement has resulted from emigration and foreign investment by Kazakhstan, and from foreign investment by Azerbaijan. But that inflow has enhanced those states' dependence on hydrocarbons and the danger of a "Dutch disease." In all six states, authoritarian and corruption-prone governance inhibit foreign investment, though in two, Azerbaijan and Kazakhstan, state funds have been established so that eventual income from fixed assets replace that from depleting hydrocarbon deposits.

Six Labor-Surplus and Trade-Dependent Economies

The six Soviet successor states from the Caspian Sea to the Hindu Kush Mountains quit the USSR in 1991 amid economic and social conditions that differed profoundly from those of their Slavic partners. Whereas the Soviet economic decline, that was an important factor in the break-up of the Union, could partly be attributed to the shortage of mass labor inputs on which Stalinist-type growth had relied, the six successor states had surpluses for which employment had to be found if eventual social unrest were to be avoided. Thus, in 1989, the Turkmen population aged 10-14 was 357% of that aged 50-54, the Uzbek 347%, the Kyrgyz 274%, the Kazakh 192% and the Azeri 191%, whereas in Belarus, it was 112%, in Russia 110% and in the Ukraine it was as low as 98% (Timofeev and Perevedentsev 1994: 55). Moreover, the growth of labor productivity in the Central Asian states had been lower during 1970-90 than in the other ten Soviet republics, only 1.5% annually in Kyrgyzstan, 1.2% in Uzbekistan, 0.8% in Tajikistan, 0.7% in Kazakhstan, and had even been negative at –0.3% in Turkmenistan, Azerbaijan having been the lowest of the ten others at +2.6%.[1] The contrast with Russia is marked: during 1970-90 material-sector labour productivity grew annually by 2.9% and in industry by 3.4% (though it declined in agriculture).

The combination of demographic pressure and low labor productivity has marked the development of the six states during one to two centuries of incorporation within the Russian economic sphere.[2] A comparative advantage in economic terms was allied to the strategic interests of both Tsarist and Soviet policymakers. The three Caspian countries plus Uzbekistan would be primary producers for export or for supplying industry elsewhere in the vast Russian/Soviet territory. Oil was extracted in Azerbaijan (Baku) and Kazakhstan (Emba/Gur'yev) before the October Revolution; all save Kyrgyzstan were substantial cotton-growers; and Kazakhstan and Kyrgyzstan mined non-ferrous ores. The economic-strategic linkage was symbolized by four transport routes—from the Imperial period the oil pipeline from Baku to Batumi (opened in 1906) and the Trans-Caspian Railway from Krasnovodsk into Turkestan (and later to

[1] Data refer to net material product, which falls short of GDP by the value of nonproductive services and an allowance for capital depreciation.

[2] The first Azeri khanates submitted to the Tsar in 1804 and the Badakhshan region of Tajikistan was annexed in 1895.

Dushanbe, capital of Tajikistan); the TurkSib Railway, planned in Tsarist times and implemented in Stalin's First Five-year Plan; and the Orenburg-Tashkent line of the Third Five-year Plan. After the Second World War trunk pipelines were laid to carry Azeri and Central Asian oil and gas northward to Soviet users. Together, with railways along the eastern coast of the Black Sea and the western coast of the Caspian Sea, they focused the three economies onto cotton, metal, oil, and gas production for exchange against industrial products and foodstuffs with the other Soviet republics.

During Soviet times political antagonism with, and the underdevelopment of, their southern neighbors (eastern Turkey, northern Iran, and northern Afghanistan) precluded trade with nearby market economies and the growth of East-West trade in the later Soviet era benefited the Slav, rather than the Turkic, Union-republics. The trade of Kazakhstan and Kyrgyzstan fluctuated with Sino-Soviet relations, flourishing during the Soviet occupation of Sinkiang, a fair volume of trade until Khrushchev's break with Mao in 1960, and small thereafter, however, trade was inhibited by dependence on old caravan routes: the laying of a railway between Kazakhstan and China neared their frontier in 1960, but was only completed after the USSR collapsed.

The initial conditions of independence gained in 1991 were therefore high trade dependence, largely directed to the rest of the USSR. Such dependence, though with more varied partners, has persisted under transition. For four states, the trade to GDP ratio[3] in 2001 was close to that of 1990, the last full year of its Soviet period: Azerbaijan's was 31% in 2001 (35% in 1990); Kazakhstan's was 39% (43% in 1990); Turkmenistan's was 35% (44% in 1990); and Uzbekistan's was 43% (44% in 1990). In the two other republics, the trade ratio has sharply changed. In Kyrgyzstan, it fell from 50% in 1990 to 34% in 2001, mainly because Russia in the 1990s ceased buying from Kyrgyz manufactures, chiefly their defense goods. Tajikistan's ratio, by contrast, already a high 48% in 1990, leapt to 67% in 2001 because its non-farm sectors were still in tatters after a civil war which ended uneasily only in 1997. All were much more trade dependent on the eve of independence than contemporary Turkey at 21% and Iran at 6%. Trade dependence on the rest of the USSR meant that the recession of all ex-Soviet partners during the 1990s was transmitted to the Caucasus and Central Asia. For the CIS, their aggregate measured GDP declined every year from 1990 to 1998 inclusive, save for a minor 1.0% spike in 1997. The significance of focusing on "measured GDP" is considered further below.

[3] Average of imports and exports as percentage of GDP.

The rational reaction of CIS members was to seek export markets outside the group where demand was strong, and all their currencies had devalued with respect to the dollar and the west European currencies, essentially now sterling and the euro. Thus between 1995 and 2003, Azerbaijan switched its non-CIS exports from 55% to 87%, Kazakhstan from 45% to 77%, Kyrgyzstan from 34% to 65%, and Tajikistan from 66% to 83%; Uzbekistan went from 61% to 72% in 2002 (ECE, 2004: 89). On the other hand, Turkmenistan showed little change—51% in 1995 and 54% in 2003. The latter did increase the non-CIS share in its imports—from 45% in 1995 to 65% in 2003, whereas Azerbaijan and Uzbekistan showed little change—from 66% and 59% in 1995 respectively to 68% and 63% in 2003, but there were substantial rises in Kazakhstan from 30% to 53%, in Kyrgyzstan from 32% to 41%, and in Turkmenistan from 45% to 65% (ECE, 2004: 90). In the case of Azerbaijan the reorientation of exports is attributable to the loss of trade with its neighbors, an outcome of conflict in the Caucasus—it lost control of Karabakh and adjoining territory to Armenia; Georgia was torn by internal separatism; frontier traffic with the Russian Federation became disrupted—above all, across to Chechnya and Dagestan. Turnover among the three Caucasus states dropped to a mere 3% of their total turnover in 1998. Even among neighboring states of Central Asia (Kyrgyzstan, Turkmenistan and Uzbekistan), mutual turnover was only 13% of their total.

An important caveat on the percentage changes just cited for the first decade of transition is on the fluctuating and ill-measured "shuttle" trade. There had been no barriers to trade among Soviet Union-republics and with transport at irrationally-low tariffs; enterprises and individuals had taken advantage of the "shortage economy" to carry goods surplus in one region to areas of deficit and thereby reaping profit. Among state-owned enterprises, the agents, termed "pushers," arranged deals to lubricate the shortcomings and constraints of central planning; among individuals, people from Central Asia and the Caucasus supplied fruits, vegetable, and flowers to the farmers' markets of the Slav and Baltic republics. After 1990, the latter flows extended to consumer goods from foreign countries (particularly from Turkey, Singapore, India, Dubai, and China) against which a wide range of domestic goods was traded. Estimates of such trade, when it was at its peak in the mid-1990s, shows that "shuttle" imports exceeded "shuttle" exports, with the balance often settled by "flight capital" arising from declared trade, for example, by underinvoicing legitimate sales and the foreign importer paying the difference into an undercover foreign bank account. Extensive privatization, the normalization of markets, and the liberalization of payments in the region (save in Uzbekistan and

Turkmenistan) have reduced the proportion of trade carried out under shuttle procedures.

The exchange controls, which favored shuttle trade, were at their tightest when the former Soviet ruble was disarticulated among the CIS. The separation in 1993 of the Central Asian and Caucasus currencies from the ruble (save in Tajikistan, where a transitional "Tajik ruble" was established in 1995, to transmute into its somon in October 2000) also inhibited legitimately-conducted trade among themselves and with the rest of the CIS. The introduction of its own currencies in 1993 furthermore sparked a second round of post-independence inflation. The Turkmen government, as also in 1996 the Uzbek, took refuge in protectionist policies, signaled by standoff with the IMF and other IFIs (though in the Uzbek case, not with the World Bank, EBRD, and ADB); when the Uzbek government, after a series of unfulfilled promises, introduced current account convertibility in 2003, Turkmenistan was alone among the six not to have adopted the neoliberal economic strategy advocated by the IMF.

None, however, will have accompanied that step in liberalization with the competitive entrepreneurial structure for which that strategy is intended. Kazakhstan is the salient example because it has had the highest inflow of FDI since independence—$15.7 billion, or $1094 per capita, against the $7.2 billion flow into Azerbaijan ($873 per capita) and $1.6 billion into Turkmenistan ($269 per capita). In those three states the bulk of the external capital went into hydrocarbon development; in Kazakhstan, however, much of it went into metals. Scant inflow has yet come in to other branches to support the diversification strategy that governments have officially embraced. Carlsberg's purchase of a major Kazakh brewing company, IRBIS, in 2002, was a significant demonstration of movement along that path. FDI into the other three countries has been modest. A disturbed security record and lack of investible projects put the per capita FDI for Tajikistan at the bottom of the list of the 27 transition countries for which the EBRD compiles statistics, a mere $34 per capita over 1989-2003; Uzbekistan's protectionism, however, runs it close to $35. Kyrgyzstan is open to offers, but its cumulative $85 per head is less than one third the average for the CIS, at $292, itself far below the $2112 per capita shown by the eight transition states admitted into the EU in 2004.[4]

Close Regional Ranking on the Human Development Index

The UNDP's Human Development Index (HDI) (UNDP 2002) embraces indicators of human capital—life expectancy, adult literacy, and

[4] FDI data from EBRD 2004: Table A.2.8.

education—as well as of economic provision. The values which it shows for Central Asia and the Caucasus reflect the apposition of a substantial inherited stock of human capital, with a meager flow of current income. The effect of divergence between the two sets of indicators may be illustrated by comparing Kazakhstan with another major oil producer, Saudi Arabia. The HDI for the year 2000 shows them ranking exactly equal at 0.75; the scale runs from Norway's 0.95 down to Sierra Leone's 0.28. But on the economic indicator, the Saudi GDP per capita at an exchange rate, calculated as purchasing power parity (PPP),[5] is just the double of the Kazakh. Both values for 2000 reflect severe declines from earlier performance, at least in terms of measured GDP: the per capita value of $5,871 for Kazakhstan is below the $8,127 shown for 1989 and that for Saudi Arabia in is $11,367 against $19,525 in 1980. Thus, the overall level-pegging on the HDI arises from much better social development where Nursultan Nazarbayev is President than where Fahd is King. On the political score, democracy may be imperfect under the President, but it is totally lacking under the King. On an economic score, only 8% of Saudi exports are non-oil, but 50% of Kazakh exports are non-oil, although another 27% are products of mineral resources. There are no statistics from Saudi Arabia on income inequality, but the Gini coefficient for Kazakhstan is almost the same as for the United Kingdom: 35.4 against the UK's 36.8, where zero represents perfect equality and 100 represents all income going to one person.

Kazakhstan thus comes up on this composite measure of socioeconomic development as at a "medium human development" level, as are the other five states here reviewed. They cluster around an index of 0.71 ± 0.04; Turkmenistan and Azerbaijan are the next below Kazakhstan, both at 0.74, followed by Uzbekistan at 0.73, Kyrgyzstan at 0.71, and Tajikistan at 0.67.

A Poor Social Environment

If the HDI measures a range of socioeconomic positives, there are corresponding indices of negatives. For example, on "government effectiveness" —an index of cronyism, corruption and the like compiled by the Economist Intelligence Unit—Kazakhstan is the world's worst, ranking at 93 with Honduras, the next being Nigeria at 89, far above the world mean of 59. Other indexes of generalized corruption do not, however, rank Kazakhstan quite as badly. The EBRD puts the country fifth from the bottom in its assessment of 27 transition economies, and another survey shows the share

[5] A dollar in PPP has the same purchasing power in the country concerned as a US dollar has in the United States.

of enterprise revenue going in bribes as 5%, less than the 6% reported by enterprises in Uzbekistan in an ill-run Slav country such as the Ukraine.

A profound social malaise is indicated by Kazakhstan's having the fifth highest murder rate and suicide rate in the world (Russia is worse on each score). Although other Central Asian and Caucasus states make better showings, in all, a disturbed social environment and institutionalized imperfections hinder entrepreneurship or keep it out of the tax base. An indicator of the latter is the estimated value of economic activities which escape measurement in official statistics of GDP. Two attempts have been made to calibrate the ratio of goods and services that fall outside the measurement net for transition economies, each providing estimates for a period covering the last Soviet and the first independence years (averaging 1990-93) and a later date. One estimates the unmeasured activity by imputation from a commonly-used physical input, electricity, and estimates for 1994-95 (Johnson et al. 1997, 1998), while the other employs a set of regressions to yield estimates for 2000-01 (Schneider 2002), covering four of the countries here under review. The two methods show remarkable correspondence for the base period 1990-93. The percentage of measured GDP added by unmeasured activities in Azerbaijan was put at 44% on the first (physical input) method and 45% on the second (termed DYMIMIC) method; in Kazakhstan 32% on both methods; in Kyrgyzstan at 34% and 35%, respectively; and in Uzbekistan 20% and 22%, respectively. Well into the post-independence recession, the first method showed sharp rises in the unmeasured activity to 1994-95 in Azerbaijan (to 59) and in Uzbekistan (to 28), although in Kazakhstan and Kyrgyzstan the increment was smaller (to 34% and 37%, respectively). On the longer-run series, to 2000-01, three showed substantial increments:—to 60 in Azerbaijan, to 42 in Kazakhstan, and to 33 in Uzbekistan, but in Kyrgyzstan the rise was only to 33. Schneider, author of the second method, also estimates the share of the working population (age cohort 16-65) engaged in the "shadow economy." He puts it highest in Azerbaijan at 51% and at 33% in both Kazakhstan and Uzbekistan, but lower, at 29%, in Kyrgyzstan. The high proportions of activity outside measured GDP in the later stages of transition contrast with an estimate of such unmeasured activity in the USSR—for 1989 it has been estimated at only 6% (EBRD 1997).

Fiscal Insufficiency

Partly because the large shadow economy is by definition unrevealed to the authorities—some concealed from them by corruption—fewer taxes are proportionately collected than in a market economy. A norm for the latter is furnished by Schneider, who calculates unmeasured GDP in 21 OECD states as averaging a 17% addition to measured GDP, and the labor force engaged in the shadow economy as some 15% of the total age group. Among the transition economies that have escaped armed conflict since independence, Kazakhstan and Kyrgystan exhibit the lowest general government revenue and expenditure[6] as a share of GDP, 23.0% and 24.0% in the one and 22.2% and 27.1% in the other on estimates for 2003 (EBRD, 2004: 42-3). Of the conflict-torn countries, Azerbaijan's revenue and expenditure in 2003 were 27.7% and 30.0% and Tajikistan's 17.2% and 16.3% of their respective GDPs. In Turkmenistan, insofar as the official data are comparable (for some quasi-fiscal institutions are outside the Treasury system), the shares, at 24.5% and 26.3%, are not much better; Uzbekistan, at 39.3% and 39.9%, on the other hand, is aligned with the three industrially-developed slavic states.

The UNDP in its *Human Development Reports* compares expenditures on education, health and defense as shares of GDP. Subject always to the caveat that unmeasured activity is proportionately greater in the transition states than in developed market economies and that it was much less in 1989, the share of public education as share of GDP in the Soviet period rose between 1985-87 and 1995-97 in Kazakhstan to 4.4%, but fell in Azerbaijan to 3.0%, in Uzbekistan to 7.7%, in Kyrgyzstan to 5.3% and in Tajikistan to 2.2% (UNDP 2002: Table 9), but recent swingering cuts in state education and health care services there have made 12,000 teachers and 15,000 medical staff redundant.

Resource Wealth but Industrial Recession

The six states today have a substantial base in mineral resources, predominantly oil, and an inheritance of sound social capital, which they are, however, constrained in augmenting by their inability to mobilize an appropriate tax base. Resource exploitation puts them as medium developed on a world scale in aggregate economic production, but also as insufficiently diversified for security against the "Dutch disease." The latter phenomenon characterized the Netherlands when large export surpluses due to natural gas caused an appreciation of the currency, which both disfavored exports of other goods while favoring increased imports to the detriment of domestic production. The reduction registered since independence in all six states in nonhydrocarbon outputs was not due to such a process, but if the production profile is to be diversified, the "disease" must be avoided.

[6] The general government account includes regional and local budgets, which are substantial in Kazakhstan; in 1998 and 1999, the balances of the central and the general budgets show opposite signs.

The virtual collapse of manufacturing branches after 1991 was, as already noted, a function of sharp declines in aggregate demand in all ex-Soviet states, which had until then been heavily dependent on their mutual trade. Because the industrial index includes minerals (including hydrocarbon) extraction, it understates the decline in manufacturing that took place in four of the six states. The index shows a nadir, on 1989=100 of 26.3 in 1996 in Azerbaijan, of 26.7 in 1995 in Kyrgyzstan, of 32.7 in 1997 in Tajikistan (the year the civil war ended), and of 47.7 in 1995 in Kazakhstan. In the latter, an index distinguishing output of raw materials from manufactures shows a drop for the former to just 59.8 in 1995, whereas the latter precipitously declined to a trivial 1.2 in 1998, reversing up to only 1.5 in 1999; that is a virtual elimination of the group's production (Kalyuzhnova and Andreff 2003). The Kyrgyz industrial index dipped again in 2002, due principally to a landslide, which in July stopped gold production for the remainder of the year, and to a resumed fall in manufacturing. Two governments, on the other hand, maintained their range of industrial activity through subsidies and protection: the index including resource extraction, mainly gas and gold, scarcely dropped in Uzbekistan, reaching a minimum of 96.4 in 1992, and in Turkmenistan reaching a minimum of 64.4 in 1998, each well surpassing the 1989 index by 2003 at 156.0 and 113.4 respectively (UNECE 2004: Table B.4).

Throughout the region, the textile and many engineering branches were virtually eliminated as domestic and foreign orders dried up or as their lack of competitiveness was exposed. The collapse of defense requirements followed the end of the Cold War and the procurement for the Russian armed forces preferentially from domestic factories—Kyrgyzstan, for example, was a supplier of naval equipment. Probably the town worst hit in the region was Sumgait in Azerbaijan, now a virtual rust belt. Food processing temporarily slumped as agricultural output declined, but has been recovering in all six states. In Kazakhstan the development of transnational "financial-industrial groups" with Russian enterprises seems to have allowed the gains in Russian industrial productivity and export markets to be shared with Kazakh partners.

Labor and Migration

During the first decade of post-independence, industrial employment has been protected in Turkmenistan and Uzbekistan. In Turkmenistan in 1999, the last published figure, it was 53% above the 1989 level, and in Uzbekistan, in 2001 it was 98% of 1989 level; but in three other states, it was halved, in 2003 reaching was 53% of 1989 in Azerbaijan, 54% in Kazakhstan, and 48% in Tajikistan; and in Kyrgyzstan, it fell to 42% (UNECE 2004: Table B.6). In Kazakhstan, about half the fall in aggregate employment (less drastic than in industry—2003 was 90% of 1989) was attributable to emigration and half may be accounted for by unemployment, officially 1.8% in 2003 (against 0.4% in 1992) and by withdrawal from the labor force. Against the Kazakh benchmark, the official tally of unemployment was higher in Kyrgyzstan, at 3.0% (0.1% in 1992), but was elsewhere smaller: in 2003 it was 2.4% in Tajikistan (0.4% in 1992), 1.4% in Azerbaijan (0.2% in 1992), and in 2000 0.6% in Uzbekistan (0.1% in 1992), but was not officially admitted in Turkmenistan.

Emigration was substantial in the first half of the 1990s, when some 1.5 million Slavs, 245,000 Germans, and 102,000 Jews left the five Central Asian republics (Michugina and Rakhmaninova 1996). Such settlers had followed the Tsarist troops there and into the Caucasus (there had been Jews since ancient and medieval times in Azerbaijan, Turkmenistan, and Uzbekistan) and were followed under Soviet industrialization by many more who were needed to staff enterprises and administrations. Among them, Kazakhstan was the principal destination of prisoners and forced laborers, and along with Uzbekistan, became the dumping-grounds for Koreans resettled in 1937-9 from the Far East when Japanese troops were occupying Korea and Manchuria and were invading Mongolia.[7] Nikita Khrushchev's "Virgin Lands Campaign" brought many young Slavs in 1953-6 to grain farming in northern Kazakhstan. In its impact by country, the return movement varied inversely with the intensity of more than a century of inflows. In the 1989 census, the Russian population of Kazakhstan was 37% of the total, of whom one-tenth departed; in Kyrgyzstan, it was 20% and one fifth left; and in three states where the proportion of Russians was between 7% and 9%, one fifth left Turkmenistan and Uzbekistan and almost half fled Tajikistan, as it was engulfed by civil war. Chechens, Ingush, and Crimean Tartars deported by Stalin in 1943 to Kazakhstan and Uzbekistan had been allowed to return in 1957. Such return was not permitted to the deported Germans, settled on the lower Volga and elsewhere in the south of Tsarist Russia in the eighteenth and early nineteenth centuries; they were mainly settled in Kazakhstan, but also in Kyrgyzstan and Tajikistan. Many of the quarter-million Germans who left after 1990 emigrated to Germany. There has been a much smaller flow in the reverse direction, in the repatriation of those with Central Asian and Caucasus roots under the influence of unemployment and ethnic discrimination in the Slav-populated republics,

---

[7] A parallel is the US government's internment of people of Japanese stock in the Pacific states after Pearl Harbour in 1941.

and nothing like the return of the well-educated among the respective Diaspora of Armenia, Ukraine, and the Baltic States.

Some emigration of the remaining population of non-native stock can be expected, as professional and administrative posts are taken over by those of local ethnicity and as Russian job opportunities expand—Russian GDP growth averaged over 6.7% in 1999-2003 (CIS 2004: 19). Little emigration can be anticipated among those of local ethnicities.

A consequence, as already noted at the beginning of this exposition, will be that the local job supply must match the expanding population of the working age if the social unrest from underemployment is to be avoided. Authoritarian regimes in all six states—though greatly disparate in the degree of repression of political and religious opposition—have hitherto kept discontent in check. Even in the medium term, however, the proportion of those of working age is significantly rising. Thus, between 2000 and 2015, the share is estimated at rising from 66% to 70% in Kazakhstan, from 58% to 67% in Turkmenistan, from 64% to 74% in Azerbaijan, from 59% to 69% in Uzbekistan, from 60% to 69% in Kyrgyzstan, and from 56% to 68% in Tajikistan (UNDP 2002: Table 5).

Saving for the Future

Two governments, those of Azerbaijan and of Kazakhstan, set aside revenue from hydrocarbon extraction towards investment in the capital assets, upon which the increment of the workforce can be employed. Under the IMF's counsel, both have prudentially invested against resource exhaustion by establishing National Funds that would set aside part of state hydrocarbon revenue for investment. Such saving is a step in the right direction, but as yet not a big enough step. An IMF technical study (IMF 2002: 12-28)[8] nevertheless observed that insufficient receipts from oil and gas exploitation were being paid into the Kazakh National Fund to ensure over the long term (the chosen time horizon was 2048) a capital stock to provide an income flow offsetting the depletion of oil reserves. The IMF modeled the future on a Permanent Income Hypothesis, but added that such investment was needed for two other reasons: to fund pensions as the Kazakh population ages and to counter the trend of negative capital productivity growth as exhibited over three decades.

But in four countries, the overall rate of investment is low. In the national accounts of 2002 or 2003, gross fixed investment in GDP was 13% in Kazakhstan, 11% in Tajikistan, 15% in Kazakhstan, but 26% in Uzbekistan, 33% in Azerbaijan (due to high FDI) and probably as

[8] The aim is to maintain a population of 15 million at $3,974 wealth per capita forever.

much in Turkmenistan (CIS 2004: 24). In the two latter countries, much of the high level of investment may be misdirected, for example, into internationally-uncompetitive projects in Uzbekistan and allocated to nonproductive sectors by an irrational President in Turkmenistan. More significant is the FDI flow, to which reference is made above, since it is not only mostly directed towards exploiting comparative advantage in international markets, but is accompanied by packages of complementary managerial, technical, and marketing inputs.

Another area in which future welfare is at stake concerns the environment. Pollution affects the oil fields of three states. In Kazakhstan TengizChevroil was fined $70 million in December 2002 by an Atyrau court for contravention of environmental protection regulations in stockpiling sulfur. In Kazakhstan the Semipalatinsk "polygon" remains radioactive from the early Soviet nuclear tests; in this state and in Turkmenistan, the Aral Sea region suffers from Soviet chemical weapons residue. Downwind in Uzbekistan the Tajik, aluminum smelter at Tursanzade still causes fluorosis among the young. The desiccation and pollution (from fertilizer and pesticides brought by irrigated water from cotton plantations) of the Aral Sea has not yet been reversed. The Caspian Sea has markedly suffered from a decrease of valuable fish species—notably sturgeon and hence caviar production—within a general degradation of biodiversity and the pollution, mainly by oil, of its water and bottom sediments.

A final constituent of the conditions pertaining to growth prospects is vulnerability to external shocks. The decline in import propensity may not persist. It was occasioned by demand recession and the "shake-out" of transition from command to market mechanisms, for the Soviet system was notorious for high materials input per unit of output. But for the oil-led economies, the higher the exposure to trade and, hence, to volatile world oil prices, the greater the variability in both total export earnings and in the terms of trade. The latter, if turning negative, could prejudice further favorable access to international capital markets, or, if set on a positive trend which appreciated currencies in real terms, could affect non-oil, tradable in the 'Dutch disease' syndrome.

The Prospects for Stability

The diversification of production away from the extraction of natural resources would enhance prospects for macroeconomic stability. Judged by the consumer-price index, all six economies are more stable than in the 1990s. In 2003 CPI inflation was down to single digits in Azerbaijan, 2.2%; in Kyrgyzstan, 3.1%; in Turkmenistan 6.5% and in Kazakhstan, 6.4%. But it was still double-digit in Uzbekistan, 10.3%; and in Tajikistan, 16.3%. The lifting of exchange restrictions would boost inflation in Turkmenistan but could be restrained by appropriate macroeconomic policies. The perspective for Tajikistan is poorer, notably due to its high external debt, equal to 65.0% of GDP in 2003. That ratio is worse in Kyrgyzstan (103%), but is safer everywhere else in the region—49% in Uzbekistan, 35% in Turkmenistan, 77% in Kazakhstan, and as low as 21% in Azerbaijan. The IMF and international donors are engaged in debt restructuring for Kyrgyzstan and Tajikistan, but until the countries can moderate current account deficits, inhibit capital flight, and build up foreign-exchange reserves, their external vulnerability will remain. In 2003 the ratios of external debt to exports of goods and services and of reserves to imports of goods and services were weak at 263%, but strong at 5.3 months in Kyrgyzstan and weak at 115% and 1.3 months in Tajikistan. At the safer extreme, the ratios were 41% and 10.0 months in Turkmenistan and 50% but only 2.1 months in Azerbaijan. A ratio of reserves equivalent to 3 months of imports of goods and services is seen as a necessary minimum, and Azerbaijan and Tajikistan can be considered within a danger zone. Throughout the transition period, Kazakhstan had operated below the conventional minimum—but rose above it at 3.9 months in 2003, with a debt/export ratio of 153%—and by restricting imports through exchange control from 1996 to 2003, Uzbekistan has improved its reserves/import ratio, to 6.7 months in 2003, while slightly adding to its debt/export ratio, 111% in 2003 (EBRD 2004).

All six states have much still to do to align their economic institutions and practice with those of developed market economies. The statistics on which this article focuses cannot reflect the many qualitative aspects of an economic system. Significant improvement in the economic operational environment will demand much effort by national and international agencies. Certainly it cannot be realized in the short run; and under the authoritarian regimes, which all have in varying degrees, it may not be in the interests of governing elites.

As a measure of how far the six countries are from such realization, the EBRD economic and legal secretariat has annually compiled indicative assessments on a scale 1 to 4 (EBRD 2004). Kazakhstan and Kyrgyzstan are accorded high scores (3 or 4) in institutional adaptation to the market (small and large-scale privatization, price and trade, and exchange liberalization), but make a poorer showing on enterprise governance and restructuring and competition policy (2). Their indicator of legislative and judicial progress on, for example, insolvency were ranked 'medium' by the EBRD, but Azerbaijan and Uzbekistan were 'low' and Tajikistan and Turkmenistan were 'very low'. Turkmenistan was also accorded the lowest score of 1 on implementation indicators, as was Tajikistan on reform of infrastructure. Azerbaijan and Uzbekistan still had a majority of scores of 2, but Azerbaijan ranked 4 on small privatization and on marketization.

In part, the opaqueness of local and national administrations and of corporate governance is attributable to a slow emergence from the Soviet political and economic heritage (all but two of the current Presidents held the Party leadership in late Soviet days) and in part to the government's extraction of rents from natural resources—hydrocarbons, cotton and gold—which weakens the fiscal constraints epitomized in the idea of "no taxation without representation." The leakage of government revenue to the benefit of the elites and the detriment of the majority, through a diminution of social spending, has exacerbated the widening of income differentials since independence. Household surveys of 2002-03 show the share of population living in poverty was largest in Tajikistan, 57% (IMF 2004c: 3) and high in Azerbaijan, 47% (IMF 2004a: 3) and Kyrgyzstan, 44% (IMF 2004b: 5), but more moderate in Kazakhstan, 33% (CIC 2004: 133, 398); a 1998 share for Turkmenistan is variously given as 44% (EBRD 2004: 188) and 30% (IMF 1999: 12). Prospects for domestic political evolution and for international governmental and corporate influence thereon are topics for other contributors to this volume, but inequality and poverty are strong indicators of eventual change. Political developments will determine whether the region's rich oil and gas potential will lubricate the engines of industry or merely grease itchy palms.

References

CIS

2004 *Statistical Yearbook*. Moscow: CIS Statistical Committee.

2003 *Statistical Yearbook*. Moscow: CIS Statistical Committee.

2002 *CIS Statistical Abstract 2001*. Moscow: CIS Statistical Committee.

EASTERLY, W. AND S. FISCHER

1994 "Growth Prospects for the Ex-Soviet Republics: Lessons from Soviet Historical Experience." Pp. 59-86 in Aganbegyan, A. et al. (eds.), *Economics in a Changing World*, Vol. 1, *System Transformation: Eastern and Western Assessments*. London: Macmillan.

EBRD

2004 *Transition Report 2004*. London: EBRD.

1997 *Transition Report 1997*. London: EBRD.

IMF

2004a "Azerbaijan Republic: Joint Staff Assessment of the Poverty Reduction Strategy Paper Progress Report." *Country Report* 04/323.

2004b "Kyrgyz Republic: Fourth Review under the Poverty Reduction and Growth Strategy." *Country Report* 04/16.

2004c "Republic of Tajikistan: Third Review under the Poverty Reduction and Growth Facility." *Country Report* 04/248.

2002 "Republic of Kazakhstan: Selected Issues and Statistical Appendix." *Country Report* 02/64, March. Washington: IMF.

1999 "Turkmenistan: Recent Economic Developments." *Country Report* 99/140.

JOHNSON, S., D. KAUFMAN AND A. SHLEIFER
1997 "The Unofficial Economy in Transition." *Brookings Papers on Economic Activity*. Washington DC: Brookings Institution.
JOHNSON, S., D. KAUFMAN AND P. ZOIDA-LOBOTON
1998 "Regulatory Discretion and the Unofficial Economy." *American Economic Review* 88(2): 387-92.
KALYUZHNOVA, Y. AND W. ANDREFF (EDS.)
2003 *Privatization: Lessons from the First Decade of the Transition Economies*. London: Palgrave.
MICHUGINA, A. AND M. RAKHMANINOVA
1996 "The National Composition of Migrants in the Flows of Population Between Russia and Foreign Countries." *Voprosy Statistiki*, Moscow, no. 12: 45-7.
SCHNEIDER, F.
2002 "The Size and Development of the Shadow Economies of 22 Transition and 21 OECD Countries." *IZA Discussion Papers*, no. 514. Bonn: Institute for the Study of Labour.
TIMOFEEV, T. AND V. PEREVEDENTSEV
1994 "The Economic Demography of the CIS." Pp. 48-58 in: Aganbegyan, A., Bogomolov, O. and Kaser, M. (eds.), *Economics in a Changing World:* Vol. 1, *System Transformation: Eastern and Western Assessments*. London: Macmillan.
UNDP
2002 *Human Development Report 2002, Deepening Democracy in a Fragmented World*. New York and Oxford: Oxford University Press.
UNECE
2004 *Economic Survey of Europe*, no. 2. Geneva and New York: United Nations.

VI. Gendered Transitions: The Impact of the Post-Soviet Transition on Women in Central Asia and the Caucasus

ARMINE ISHKANIAN

ABSTRACT

In this chapter, I explore the impact of the post-Soviet political and socioeconomic transitions on women in the former Soviet republics of Central Asia and the Caucasus. I review the impact of Soviet policies on gender roles and relations in order to contextualize post-Soviet developments. The central segment, which examines gender roles and relations after socialism, is divided into two sections. In the first section, I examine the impact of local political and socioeconomic transitions on gender relations and local responses to those transitions. In the second section, I discuss the impact of regional/global events and interactions on gender roles and relations. Throughout the chapter, I consider the similarities and differences of the transitions and the responses to those transitions in the post-Soviet republics of Central Asia and the Caucasus.

Introduction

Scholars writing about the post-Soviet "transitions"[1] confront issues connected with the acceptance or rejection of new models. Research therefore has delved into the complex ways in which the past and present

[1] The term "transition" has been problematized by various scholars, including Michael Buroway, Katherine Vedery, and Barbara Einhorn, who argue that "transition" implies an evolutionary development that has a single, well-defined objective and trajectory.

get expressed in the creation of national ideologies, of new identities, social classes, and non-governmental organizations (NGOs) in these societies (Akiner 1997; Berdahl et al. 2000; Bridger and Pine 1998; Buckley 1997; Burawoy and Verdery 1999; Creed and Wedel 1997; Dawisha and Parrott 1997; Dudwick 1997; Funk 1993; Gal and Kligman 2000; Sampson 1996; Verdery 1996; Wedel 2001). Yet another major aspect of post-Soviet transitions that has increasingly gained salience is the "gendered" nature of the post-Soviet transitions (Gal and Kligman 2000). In the former Soviet states of Central Asia and the Caucasus, Soviet and pre-Soviet beliefs about the nature of the family, the state, political leaders, capitalism, and community/social life affect the transition to democracy, capitalism, and the development of civil society. In these societies, the category of women was and continues to be an ideological site for political, religious, and economic projects. In this article, I explore the impact of post-Soviet political and socio-economic transitions on women in Central Asia and the Caucasus. While I make generalizations about the lives of women in Central Asia and the Caucasus, gender as a variable is one among many, including ethnicity, age, education, geographic location, and religion, which determine and affect how individuals experience the transitions. Clearly, there are variations among women living in the societies of Central Asia and the Caucasus, but in this chapter, I will identify shared tendencies and patterns that cross national borders since the women living in post-Soviet Central Asia and the Caucasus share a sufficient number of common experiences due in large part to the legacy of a shared Soviet past.

Gendered Transitions

Gail Lapidus (2000) argues that virtually no one (from the West or East) anticipated, in the initial euphoria, that gender issues would prove to be one of the most problematic aspects of the transition, and indeed that political democratization and economic reforms might significantly increase rather than attenuate the gender asymmetries (P. 102). Yet this has become the case in the countries of the Soviet Union, where women have not only become the majority of the unemployed, but have also become depoliticized and are largely left out of the government, political parties, and the official public sphere.[2] While post-Soviet developments

[2] In "Working at the Global/Local Nexus: Challenges Facing Women in Armenia's NGOs Sector," in Women in Post-Communist Transitions, edited by Carol Nechemias and Kathleen Kuehnast, Woodrow Wilson Center Press, forthcoming, I describe how Armenian women who were left out of the government, official public sphere, political parties and now make up the majority of NGO leaders and members.

in Central Asia and the Caucasus have not resulted in adequate legal, economic, and political rights yet, given women's universal literacy, the high rates of female participation in the labor force and the Communist Party and local governments during Soviet times, as well as the intervention of international NGOs and transnational feminist networks, many women in Central Asia and the Caucasus have found ways to resist and counter the adverse trends of post-Soviet developments.[3] Therefore, as Mary Buckley (1997) maintains, while the transitions have not been easy for women, it is problematic to consider women as being victims of the post-Soviet collapse. They are much more than victims; they are agents of change and reaction, who have inventively found ways of managing in the new and often difficult circumstances surrounding them (P. 7).

The editors of this volume, Mehdi Parvizi Amineh and Henk Houweling, urge us to consider how Central Eurasia is an important region that is both implicated in global processes and yet plays it own role in them. By examining how women in Central Asia and the Caucasus have been affected by the local socioeconomic and political developments and their activism in transnational movements (e.g., NGO networks), we can begin to understand how this region is implicated in global processes and how citizens of the region are responding to the changes brought about by the post-Soviet transition and the current period of globalization.

I begin the chapter with a brief overview of the impact of Soviet policies on women since these policies continue to influence gender relations and responses in the post-Soviet period. Following an overview of Soviet policies on the "woman question," I will discuss the impact of the post-Soviet transitions on women in Central Asia and the Caucasus.

"Breaking the Cake of Custom": Soviet Policies Regarding the "Woman Question"

The leaders of the new Soviet state who came to power in the 1917 promised a radical reconstitution of society, which would fundamentally transform economic, social, and political institutions and relations. The few Communist Party leaders who had given some consideration to the role of women under socialism believed that the alteration of women's roles in society was a function of the economic and political reconstitution of the larger society (Lapidus 1978: 55). These leaders argued that as women took on larger roles in the economic, social, and political

[3] Transnational advocacy networks, according to Keck and Sikkink (1998) are "Networks that are organized to promote causes, principled ideas, and norms, and they often involve individuals advocating policy changes that cannot be easily linked to a rationalist understanding of their 'interests'" (Pp. 8-9).

life of the Soviet Union, they would emerge from the confines of the household into the wider public arena. This entry into social production and political life would, they maintained, have profound consequences on male-female relationships (Lapidus 1978: 55). In order to stimulate this transformation, the Soviet government adopted legislation during the early 1920s, establishing civil marriage, easy divorce, abortion services, maternity pay, and childcare facilities. All restrictions on women's freedom of movement were abolished, while laws were adopted that gave women equal rights to hold land, to act as heads of households, to participate as full members in rural communes, and later to be paid as individuals rather than as part of a household for collective farm labor (Lapidus 1978: 60). Although the principles of equal pay for equal work were enshrined in law, new laws and policies were no guarantee that women would enjoy an increased role in social production (Buckley 1989: 19). Formal legal equality did not translate into real equality in gender relations in the private or public spheres as women, laboring under the "double burden," continued to be responsible for nearly all domestic chores in addition to working outside the home.

The Zhenotdel and Women in Central Asia and the Caucasus

In order to expand the influence of the party over a large number of working-class and peasant women, in 1919 the state created the *zhenskii otdel* (abbreviated as Zhenotdel), Women's Department of the Central Committee Secretariat. Under the leadership of Inessa Armand and later Alexandra Kollontai, the Zhenotdel was charged with spreading the message of the Party to the unorganized women in factories and villages throughout the Soviet Union. Zhenotdel representatives were sent to the Central Asian and Caucasus republics to teach, mobilize, and politicize local women, to draw them into the Party, trade unions, cooperative organizations and the soviets, and to promote literacy (Lapidus 1978: 66). The peasant household, in particular, was seen as the very embodiment of tradition and backwardness and the bearer of counterrevolutionary values. According to Soviet ideology, Muslim women in Central Asia and the Caucasus endured a triple oppression of class, nationality, and family. The assimilation of the Muslim societies of Central Asia into the Soviet Union presented special difficulties as various Party organizations failed to penetrate or destroy traditional associational networks by a direct assault on local elites. Hence, to "break the cake of custom" (Matossian 1961: 61), Soviet social engineers developed numerous methods of agitation and recruitment in order to inform women of their new rights and responsibilities to the Party and the state.

Gregory Massell's (1975) study, *The Surrogate Proletariat: Moslem Women and Revolutionary Strategies in Soviet Central Asia, 1919-1929*, illustrates the process through which the Soviet government replaced social class with sex as the decisive lever for effecting social change in Central Asia. Massell argues that the Communist Party came to experiment with a number of approaches in drawing in the Central Asian republics into the fold of the system. One of these approaches was the "in depth" approach, which was aimed at undermining the traditional social order in order to destroy family structures and the kinship system. Soviet social engineers believed that this could most speedily be achieved through the mobilization of women. Massell explains, "It may be said, then, that Moslem women came to constitute in Soviet political imagination, a structural weak point in the traditional order: a potentially deviant and hence subversive stratum susceptible to militant appeal—in effect, a surrogate proletariat where no proletariat in the real Marxist sense existed" (1975: xxiii).

The authorities believed that, if they could engender conflict within the traditional family structures, this would provide them with leverage for the disintegration of those structures and their subsequent reconstitution. Throughout the Soviet republics of Central Asia and the Caucasus, local Zhenotdels established women's clubs, workers' clubs, tea-houses (*chaykhana*), workshops (*artel*), and evening schools for adult literacy or "illiteracy liquidation centers" (*likpunkty*) to integrate women into the Soviet system. The overarching goal of the Soviet state was not so much to liberate women, but to organize them as a political and economic force that would become workers in the industrialized economy.

One such club that was established during this period was the Ali Bayramov Club in Baku, Azerbaijan. This club was created by Dzheiran Bairmova in 1920. By year's end, the club had 100 members. Its press organ, *Sharq Qadini* (Woman of the East), played an important role in the legal and political education of women in Azerbaijan. In Armenia, which is a Christian country, combating the influence of Islam was not the main concern of the Communist Party. Instead, the party in Armenia identified the traditional Armenian family as a "backward" institution and sought to transform it by dismantling family loyalties. To do this the Soviet leadership created the *Kinbazhin*, the Armenian chapter of the *Zhenotdel*. During the 1920s, Kinbazhin workers would select representatives (*delegatkii*) who would visit homes and give women "scientific" advice on how to raise children and on simple rules of hygiene. These delegatkii would also try to establish rapport with the children of the household and encourage them to report cases of child beating, wife beating, and forced marriages, which Mary Matossian (1961) argues, had "immense potentialities for disrupting traditional family patterns" (P. 66). In addition to Kinbazhin, the Commission for the Improvement of the Way of Life of Women (*Kanants Kentsaghe Barelavogh Hantznazhogove*) was created in 1923 to "advise government organs, conduct propaganda campaigns, offer legal advice to women, and provide an 'inspection service' to see that Soviet legislation regarding the family and traditional offenses was put into effect" (Matossian 1961: 67). These and other intrusive Soviet institutions and practices were resented and resisted throughout the Central Asian and Caucasian republics and they had the paradoxical effect of strengthening family and kinship networks. Opponents in some of the Central Asian republics even referred to the Zhenotdel as the *jinotdel* (the department of bad spirits) (Tadjbakhsh 1998: 169). Hence, in these societies, the cake of custom was only partially broken as the family became not only a mode of resistance to the state, but also remained as the primary means of identification, support, and advancement throughout the Soviet period.

By 1926, the Soviet authorities realized that legislative reforms alone were insufficient. A more interventionist and aggressive campaign called *Hujum,* or intractable assault on customs and traditions was launched in Central Asia (Tadjbakhsh 1998: 169). Mass unveilings were part of this movement and women were encouraged to publicly shed and burn their veils. On March 8, 1927, the International Day of Women, women were ordered to appear in parades throughout the Soviet Union. That evening, a simultaneous mass "un-veiling" was ordered in the largest cities of Central Asia and apparently, on that day, over 10,000 women burned their veils in Uzbekistan alone. Shirin Akiner (1997) maintains that, through the Zhenotdel's sustained efforts, traditional culture was either destroyed or rendered invisible, confined to the most intimate and private sphere, whereas in the public arena new identities were created and society was secularized (P. 261).

The Zhenotdel continued to operate until 1930 in spite of accusations that it had "feminist tendencies," which "under the banner of improving the women's way of life, actually could lead to the female contingent of the labor force breaking away from the common class struggle" (Lapidus 1978: 71). In 1930, with Stalin's consolidation of power within the Party nearly complete, he ordered a general reorganization of the Central Committee. This led to the formal abolition of the Zhenotdel and to the declaration that the "woman question" had been solved.

The Woman Question Solved?

After 1930, Zhenotdel activities were assigned to regular Party organizations and to the Commissions for the Improvement of the Working and Living Conditions of Women that were attached to local, provincial, and republic executive committees. This, Lapidus (1978) argues, was indicative of the growing predominance of views that were hostile to any further strain in family and communal relations; instead, under Stalin the status quo was to subordinate a broader definition of liberation to the need for social stability, control, and productivity for harnessing the energies of men and women alike to the common cause of socialist construction (P. 94). Over the next twenty years serious discussion about women's issues was largely silenced, and discourses about production and output displaced the earlier themes of liberation, equality, and domestic labor (Buckley 1989: 13). As a result of this shift in state ideology, in the 1930s and 1940s women were portrayed as virtuous, self-sacrificing *Stakhanovite*, shock workers, who overfilled production quotas as part of the Soviet project of constructing socialism. During and immediately after World War II, the state also began to encourage couples to have many children (four or more) and rewarded them with subsidies such as free milk, living stipends, and better homes. Women with ten or more children, besides all the economic benefits, were also awarded a medal of honor and given the title of Heroine Mother of the Soviet Union (Matossian 1961: 182).

This absence of debate about women's roles in society lasted until the mid-1950s, when, following Stalin's death in 1953, women's issues returned to public debate as part of the general thaw in Soviet society. During the 20th Party Congress in 1956, Khrushchev expressed regret for the relative absence of women from prominent positions in the state and Party. In his view, different groups had different needs and should be treated differently. Women constituted one such group in Khrushchev's view (Buckley 1989: 14). Because of this, he supported the creation of women's organizations called *zhenskii sovety* (*abbreviated as Zhensovety*), or women's councils to cater to the needs and interests of women. The political activities of the Zhensovety were aimed at developing the "New Communist Women," who were educated in the spirit of the "high moral principles needed for building communism" (Browning 1986: 87). As before, the Zhensovety delegates were charged with combating the influence of religion in the Muslim regions of Central Asia, but now, they were also responsible for monitoring the spread and popularity of religion in the Catholic regions of the Baltic republics and in the central Russian areas where the Orthodox Church retained a strong foothold. In general, the Zhensovety failed to produce female leaders in the Communist Party, even though they acted as autonomous consciousness raising groups among women (Tadjbakhsh 1998: 171).

Achievements and Failures

On the surface, it appears that the women of the Soviet Union had accomplished an impressive level of emancipation by the late 1980s. Education is one area in particular in which the state socialism of the Soviet Union deserves undeniable credit (Tohidi 1998: 142). Free universal primary education was introduced in the 1930s, universal eight-year education in the 1950s, and universal secondary education in the 1970s.

There was universal literacy among women in the Soviet Union, and women were generally slightly better educated than men. In the late 1980s, women constituted 61% of specialists with higher or secondary specialized education and 54% of students in higher educational establishments (Pilkington 1992: 182). Although there was gender-based occupational segregation, women comprised 50.9% of workers in the Soviet economy and their wages were between 70 to 85% of men's wages (Pilkington 1992: 183). In spite of these educational and professional achievements, however, women continued to endure the dreaded double burden as they fulfilled quotas in the factories and farms and tended to their families and homes. In Central Asia in particular, the double burden was more onerous owing to the prevalence of large families, relatively low provisions of communal amenities such as crèches, canteens, and laundries, and outside the main cities, the chronic scarcity of labor-saving devices such as washing machines and vacuum cleaners (Akiner 1997: 281).

In terms of health, women's life expectancy was nearly ten years longer (73.9) than men's life expectancy (64.8), and the level of primary health care and the overall health status of people, including that of women and children, seemed to be comparable to that of many developed countries. The reality of women's health, however, given the unreliability of Soviet statistics, was less encouraging. In Central Asia, as throughout the Soviet Union, the state's pro-natal policy, the lack of family planning, and the minimal access to contraceptives rendered abortion the primary means of birth control (Tohidi 1998: 142). The prevalence of abortion was a major health care issue that was neglected by Soviet authorities.

In the realm of politics, a quota system was instituted by the authorities to maintain a proportional level of "representation" at all levels of state and Party governance. Women, however, rarely appeared in the leadership positions in the higher echelons of power. The contrived nature of this system of representation became quite clear following the collapse of the Soviet Union and the nearly absolute withdrawal of women from political life throughout the former Soviet states. This decline in women's representation began during the period of *perestroika* (restructuring) as women lost their one-third representation in the local soviets and the Parliament and intensified immediately following independence (Posadskaya 1993: 163).

Women in Central Asia and the Caucasus after Socialism: Between Tradition and Modernity

Socio-Economic Developments

Women in the former Soviet Union had entered the working world in vast numbers after Sovietization and enjoyed the benefits of a socioeconomic safety net. In the post-Soviet period, they have suffered most from the difficult economic transition. The transition to a market economy has been costly in terms of real income and output decline, disproportional unemployment and underemployment among women, widespread impoverishment, a rapid deterioration of living standards and social safety nets, the loss of maternal and childcare benefits, deepening gender inequalities, and the decreasing presence of women in the government and formal political parties. In 2002, ten years after independence, all of the former Soviet republics in Central Asia and the Caucasus are listed in the "Medium Human Development" category on the United Nations Development Program Human Development Index (HDI). As such, these countries are considered "developing" countries. Only the Baltic states of Estonia, Lithuania, and Latvia are included in the "High Human Development" category on the HDI, ranking 42nd, 49th, and 53rd, respectively. This "de-development," de-modernization, or reprimitivization, as it has been variously called, has created a great deal of anxiety, anger, and frustration for the citizens of the former Soviet Union (Bridger and Pine 1998; Burawoy and Verdery 1998; Creed and Wedel 1997; Ishkanian 2000; Platz 2000; Wedel 2001). International aid and development agencies, including the United Nations Development Program (UNDP) and the World Bank, among others, have begun to address the de-development in the former socialist states, and their policies and rhetoric have shifted, moving from helping these countries to make or effect the transition from communism to capitalism/democracy to helping them make the transition from poverty to sustainable development. This shift is evident in the types of publications that are released by these agencies and the grants that are apportioned.

In 2000, the World Bank published a report on poverty in the Commonwealth of Independent States (CIS): *Making Transition Work for Everyone: Poverty and Inequality in Europe and Central Asia.* According to this report, in 1998 an estimated one out of every five person in the transition countries of the Europe and Central Asia survived on less than $2.15 USD a day (2000: 1). The World Bank authors acknowledge the fact that the transitions have brought new economic opportunities for some, but for many others, it has meant unaccustomed material hardship and loss of security, including loss of jobs, prolonged non-payment of salaries, hyperinflation and loss of savings, and the drastic erosion of accustomed supports such as low-cost or free social services and subsidies. The high rates of absolute poverty, the lack of economic growth, growing social polarization, and crumbling infrastructure have had a serious impact at the micro-level where people have lost their jobs and status in society, and are faced with harsh material conditions. As Valentine Moghadam (2000) argues,

> From a gender perspective, the market reforms that have been adopted in Central Asia and the Caucasus, prescribed and underwritten by the international financial institutions, and endorsed by neoliberal thinkers and policy makers in Western countries have been deeply flawed. The social costs of the transition have been far higher than expected, and the burden borne by women has been especially onerous. (P. 32)

She asks, "Why should women be more vulnerable than men?" Moghadam maintains that the reasons for the gender-differentiated impact of the market reforms in the former Soviet republics are both cultural and economic (P. 31). They are cultural in the sense that traditional gender ideology regarding men's and women's roles shapes the types of work that men and women can engage in, and it is economic in the sense that the nature of the reforms themselves and the assumptions in the neoliberal economic thinking that inform them are not gender-neutral. Neoliberal economic policies privilege the operation of a free market unencumbered by state interventions and tend to favor the accumulation of capital over and above the well-being of labor. This does not mean that social safety nets or policies targeted at poor households are not considered; indeed, as Moghadam points out, the neoliberal approach to welfare has a preference for means-tested and targeted programs rather than universal programs based on notions of solidarity and equality. According to neoliberal policymakers and the agencies such as the World Bank that push such reforms, the pains of adjustment are both transitory and necessary to achieve macroeconomic stabilization and respectable rates of growth. In the post-Soviet states, however, these adjustments and reforms have been long in coming, and millions of people continue to live in abject poverty with little hope for a better future. While poverty affects all citizens, women have tended to suffer more from these adjustment policies that eliminate subsidies for food, utilities, transportation, childcare, and other necessities. [4]

The transition to a market economy has not only, in many instances, failed to remove the disadvantages for women in the Soviet system, but in most cases it has actually intensified the gender asymmetry and inequalities. Simultaneously, prior advantages and benefits that women enjoyed have been jeopardized. Unlike many developing countries, however, gender gaps in literacy and educational attainment are not wide in the post-Soviet Central Asia and the Caucasus states (Moghadam 2000: 26). Women had relatively easy access to education and employment, but the

[4] There is a well-known formula that is frequently referred to by development workers. According to this formula, women are one-half of the world's population, they perform two-thirds of the world's work, and market over three-fifths of the world's food. Yet, they represent three-fifths of the world's illiterate, receive one-tenth of the world's income, and own less than one-hundredth of the world's property.

deterioration in wages and other social benefits has meant that they now endure worsening working conditions in the new market economies and that they are less likely to set up their own businesses. Throughout the former Soviet republics, a very small proportion of women's businesses have been able to survive the vicissitudes of the market system and the challenges associated with doing business in circumstances that involve paying "tributes" (Humphrey 2002: 144) to both racketeers and corrupt public officials. For these reasons, most women in Central Asia and the Caucasus have tended to work in the less profitable and less high profile sectors of the market economy, including working as small-scale shuttle traders or as merchants in local markets and open-air bazaars (*yarmarka*). Throughout the countries of Central Asia and the Caucasus, "suitcase" or shuttle trading is booming and, as Seteney Shami (2000) argues, it is the sole province of women (P. 326). Women from these countries travel to Turkey, the United Arab Emirates, Poland, and other destinations to bring back clothing, shoes, and various household items that they then sell in local bazaars for varying margins of profit.

Cynthia Werner (2001) discusses the emergence of a New Silk Road in Central Asia and argues that one of the more striking aspects of this trade is the predominance of women in the marketplace. In particular, Werner claims, women dominate the exchange of cloth and clothing that are exchanged as gifts, and food products that are used to feed families and honor guests. As Werner observes, "The high visibility of merchant women, including Muslim women, stands in contrast to popular stereotypes of secluded Muslim women. Not only do these Muslim women work in public places, they also travel to markets in distant towns where their activities are less likely to be observed by kinsmen and neighbors" (P. 2). Werner further argues that women's active participation in the markets is due to the growing unemployment and the factors that constrain women's opportunities to work (P. 4).

In several of the countries of Central Asia and the Caucasus, including Armenia, Azerbaijan, Georgia, and Tajikistan, the problems of the "gendered" nature of the transition have also been exacerbated by civil strife and conflict. Of these countries, Tajikistan has fared the worst; even during the Soviet period, Tajikistan was among the most rural and least developed of the republics. The civil war that plagued the country from 1992-1993 ruined the social and physical infrastructure and has had adverse effects on women, children, and the elderly (Moghadam 2000: 27). There, the fear of abduction and rape, particularly of girls and young women, is strong enough to affect their freedom of movement. In Azerbaijan, women again constitute the majority of the unemployed and experience insecurity in an evermore-violent social atmosphere (Tohidi 1998: 144).

Religion

During the Soviet period, Islam was a matter of private life that was preserved and practiced by women. Increasingly, however, it is becoming politicized in the hands of men at the national and regional levels where it is being manipulated in accordance with or in reaction to new capitalist realities or the old gender arrangements (Tohidi 1998: 157). In Tajikistan, Collette Harris (2000) argues that the Islamization has not been without conflict. She explains how during the coalition Government of National Recovery (May-November 1992), pressure was put on urban girls and women to abandon their European clothes in favor of native costumes, or in other words to "desovietize" themselves. This, she explains, was experienced by many as incipient Islamization and as the first step towards the reestablishment of pre-Soviet gender identities. As such, it became a major factor producing support for the communists (Harris 2000: 212). Since the opposition's return from exile, it is now seen as entirely possible that there may be a serious attempt in the near future to establish strongly Islamized gender identities. This, Harris contends, is seen as undesirable by women at all levels of society and influences attitudes towards Islam, which after the suppression of the Soviet years, has once more come into the open (P. 212). Akiner (1997) claims that, although most Central Asians welcomed the reintroduction of Islam into the public space, the majority do not want it to assume a regulative function, adding, "they still feel strongly that religion and the state should be separate" (P. 285). Paula Michaels (1998) identifies a similar position among Kazak women who continue to play high profile roles in the public sphere, such as entering new university departments and creating and joining NGOs. Michaels writes,

> Economically, socially, and politically, Kazak women's lives lie at a crossroads. Pre-Soviet and Soviet influences have given rise to the multi-faceted roles women play in present day Kazak society. For now, the renewed interest in Islam seems to play only a modest role, and other factors, such as the rise of the market economy, show signs of exerting a more dominant influence on the position of women. (P. 199)

Public veiling is also not increasing significantly in many of these former Soviet Muslim republics. While unveilings were politicized during the early Soviet years, the return to the veil is currently regarded as a symbol of personal commitment to Islam. It has not, as Akiner (1997) maintains, become politicized, as has been the case elsewhere (P. 286).

The legacy of Soviet emancipation and the politicization of Islam today has meant that for most, women in Central Asia are caught between competing impulses. Some, as Akiner argues, feel the need to return to their "authentic" roots, while others wish to continue along the road to greater personal independence and freedom of choice (P. 263). The situation remains in flux and will be shaped by national, regional, and global political, economic, and cultural developments in the coming years.

Politics

Scholars of nationalism have demonstrated the gendered nature of nation-state building and nationalism (Jayawardena 1986; Pateman 1988; Chatterjee 1993). In the current post-Soviet period, nation-state building processes in each of the former republics have led to the elimination of certain previously held rights as women have lost representation in local and national governments. Although women had been crucial in the independence movements, immediately after the collapse of the Soviet Union, women in all of the former Soviet states found themselves excluded from the new governments. In Armenia, for example, the removal of the quota system led to a significant decrease in political representation among women: in 1985, 121 of 219 members of parliament were women, while the number of female parliamentarians dropped to 8 following the 1991 National Assembly elections. Removal of the quota system of representation used in the Soviet Union, however, can explain only part of the decline in female representation. Other factors, including the strain of the double burden, gender role socialization, the commonly shared belief that politics is "men's work" and is inherently corrupt and dirty have contributed to the small number of women in public office and to the low levels of women's participation in political parties in all of the former Soviet republics.

Women's inadequate representation and small share in political power is a worldwide problem and not unique to the post-Soviet states of Central Asia and the Caucasus. However, given the Soviet legacy of egalitarian laws and rhetoric, the post-Soviet political activism of women in national-liberation movements, universal literacy, and women's high level of economic participation, one would expect a much stronger representation of women in political power. While there has been a decline in women's representation at the formal (national and local) government levels and in political parties, there is an unprecedented increase in women's participation in NGOs, as I will explain in the next section. By choosing NGOs, women reaffirm the ascribed gender roles and gender-based divisions of labor and avoid the criticisms that they would face if they enter political parties or government, but are still able to work in and through the public sector to achieve their personal and community objectives.

Women after Socialism Part II: Regional and Global Developments

NGOs, Donors, and Transnational Advocacy Networks

World politics currently involves, alongside states, many non-state actors, including non-governmental organizations (NGOs) that interact with each other, with states, and with international organizations (Keck and Sikkink 1998: 1). Whether NGOs are seen as latter-day evangelists or as manifestations of grassroots democracy, it is clear that in the last twenty years there has been tremendous growth in their number and influence.[5] This is the result of many factors, including Western donors' disillusion with the national governments of developing countries, their belief that civil society is an important part of democratization, and the idea that a "global civil society" will promote more equitable and just economic development and progress.

In spite of the problems of defining and locating civil society in the West, in the 1990s the idea became a central part of Western aid programs to Eastern Europe and the former Soviet Union. US agencies and other first world donors embraced the idea of civil society development as critical to democratization and "successful transition." This became a new mantra in both aid and diplomatic circles (Ottaway and Carothers 1998: 6), as the West readily made funding available to the former Soviet states and countries of Eastern Europe. Policymakers and international development organizations hailed NGOs as "stakeholders in the transition and development of these [post-socialist] countries" (World Bank 2001) and the "connective tissue of democratic political culture" (Wedel, cited in Hann 1996: 1). While the link between civil society, democratization, and NGOs is a late twentieth-century phenomenon and one that should be understood in the context of deregulated and increasingly globalized economies, it is very significant because it has led to the phenomenal growth in the number of NGOs in the countries of the former Soviet

[5] NGOs can be charitable, religious, research, human rights, and environmental organizations, and they can range from loosely organized groups with a few unpaid staff members to organizations with multimillion-dollar budgets, employing hundreds of people all over the world. By this definition, a local bird-watching society, the Ku Klux Klan, and Amnesty International are all NGOs. The only agreement most scholars have about NGOs is that they (ideally) exist outside both the state (hence, non-governmental) and the market. Sometimes they are called "third sector organizations," the third sector being that which lies between the first (governmental) and second (market) sectors. Some NGOs are voluntary groups with no governmental affiliation or support; others are created and maintained by, and loosely linked to, governments. This has led to a proliferation of acronyms, for example GONGOs (governmental non-governmental organization) and QUANGOS (quasi-governmental NGOs).

Union where democratization and a vibrant civil society have been directly linked to the presence of NGOs (Hann 1996: 7). One of the largest donors of NGOs in the former Soviet Union, the Eurasia Foundation, describes NGO development as "help[ing] build democracy by providing citizens with a formula for collectively voicing their views and lessen[ing] the pain of economic transformation by providing alternative vehicles for the delivery of critical social services" (Eurasia Foundation 1998: 2).

In Armenia, Azerbaijan, Uzbekistan, and in many of the other states of the former Soviet Union and Eastern Europe, NGOs are overwhelmingly led by Soviet era elites, either intellectuals or former Communist Party apparatchiks who were quick to recognize the potentials offered by NGO sector participation and to make the transition from state or Communist Party structures into NGOs (Abramson 1999; Ishkanian 2000; Hemment 2000; Phillips 2000; Sampson 1996). For example, in Armenia the Soviet era Women's Council (*zhensovet*) became the Women's Republican Council in the post-Soviet period with its leader and hierarchical structures intact. Few working-class people or rural residents, or even intellectuals who were not part of the former structures of power, were able to make this transition. These Soviet elites, in addition to possessing the organizing and language skills, also had an advantage of belonging to social networks that put them in contact with the Westerners who control or influence the distribution of grants.

Another similarity shared by NGOs operating in post-socialist countries is that women overwhelmingly run them. Various scholars have examined the reasons for this feminization of the NGO sector in the post-Soviet countries and have identified several factors contributing to this feminization, including 1) women's exclusion from the spheres of government and business; 2) women's networking and linguistic skills, particularly important in establishing ties with foreign donors; 3) women's traditional interest in and responsibility for social problems, including disabilities, health, and children's issues; 4) women's desire to avoid the taint of corruption, a problem more associated with the formal political arena than the new informal, civic arena; 5) women's secondary status, a situation that leaves avenues of participation that lack monetary award or "prospects" open to women due to men's lack of interest and their preference for the worlds of business and formal politics; and finally and most importantly, 6) a preference among donors in supporting women's initiatives and empowerment.

The number of women's organizations particularly grew after the 1995 UN Fourth World Conference on Women in Beijing (Berg forthcoming; Ishkanian forthcoming; Olson 2001). Although international conferences, such as the Beijing conference, have not necessarily led to the creation of women's networks or NGOs, they have legitimized the issues addressed by women's NGOs, and they have brought together unprecedented numbers of women around the world (Keck and Sikkink 1998: 169). The Beijing conference in particular not only provided women in the former Soviet states with an introduction to the international world of NGOs (i.e., global civil society) but also stimulated greater funding and interest in the role of women in development. Donors began to claim that women were more "cost-effective" as beneficiaries of development and civil society aid (Buvinic et al. 1996: 13). The Beijing conference was the defining moment in the development of women's NGOs in Armenia because the women who attended it, either as members of the government delegation or as NGO members, returned to Armenia informed and educated about global gender discourses, issues, and concerns, which they proceeded to translate into the local Armenian context.[6]

Donors' focus on women began in the 1970s when international development agencies began to make "women" visible as a category in development and research policy (Kabeer 1994: xi). This came to be known as the Women in Development (WID) approach. The thinking went, if policymakers, donors, and planners could be made to see women's concrete and valuable contributions to their economies, then women would no longer be marginalized in the development process. This trend grew in the 1990s and continues today as many of the largest donors, including the World Bank, US Agency for International Development (USAID), and various UN agencies, all have departments focused on gender issues intended to promote gender equality in development or GID (gender in development). Women's NGOs in Central Asia and the Caucasus recognize the ascendancy of the GID approach and have become quite adept at employing the appropriate discourses. "Talking gender" has become an important factor in winning grants.

In Azerbaijan, a new civil society is emerging, Tohidi (1998) claims, with an increasing number of informal and non-governmental women's organizations. Whether the women who are involved in NGOs will have

---

[6] Scholars working in Asia and Latin America have also documented the impact of UN conferences on local women's organizing and discourses. For instance, according to Mary Fainsod Katzenstein, anticipation of the 1975 UN conference in Mexico City on women played a "catalytic role in the emergence of the contemporary women's movement in India" (Katzenstein, Mary Fainsod cited in Narayan 1997: 91). Sonia E. Alvarez (1998) describes the Beijing conference as an effusive celebration of "global sisterhood," adding that it was the site where professionalized, thematically specialized, and transnationalized feminist NGOs focused their energies on influencing the International Platform for Action and in helping articulate the "global women's lobby." Alvarez refers to this professionalization and specialization of women's groups as the "NGOization" of the Latin American women's movements (Pp. 293-296).

greater opportunities than previously to engage in decision making as free agents of change with genuine representation is not yet clear, however (Tohidi 1998: 144). In Tajikistan, a network of women's organizations has also been established whose members regularly meet in Dushanbe and exchange information and build strategies for mutual cooperation (Harris 2000: 224). In addition to national networks created by local NGOs operating in the post-Soviet sphere, there are transnational advocacy networks that have been created to coordinate the efforts of NGOs in the various Central Asian and Caucasian republics. An example of such a transnational advocacy network is the "Working Together—Networking Women in the Caucasus," which is sponsored by the Institute for Democracy in Eastern Europe (IDEE).[7] Created with funding from the Bureau of Educational and Cultural Affairs of the US Department of State, "Working Together" is a program for women leaders in Armenia, Azerbaijan, and Georgia. In 1999, IDEE launched the "Working Together" project "in response to the needs of women NGO activists for greater cross-border networking and NGO development in a historically and ethnically divided region, and to the need for promoting and advancing women in societies where men have traditionally played dominant roles in the community" (*http://www.idee.org/*). Building on the success of the Caucasus program, IDEE launched a similar program in Central Asia, titled "Civic Bridges—Networking Women in Central Asia" for women leaders in Tajikistan, Turkmenistan, and Uzbekistan.

Through a range of training, civic education, NGO development, and cross-border networking activities, the IDEE programs attempt to enhance the leadership abilities and capacity of women leaders and their NGOs, to advance women's participation in public life, and to create a strong regional network of women's organizations with ties to NGOs in Central and Eastern Europe and the former Soviet Union. During the first phase of the Caucasus program, several issues common to the entire region were identified, including, civic education, reconciliation, working with refugees

[7] The Institute for Democracy in Eastern Europe (IDEE) is a not-for-profit tax-exempt corporation begun in 1986 to support the growing opposition movements in Eastern Europe, seeking democratic change and an end to communism. Since 1989, IDEE has helped democrats in the region to overcome communism's harsh and oppressive legacy and TO rebuild—or build anew—institutions of a democratic political system and a plural and open society. IDEE has administered over $10 million in assistance, training, internship, exchange and education programs to more than 2,500 publications, civic and human rights organizations, political groups and opposition movements in the following countries: Armenia, Azerbaijan, Belarus, Bosnia and Herzegovina, Bulgaria, Chechnya, Crimea, Croatia, Cuba, the Czech Republic, Estonia, Georgia, Hungary, Kosovo, Kyrgyzstan, Latvia, Lithuania, Macedonia, Mongolia, Montenegro, Poland, Romania, Russia, Serbia, Slovakia, Ukraine, and Uzbekistan. Retrieved August 20, 2002 (*http://www.idee.org/*).

and internally displaced persons, and gender-based violence, that could be more effectively addressed by a regional coalition. In the program's second phase, which culminated in an advanced training seminar and regional NGO assembly, sixty participants from the region came together to share experiences and make new contacts, suggest solutions to common problems, plan joint projects, and evaluate the program's second year.

In 2002, the "Working Together" NGOs from Armenia, Azerbaijan, and Georgia published four NGO newsletters highlighting their activities, achievements, and plans. The newsletter is distributed primarily by e-mail and is posted in Armenian, Azeri, Georgian, and Russian on the Internet. Print copies are also available in the regional languages and are distributed free of charge to those without Internet access. In addition to the trainings and publications, IDEE has provided financial and material assistance to NGOs for projects promoting greater cooperation among participants and the transfer of skills and knowledge to their communities. The expanded small grants competition for 2001-2002 allocated money projects that promote cross-border or cross-regional cooperation (*http://www.idee.org/*).

NGOs in the Central Asia and the Caucasus countries are constantly working between the local, regional, and global levels. As intermediaries, NGO members benefit from the Western aid because it provides them with increased leverage and autonomy at the local level and an ability to continue working in respectable jobs instead of having to do menial, humiliating (by local standards) work. Aid, however is a double-edged sword, and while it provides NGOs with funding and support, it also exposes them to foreign direction and control. This dependency of local NGOs on the "uncertain largess of donors," as William Fisher (1995) calls it, has direct and indirect effects. He describes these as, a) redirecting the accountability toward funders and away from the group's grassroots constituencies and, b) transforming NGOs into contractors, constituencies into customers, and members into clients (P. 454). This criticism exposes NGOs to attacks within their own countries, raising questions about whether they truly represent their constituents and is one of the most difficult challenges facing NGOs in Central Asia and the Caucasus if they hope to win legitimacy from among their own populations. If they sacrifice the local for the global, then they are betraying their mission as local organizations. If they ignore the needs and wants of international donors, they risk losing funding that is critical to their survival and success.

In the post-Soviet period, many of the former republics are now identified as "developing" countries; women in these countries have had "development encounters" of their own with Western development workers, consultants, and "experts." The implications of the asymmetrical relations between the global and local actors engaged in development encounters cannot be overlooked, given that the power inequalities inherent in these encounters affect the production of knowledge, the circulation of information, decision-making, and the outcomes of development or transition projects.[8]

Conclusion

In this chapter, I have addressed the impact of Soviet policies on women in Central Asia and the Caucasus. In writing about the "gendered transitions" of the post-Soviet period, I discussed the shared tendencies and patterns that cross national borders. Women's lives in these countries have been adversely affected by the economic and political developments of the transition, but it is also true that women have taken an active part in promoting social change in their societies. The legacy of a shared Soviet past and the current widespread impoverishment have meant that women in Central Asia and the Caucasus often face similar dilemmas and that, at times, they have found similar coping and survival strategies. For instance, through their participation in NGOs, women have not only been able to maintain a modest existence, but more importantly, they have gained the knowledge, skills, and social connections needed to promote progressive developments in their countries. Yet participation in NGOs and transnational advocacy networks is not without struggles and problems. On the contrary, most women in Central Asia and the Caucasus have very high levels of literacy, professional experience and training, and knowledge of the world, and they resent the patronizing attitude of Western consultants who propose programs that have little relevance to local conditions, culture, history, and traditions. While it remains to be seen how future developments will impact gender relations and roles, it is clear that women are not simply the "victims" of the transition. On the contrary, they are agents of change who adapt, resist, manipulate, and accommodate the developments of the transition period and globalization in their lives.

References

ABRAMSON, D.M.
1999 "A Critical Look at NGOs and Civil Society as Means to an End in Uzbekistan." *Human Organization* 58(3): 240-250.

AKINER, S.
1997 "Between Tradition and Modernity: the Dilemma Facing Contemporary Central Asian Women." Pp. 261-304 in Buckley, M. (ed.), *Post-Soviet Women: From the Baltic to Central Asia.* Cambridge: Cambridge University Press.

---

[8] I discuss these issues in greater detail in "Importing Civil Society?: The Emergence of Armenia's NGO Sector and the Impact of Western Aid on Its Development," in *Armenian Forum: A Journal of Contemporary Affairs*, forthcoming.

ALVAREZ, S.E.
1998 "Latin American Feminisms 'Go Global': Trends of the 1990s and Challenges for the New Millennium." Pp. 292-317 in Alvarez, S.E., Dagnino, E. and Escobar, A. (eds.), *Politics of Culture, Culture of Politics*. Boulder, CO: Westview Press.

BERDAHL, D. ET AL.
2000 *Altering States: Ethnographies of Transition in Eastern Europe and the Former Soviet Union*. Ann Arbor: University of Michigan Press.

BERG, A.
Forthcoming "Two Worlds Apart: The Lack of Integration Between Women's Informal Networks and Non-Governmental Organizations in Uzbekistan," in Kuehnast, K. and Nechemias, C. (eds.), *Post-Soviet Women Encountering Transition: Nation-Building, Economic Survival, and Civic Activism*. Washington, DC: Woodrow Wilson Center Press.

BRIDGER, S. AND F. PINE
1998 Surviving *Post-Socialism: Local Strategies and Regional Responses in Eastern Europe and the Former Soviet Union*. London: Routledge.

BROWNING, G.K.
1986 *Women and Politics in the USSR: Consciousness Raising and Soviet Women's Groups*. New York: St. Martin's Press.

BUCKLEY, M.
1997 *Post-Soviet Women: From the Baltic to Central Asia*. Cambridge: Cambridge University Press.
1989 *Women and Ideology in the Soviet Union*. New York: Harvester Wheatsheaf.

BURAWOY, M. AND K. VERDERY
1999 *Uncertain Transitions: Ethnographies of Change in the Post-Socialist World*. Lanham: Rowan & Littlefield.

BUVINIC, M. ET AL.
1996 *Investing in Women: Progress and Prospects for the World Bank*. Washington DC: Overseas Development Council in Cooperation with the International Center for Research on Women.

CHATTERJEE, P.
1993 *The Nation and Its Fragments*. Princeton: Princeton University Press.

CREED, G.W. AND J.R. WEDEL
1997 "*Second Thoughts from the Second World: Interpreting Aid in Post-Communist Eastern Europe*." *Human Organization* 56(3): 253-264.

DAWISHA, K. AND B. PARROTT
1997 *Conflict, Cleavage, and Change in Central Asia and the Caucasus*. Cambridge: Cambridge University Press.

DUDWICK, N.
1997 "Political Transformations in Postcommunist Armenia: Images and Realities." Pp. 69-109 in Dawisha, K. and Parrott, B. (eds.), *Conflict, Cleavage, and Change in Central Asia and the Caucasus*. Cambridge: Cambridge University Press.

EURASIA FOUNDATION
1998 Press release. Yerevan, Armenia.

FISHER, W.
1995 "Doing Good? The Politics and Anti-Politics of NGO Practices." *Annual Reviews in Anthropology* (26): 439-464.

FUNK, N.
"Feminism East and West." Pp. 1-14 in Funk, N. and Mueller, M. (eds.), *Gender Politics and Post-Communism*. London: Routledge.

GAL, S. AND G. KLIGMAN
2000 *The Politics of Gender After Socialism*. Princeton: Princeton University Press.

HANN, C. AND E. DUNN
1996 *Civil Society: Challenging Western Models*. London: Routledge.

HARRIS, C.
2000 "The Changing Identity of Women in Tajikistan in the Post-Soviet Period." Pp. 205-228 in Acar, F. and Gunes-Ayata, A. (eds.), *Gender and Identity Construction: Women in Central Asia, the Caucasus, and Turkey*. Boston (Brill).

HEMMENT, J.
2000 "The Price of Partnership: The NGO, the State, the Foundation, and its Lovers in Post-Communist Russia." *Anthropology of East Europe Review* 178(1): 33-36.
HUMPHREY, C.
2002 *The Unmaking of Soviet Life*. Ithaca: Cornell University Press.
ISHKANIAN, A.
2000 *Hearths and Modernity: The Role of Women in NGOs in Post-Soviet Armenia*. Ph.D. dissertation, Department of Anthropology, University of California, San Diego.
Forthcoming "Working at the Global-Local Intersection: The Challenges Facing Women in Armenia's NGOs Sector," in Nechemias, C. and Kuehnast, K. (eds.), *Post-Soviet Women Encountering Transition: Nation-Building, Economic Survival, and Civic Activism*. Washington, DC: Woodrow Wilson International Center Press.
JAYAWARDENA, K.
1986 *Feminism and Nationalism in the Third World*. London: Zed Press.
KABEER, N.
1994 *Reversed Realities: Gender Hierarchies in Development Thought*. London: Verso Press.
KECK, M.E. AND K. SIKKINK
1998 *Activists Beyond Borders: Advocacy Networks in International Politics*. Ithaca: Cornell University Press.
LAPIDUS, G.W.
1978 *Women in Soviet Society: Equality, Development, and Social Change*. Berkeley: University of California Press.
2000 "Discussant's Comments." Pp. 102-106 in Lazreg, M. (ed.), *Making the Transition Work for Women in Europe and Central Asia*. Washington, DC: The World Bank.
MASSELL, G.
1975 *The Surrogate Proletariat: Moslem Women and Revolutionary Strategies in Soviet Central Asia, 1919-1929*. Princeton: Princeton University Press.
MATOSSIAN, M.
1961 *The Impact of Soviet Politics in Armenia*. Netherlands: Leinden Press.
MICHAELS, P.
"Kazak Women: Living the Heritage of a Unique Past." Pp. 187-202 in Bodman, H.L. and Tohidi, N. (eds.), *Women in Muslim Societies: Diversity Within Unity*. Boulder, CO: Lynne Rienner Press.
MOGHADAM, V.
2000 "Gender and Economic Reforms: A Framework for Analysis and Evidence from Central Asia, the Caucasus, and Turkey," in Acar, F. and Gunes-Ayata, A. (eds.), *Gender and Identity Construction: Women in Central Asia, the Caucasus, and Turkey*. Boston: Brill.
NARAYAN, U.
1997 *Dislocating Cultures*. New York: Routledge.
OLSON, L.
2001 "Women and NGOs: Views from Some Conflict Areas in the Caucasus." Occasional Paper from the Women in International Security "New Bridges to Peace" Workshop. Retrieved August 21, 2002 (*www.wiis.org*).
OTTAWAY, M. AND T. CAROTHERS
1998 *Funding Virtue: Civil Society Aid and Democracy Promotion*. Washington DC: Carnegie Endowment.
PATEMAN, C.
1998. *The Sexual Contract*. Palo Alto: Stanford University Press.
PHILLIPS, S.D.
"NGOs in Ukraine: the Makings of a 'Women's Space?'" *Anthropology of East Europe Review* 178(2): 23-29.
PILKINGTON, H.
1992 "Russia and the former Soviet Republics: Behind the Mask of Soviet Unity: Realities of Women's Lives." Pp. 180-235 in Corrin, C. (ed.), *Superwoman and the Double Burden: Women's Experiences of Change in Central and Eastern Europe and the Former Soviet Union*. Toronto: Second Story Press.
PLATZ, S.
2000 "The Shape of National Time: Daily Life, History, and Identity during Armenia's Transition to Independence, 1991-1994." Pp. 114-138 in Berdahl,

D., Bunzl, M. and Lampland, M. (eds.), *Altering States: Ethnographies of Transition in Eastern Europe and the Former Soviet Union.* Ann Arbor: University of Michigan Press.

POSADSKAYA, A.
"Changes in Gender Discourses and Policies in the Former Soviet Union." Pp. 162-179 in Moghadam, V. (ed.), *Democratic Reform and the Position of Women in Transitional Economies.* Oxford: Clarendon Press.

SAMPSON, S.
1996 "The Social Life of Projects: Importing Civil Society to Albania." Pp. 121-142 in Hahn, C. and Dunn, E. (eds.), *Civil Society: Challenging Western Models.* London: Routledge.

SHAMI, S.
2000 "Engendering Social Memory: Domestic Rituals, Resistance, and Identity in the North Caucasus." Pp. 305-331 in Acar, F. and Gunes-Ayata, A. (eds.), *Gender and Identity Construction: Women in Central Asia, the Caucasus, and Turkey.* Boston: Brill.

TADJBAKHSH, S.
1998 "Between Lenin and Allah: Women and Ideology in Tajikistan." Pp. 163-185 in Bodman, H.L. and Tohidi, N. (eds.), *Women in Muslim Societies: Diversity Within Unity.* Boulder, CO: Lynne Rienner Press.

TOHIDI, N.
1998 "Guardians of the Nation: Women, Islam, and the Soviet Legacy of Modernization in Azerbaijan." Pp. 137-161 in Bodman, H.L. and Tohidi, N. (eds.), *Women in Muslim Societies: Diversity Within Unity.* Boulder, CO: Lynne Rienner Press.

UNITED NATIONS DEVELOPMENT PROGRAMME (UNDP)
2002 *Human Development Report 2002: Deepening Democracy in a Fragmented World.* New York City: United Nations Press.

VERDERY, K.
1996 *What Was Socialism and What Comes Next?* Princeton: Princeton University Press.

WEDEL, J.R.
2001 *Collision and Collusion: The Strange Case of Western Aid to Eastern Europe 1989-1998.* New York: St. Martin's Press.

WERNER, C.
2001 "Between Market and Family: Women Traders on the New Silk Road." Paper presented at the Women in Post-Communist Societies Working Group Meeting, Kennan Institute, Woodrow Wilson Center for Scholars, Washington, DC.

WORLD BANK
Armenia Country Assistance Strategy Report. Retrieved August 20, 2002 (*www.wbln0018.worldbank.org/eca/eca.nsf/*).

PART THREE

INTERACTIONS BETWEEN OUTSIDERS, NEIGHBORS AND CENTRAL EURASIAN REPUBLICS

VII. Sino-Indian Relations: Security Dilemma, Ideological Polarization, or Cooperation Based on 'Comprehensive Security'?*

KURT RADTKE

ABSTRACT

Geopolitics in Central Asia are not wholly determined by its giant neighbors China, India, and Russia, but the strategic approaches adopted by these three countries have a major impact on the dynamics of Central Asia. This contribution aims to throw more light on the nature of the strategic and security discourse between China and India as one way to increase our understanding of the context in which Central Asian states operate. Despite globalization, Asian governments tend to cling to statist approaches. China in particular emphasizes the role of "large powers" (daguo) in determining the global structure, and regards itself as one of those large powers. Cooperation with other powers demands a minimum level of agreement on common goals for the future global system, but recent emphasis on moral, and thus ideological elements in US global strategies has the potential to reimpose ideological polarization on the global system. Countries in Southeast or Central Asia tend to adopt policies of diversification by

* I should like to express my gratitude to Dr. Raymond Feddema (University of Amsterdam), with whom I discussed on numerous occasions many issues dealt with in this article. The contents of this paper are, of course, my sole responsibility. Together with Dr. Feddema, I edited a book entitled *Comprehensive Security*. Our book also contains numerous references to sources dealing with issues such as the role of "comprehensive security" in the Council on Security Co-operation in the Asia-Pacific (CSCAP) a track II dialogue. The reader is advised to consult the introduction to the book for further details.

strengthening their links with all major global powers, including the United States, hoping to avoid polarization while at the same time staying clear of bandwagoning. This is one of the reasons why the New Great Game cannot simply be described in terms of Great Powers that engage among themselves in maneuvers of bandwagoning and balancing. Sharing concepts such as "comprehensive security" may provide greater leeway for policymakers who do not wish to become prisoners of man-made dilemmas. Nineteenth-century concepts of balance of power seem no shining beacon for policymakers of Eurasia and the United States in the twenty-first century.

Introduction

Geo-politics in Central Asia are not wholly determined by its giant neighbors China, India, and Russia, but the strategic approaches adopted by these three countries have a major impact on the dynamics of Central Asia. This contribution aims to throw more light on the nature of the strategic and security discourse between China and India as one way to increase our understanding of the context in which Central Asian states operate. Globalization is forcing the United States, the European Union, and Japan to develop new approaches towards international order beyond the ideals of traditional concepts of a system populated by sovereign nation states. Especially, younger states in Asia and Africa, however, tend to cling to static approaches (Ayoob 2002; Barnett 2002). Although usually not spelled out explicitly, China in particular emphasizes the role of "large powers" (daguo) in determining the global structure, and regards itself as one of those large powers. Since the United States has overwhelming military and economic power in this multipolar system, Chinese strategists tend to devise strategies to redress power imbalance by seeking a counterweight to the United States along the lines of nineteenth-century balance of power concepts. There are big question marks as to whether this is a realistic and feasible approach. For countries to share global strategies they need to agree on common goals for the global system. The basic question is, first of all, whether powers other than the United States are interested, willing, and able to agree on such a common vision for global order that makes them align against the United States, and second, whether they have more to gain than to loose by adopting that stance. The question arises also as to what China can offer Russia, Japan, India, Central Asian, and European powers to make them move closer towards China, distancing themselves from the United States. In Asia, China has so far failed to persuade Japan (or the United States) that it has the means and the will to make specific contributions to solving the North

Korean issue. China's recent defense agreement with Bangladesh is unlikely to gain China many friends in India. Other Asian countries, whether in Southeast or Central Asia, tend to adopt policies of diversification by strengthening their links with all major global powers, including the United States, hoping to avoid polarization. A case in point is Myanmar's opening towards India. The New Great Game cannot simply be described in terms of Great Powers that engage amongst themselves in maneuvers of bandwagoning and balancing.

Although the growth of Indian economic and military strength is not as fast as that of China, India has the potential to catch up with China during the next two decades. Even before the Bush administration came to power, but now increasingly so, the United States wishes to strengthen its global security by fighting terrorism and "rogue states" ("regime change"). At the same time, it is actively engaged in pushing countries around the globe onto the road of becoming "genuine" market democracies, and that strategy includes China. Much will depend on whether China can maintain its march towards a stable market economy while transforming its political system. Since India is already a parliamentary multiparty democracy, US specialists seem to agree that it can cope better with political instability in the wake of economic transformation. If India manages to achieve projected growth, it will become a serious competitor for China within the next two decades, not only economically, but also as a politically as a rival, and both countries may willy-nilly find themselves engaged in a future arms race.

Military insecurity is not only a threat to bilateral relations, but to regional and global stability as well. Sudden changes in exchange rates, collapse of the stock market, outbreaks of infectious disease, and many more non-military crises have increasingly drawn the attention of governments and security planners. For decades, there has been a keen awareness of the linkages between military security and social, political, and economic stability, as the study of documents related to the occupation of Japan and Germany shows. [1] During the sixties, various newly independent countries in Southeast Asia developed comprehensive concepts for national stability, such as "national resilience" (*Ketahanan nasional*, see below). Government-sponsored discussions in Japan conducted mainly between 1979 and 1981 developed the concept of "comprehensive security" further (*soogoo anzen hoshoo*). [2] Although definitions of the term differ with time and

[1] US occupation policies towards Germany and Japan were based on the concept of indivisibility between peace and economic prosperity. Moreover, political unity without economic unity would not be acceptable.

[2] The term "comprehensive security" is also being used in contexts such as maritime security without reference to the Asia-specific concept. The Concept was also discussed

place, "comprehensive security" is usually associated with discussions on security that consciously avoid references to hypothetical enemies in order to facilitate dialogue. The track two CSCAP group established a committee focusing on this concept (Rolfe 1997; Dickens 1997), and in recent years the term has gained the attention of Chinese security specialists who credit Japan with creating the concept.

Both Japan and India look back on a long and strong history of "pacifism" that made it difficult, if not impossible, to conduct discourses on military security, in particular when these involved clear distinctions between friends and enemies, balance of power, or bandwagoning. Although both China and Japan stressed discourse on "pacifism," the Chinese discourse has to be understood as part of a communist-controlled "peace movement" designed to promote China's strategic influence. Until most recently, Japan's "pacifist" discourse forsook great power ambitions:

> Japan's own judgment, as presented by the central decision-makers as well as the general public, has been that it has ceased to be a relevant independent actor in postwar international and regional security due to various constraints stemming from its wartime aggression. Much of Japan's postwar security policy has indeed been premised on its national determination not to reemerge as a traditional great power in the game of regional and international security. (Soeya 1998)

In such a context, the adoption of "comprehensive security" in Japan facilitates public discourse on security issues. First intended to conceptualize security within national boundaries, the concept has more recently also been applied to discussions at the wider regional level. It is no panacea for peace, nor is it meant to replace military security with pacifist approaches. The nature of this discourse has definitely facilitated the creation of an epistemic security community among ASEAN countries, and is thus of more than purely "academic" interest (Leifer 1996). Also, it increases the awareness that overspending on military "security" will dangerously erode the capacity to deal successfully with other issues affecting national security, such as increasing demands for energy and water, to mention just two examples. Despite the success of neorealism and its derivatives in penetrating and changing the nature of discourses on security in virtually all Asian countries, this should not lead us to assume that thinking on security in Asia is now following the framework determined basically

at "The Third ASEAN Regional Forum (ARF) Track Two Conference on Preventive Diplomacy," jointly sponsored by the Institute of Defense and Strategic Studies (IDSS) Singapore, and the International Institute for Strategic Studies (IISS) in the United Kingdom, held in Singapore from 9-11 September 1997.

by the US security community.[3] A surprisingly great diversity of recent approaches towards security strategy is found in China (the bibliography includes only a fraction), and that includes the relatively new field of geopolitics (Shen 2001). Security concepts "made in Asia," such as Japanese style comprehensive security, if discussed at all, tend to be classified as "alternative" concepts. This essay will first present an overview of some security issues affecting Sino-Indian relations, and then proceed to suggest how the notion of comprehensive security, or similarly related ones, may play a positive role in the security discourse.

China and India: Strategic Confrontation or Non-Adversary Competition?

Despite a long history spanning several decades of border disputes, including armed clashes, and mutual suspicions concerning the build-up of nuclear-armed forces, the number of academic specialists in China and India publishing on issues related to mutual security remained surprisingly low until roughly the middle of the nineties.[4] In contrast, there has been a spate of publications since then.[5] Juli A. MacDonald (2003) has aptly summarized the background to recent interest in the region:

> By 2015, India emerges as the second largest economy in Asia behind China.... India is building a strategic relationship with Iran that has the potential to reshape Central and South Asia.... Indians view Central Asia as a conduit to the United States and as a region in which India can balance U.S. influence.... Central Asian states view India not only as an economic partner but also as a stabilizing peer at a time when the United States, Russia, China, Japan, and the European Union are busy vying for influence in the region. The long-term competition to watch in Central Asia and the Caspian region is between India and China, not India and Pakistan....

[3] There is a considerable time lag between the creation of new concepts in the US and their reception in Japan or China, to mention only a few countries. While specialist journals in both countries have made strenuous efforts to catch up with recent developments during the past few years, new concepts such as the "New Strategic Framework" are still not very well known in the Asian academic establishment.

[4] This does not necessarily reflect the amount of research by military professionals in both countries, but here again, it seems safe to say that, relatively speaking, Indian military specialists on China have been more active than their Chinese counterparts. I became first acquainted with Indian research on China in academic institutions when I had the opportunity for field research in India in the winter of 1993 as part of the Indo-Dutch Project for Alternative Development (IDPAD). Throughout the past decade, I have kept track of Chinese publications as part of my ongoing research on the conceptualization of security issues in Japan and China.

[5] There are few obstacles to locating Indian publications in English on this topic. English translations of Chinese research tend to focus on articles in journals and newspapers and are available through translation series such as *FBIS*.

> Indo-U.S. relations converge on many levels in Central Asia, and it is likely that most American governments will view India as more geostrategically neutral—in contrast to Iran, China, or Russia. (MacDonald 2003)

A RAND study coauthored by Zalmay Khalilzad drew similar conclusions:

> Thus, the study predicted, "if India's economic and technological development can be sustained and accelerated, India should be in a position to claim a larger role for itself in world affairs." India's wary posture towards China will continue, the study predicted, but said, "Whether this posture will degenerate into outright political-military competition is less clear." But it believed that "India will most likely continue to develop its nuclear deterrent capability vis-à-vis China"—and while the Chinese will not like this, their options for dealing with it would appear to be limited.

The study advised that the US nurture a balance-of-power structure involving China, India and Russia to deter them from threatening regional security, dominating one another or coalescing against the US "Washington should seek strengthened political, economic and military relations with all, but especially those least likely to challenge US strategic interests" (cited in Khalizad, Haniffa, Aziz 2001).

It was India's explosion of nuclear bombs in May 1998 that aroused major concern, particularly in China. The following excerpt from a contemporary book represents a fairly restrained Chinese reaction; others were much more vociferous and emotional. It points out that some Indians use references to China's friendship with Pakistan in such a way as to create problems for relations between China and India:

> At the moment, the question of how to deal with India's development of nuclear weapons has already become a practical issue influencing Sino-Indian relations. In fact, the development of nuclear weapons itself by India is irresponsible.... [in particular since references to a] "latent threat" from China was given as one of the pretexts.... This year, on 3 May, India's defense Minister George Fernandes publicly declared that China is "India's "number one latent threat", raising a demand for "decisive" talks with China [on the issue]. Later it turned out that [his reference to] "decisive talks" was [made with an eye to the] development of nuclear weapons, intending to enter into a nuclear arms race with China. Concerning the question of India developing nuclear weapons Jiang Zemin on 17 June in an interview with American correspondents pointed out that India had been first to start the South Asian crisis: both India and Pakistan ought to abandon their plans for nuclear weapons, and unconditionally sign the NPT and the CTBT. If India indeed manages to calm down, abandon its own nuclear armaments, and unconditionally sign both treaties, India will receive China's understanding, so there won't be any influence on the development of Sino-Indian relations: On the other hand, if India one-sidedly decides to go ahead, going further down the road that is against the global current, China will have to raise its vigilance. (Wang 1998: 349)

Other articles were not so reticent and went to great length to explain India's chauvinistic great power dreams on the basis of India's expansionist history (Wang 1998: 268ff).[6] Indian and other authors have analyzed in detail how Chinese reactions to India's build-up of nuclear forces changed over the past few years from an attempt to roll back the nuclearization of India and Pakistan, to an acceptance of the new situation (Rappai 1999). While not representing an official governmental position, Shiping Tang (2002) has recently published an analysis in the widely read journal *Strategy and Management* that suggests that more emphasis should be placed on the need for cooperation between China and India, even if India should tilt more to the United States: "India's [strategy lies in] using US and Russia to balance China and maintaining a balance of not too bad relations with China. In that way, India can receive profit from the US and Russia, without too much offending China. Of course, India may to some extent tend more towards the US" (P. 38). However, he added that "Like China, India's challenges are mainly domestic ones for a considerable period to come" (P. 38).

Shiping accepts that both India and China are in need of maintaining balanced relations with all major players to ensure that they as weaker powers maintain access to resources controlled by the larger powers. Although he does not refer explicitly to "comprehensive security," his setting of priorities is characteristic for that concept.

Tang Shiping's reference to "balanced relations" should not be confused with power balancing neorealist style. Its ultimate aim lies in promoting a dense network of bilateral relations among a multitude of players, the best guarantee against "bandwagoning" and polarization that appears to be the inevitable outcome of a neorealist approach. It may be added that India's novel approach to nuclear strategy ("recessed deterrence") with its emphasis on nonthreatening deterrence may also contribute to a more relaxed Chinese response (Haniffa 2001). Also, India's nuclearization has not prevented both countries from negotiating successfully on confidence building measures (*The Hindu*, 24 September 2002).

A RAND study published in 2000 advocates a similar multilateral approach in support of stable relations, but eventually emphasizes the long-term prospect of rivalry between India and China:

> By 2015, India emerges as the second largest economy in Asia behind China. Second, India is building a strategic relationship with Iran that has the potential to reshape Central and South Asia. For India, Iran represents a critical balance to Chinese influence in Central Asia generally and Pakistan specifically. Historically, Iran has been India's gateway to Central Asia, and

[6] Government control of the media in China varies with time and place. Although Chinese media stand no comparison to the enormous breadth of conflicting opinions published in India, it would be wrong to interpret all publications as having been explicitly approved by the Chinese censors.

> India's recent activities suggest that it seeks to revive the links of the ancient Silk Road.... Central Asian states view India not only as an economic partner but also as a stabilizing player at a time when the United States, Russia, China, Japan, and the European Union are busy vying for influence in the region... The long-term competition to watch in Central Asia and the Caspian region is between India and China, not India and Pakistan.... Indo-U.S. relations converge on many levels in Central Asia, and it is likely that most American governments will view India as more geostrategically neutral—in contrast to Iran, China, or Russia. We should be alert for active cooperation in Central Asia and the Caspian between the United States and India, especially as American zeal for remaining in Central Asia wanes or as its interests pull it elsewhere. (MacDonald 2003)

It is fair to say that both China and India exercise extreme care to avoid direct confrontation. The introduction of CBMs in Sino-Indian relations is one of the first instances in which major Asian powers successfully manage a conflictuous security relationship without Western mediation (Kanwal 1999). On the other hand, there are Indian reports of a build-up of Chinese (tactical, if not strategic) forces in West China, and Chinese reports of the acquisition by India of hardware whose only potential target can be China. From a long-term perspective, China may have come to regret its previous strong support for Pakistan and North Korea, irrespective of whether this involved assisting both countries in becoming able to launch (nuclear tipped) missiles.[7] More than anything else, it was the Pakistan factor that made India decide to pursue the build-up of its own nuclear armed forces. Once that step was taken India of necessity had to structure its nascent nuclear forces, taking into account the strategic forces of its neighbor China. India now seems to proceed towards a first-strike ability versus Pakistan. This explains India's support for the US-led initiative for a multilateral antiballistic missile system, and India's negotiations for or purchase of systems from Russia such as the aircraft carrier *Admiral Gorshkov*, the Akula 2 nuclear powered attack submarines and the TU 22 M3 strategic bomber, and agreements with France to purchase submarines and guided missiles (Fu 2003: 45). Despite the earlier mentioned conclusion of CBMs, Chinese and Indian approaches to nuclear posture display a huge gap (Wang 1998: 349ff; Subrahmanyam 2000). It remains to be seen whether India and China can indeed avoid a costly and unproductive arms race.

China's Strategies—Defensive Moves Outweigh Offensive Stratagems

The position of China in the global and Asian international system does not depend only on its relations with India and its other Asian neighbors.

[7] This has been consistently denied by China.

Foreign policy and security strategies of the United States and China, as well as their perceptions in both countries and the international community at large, will have an important, if not decisive, impact on global politics. Despite initial signs after India's nuclear explosions that the United State and China would work together to either roll back or limit India's nascent nuclear force, the United States increasingly accepts (or even supports) a nuclear armed India, possibly even as a counterweight to China's position in Asia (Haniffa 2001).

There are numerous indications that, as early as the eighties, China began quietly adjusting its relations with its former close allies, such as North Korea and Pakistan. One reason was the negative influence those relations had on China's overall strategy of assuring stable relations, primarily with the United States, at a time when China had to weigh the possibility of its Asian neighbors, Japan and India, becoming more independent players in regional and global strategy. Although the United State and China had for some time cooperated in supporting mujaheddin to bleed the Soviets white in Afghanistan, China's concern that this support might backfire became acute years before the United States ended its cooperation with the mujaheddin and the Taleban:

> From the end of the Second World War until the beginning of the nineties, local wars were time and again mainly restrained by the Cold War between the two hegemonies, the US and the Soviet Union. After the conclusion of the Cold War, regional clashes and local wars were increasingly influenced by religious factors. This is particularly evident in the case of the influence of Islam. After the dissolution of the Soviet Union, Iranian fundamentalism constantly expanded and penetrated deep in the five Central Asian states. In the middle of February 1992 a conference attended by the heads of Iran, Turkey, Pakistan and the five central Asian states was opened in Tehran, providing convenient conditions for Iran and the others to interfere in the affairs of the five central Asian states. In Tajikistan, Afghanistan and Pakistan the external forces had close links with Islam.... After the Persian Gulf War Iran not only frequently interfered in Middle Eastern Affairs, but also actively and greatly penetrated the central Asian region, aiming to establish a "Greater Islamic Sphere." Particularly noteworthy is that at the moment Iran's influence in the Middle East exceeds that of other countries (such as Egypt, Saudi Arabia, Syria, Iraq), and there is a great possibility that Iran may become a regional superpower.... In the South Asian region, India is the No. 1 regional power. India not only interferes in Sri Lankan affairs, but also stretches out in the South China Sea, aiming to becomes the fifth pole in the Asia Pacific and become a permanent member of the Security Council. In the East Asia region Japan is about to replace the strategic position of the United States in the field of economics, technology, foreign affairs, military and culture. (Xi 1996: 276-7)

This is, of course, also the background to early Chinese initiatives to fight terrorism in Central, West, and South Asia. Much attention was given to the fight against terrorism long before the attack on the WTC (Jin

2001; Fankongbu 2001).[8] It should not surprise us that China and the United States have different views regarding which states sponsor terrorism. Ahmed Rashid (2000) warns that China may be forced to expand its role by strengthening its links with Iran, still listed as a member of the "axis of evil" by the United States: "China's regional partner in this drive for security is likely to be Iran, rather than its long time ally Pakistan, Russia or the United States. An increased Chinese military and political presence in the region will further complicate the Great Game of influence in Central Asia" (Rashid 2000). Considering China's concern about foreign-sponsored support for ethnic separatism mainly in Xinjiang, less so in Tibet, there is no reason to doubt the genuineness of China's fight against terrorism. Some Chinese writers exhorted the United States to take more effective action against Osama bin Laden. This was also a major reason for China to change its previous support for the Pakistani position on Kashmir from the beginning of the nineties, when it adopted an increasingly neutral stance (Kumar 2002; Zhu 2002: 331, 344 ff.):

> When the United States concluded the critical Clinton-Sharif deal of July 4, 1999, may have finally facilitated an honorable retreat for the Pakistani armed forces, yet, looking at the factors that actually made this deal possible, it was clearly China's continued posture of neutrality that provided the most decisive input in convincing the Pakistani leadership of the futility of continuing to back up its losing armed forces as also of seeking to internationalize the Kashmir issue in the face of Pakistan's growing global diplomatic isolation. (Singh 1999)

At the same time, China feels forced to maintain links and bonds with countries whose overall policies it may not support, such as Iran, but also Pakistan and North Korea. Likewise, its very recent defense cooperation with Bangladesh raised more than a few eyebrows in India (Kapila 2001).

Re-Polarization in the Form of Alliances, or Diversification Through Establishing Complex Networks of International Relations?

The end of the Cold War and the acceleration of globalization did not by themselves bring Asian neighbors closer together. As Nakanishi (1998) emphasizes, even more than a decade after the collapse of the Soviet Union, there is no common strategic vision. He argued that comprehensive security might offer such a vision. There is the lack of a common vision among Western powers as well; put bluntly, the United States acts as

---

[8] Among the numerous books in Chinese on topics related to this essay used in writing this essay, the following were not cited in detail: Ma 2002; Zhou 2002; Mi 2001; Yang 1999; Xue 1999; Zhao 2000; Sun 2001; Wang 1997; Yan 1991; Zhou Bolin 2002; Luncong 2001; Xu 2001; Sun 2001; Xia 2002.

the major partner in deconstructing regimes judged to be a security threat, while US allies are asked to contribute in the reconstruction.[9] Apparently, the core issue in global politics now is neither unilateralism nor multilateralism, nor is it the role of the United Nations in shaping global order. At the root is a US reemphasis of the decisive role of values in judging the legitimacy of governments and states. This, in turn, has caused a revival of the function of ideology and, with it, polarization. Condolezza Rice, the National Security Advisor in the Bush administration, provides the following explanation:

> There is an old argument between the so-called "realistic" school of foreign affairs and the "idealistic" school. To oversimplify, realists downplay the importance of values and the internal structures of states, emphasizing instead the balance of power as the key to stability and peace. Idealists emphasize the primacy of values, such as freedom and democracy and human rights in ensuring that just political order is obtained. As a professor, I recognize that this debate has won tenure for and sustained the careers of many generations of scholars. As a policymaker, I can tell you that these categories obscure reality. In real life, power and values are married completely. Power matters in the conduct of world affairs. Great powers matter a great deal—they have the ability to influence the lives of millions and change history. And the values of great powers matter as well. (Rice 2002)

From a strategic point of view, the ultimate question is whether the logic inherent in the quest for military security and power—the security dilemma—will once more be linked to ideological polarization. The answer to this question will determine the structure of international relations in Asia, unless countervailed by the power of a common awareness of shared weaknesses that fosters regional cooperation. Stressing the notion of comprehensive security may be one way to reduce the primacy of the security dilemma, as well as softening the growing ideological divide.

Internal weaknesses, including a fragile social, economic and political infrastructure, and underdevelopment have also been conducive to authoritarian and corrupt regimes in Central, West, and South Asia. These regimes are usually aware that they cannot afford to become involved in major military expenditure that would completely erode the very basis of state power. They also remain suspicious of being merely the objects to be used in power games by their larger neighbors. For this reason, they aim to reduce to some extent possibilities for the powers to engage in mutual "balancing" and "bandwagoning" games. Awareness of internal weakness, and not pacifism, is the driving motif. So far, governments in Central

[9] Ignacio Ramonet (2003) argued that there is a division of labor between the United States focusing on military and mediatory aspects, while France and Germany emphasize politics.

Asian states have not been able (or willing?) to cooperate in one regional organization, but have opted to participate in a multitude of regional organizations and institutions in which Russia, the United States, and also NATO and Japan take part. They remain apprehensive about each of the great powers: [10]

> None of the attempts made within the framework of the Commonwealth of Independent States (CIS), the Central Asian Union, the Shanghai Cooperation Organization and other vehicles of integration to put in place a regional security system that would guarantee the military-political and socio-economic stability of the region has yet produced a result. (Esenov 2001: 244-245)

Japan also cooperates with the OSCE in an attempt to promote common Western interests in Central Asia, as for instance at the conference between Japan and the OSCE on comprehensive security held in Tokyo on December 11 and 12, 2002. Although the Chinese government was not an official participant, it was present in its capacity as a member of the Shanghai Forum. [11]

The great powers, on the other hand, are careful not to provoke each other by attempting to exclude the other parties from access, a response resembling similarly motivated caution in the so-called Open Door policy of the powers in China a century ago. [12] For better or worse, this works at least for the time being as a factor preventing Central Asian countries from taking part in polarizing alliances. Although we have seen a plethora of attempts to establish institutions for regional cooperation in areas such as the economy or security (antiterrorism), so far none of these attempts have created the basis for stable regionalism. Rather than "balancing" and "bandwagoning" most actors seem to be keen to prevent all others from bandwagoning against themselves by diversifying their international relations. In Southeast Asia, the latest example is Myanmar, which decided to strengthen its ties with India, after having been suspected for a long time of being a covert ally of China in an attempt to challenge India's

[10] On apprehensions by Central Asian states concerning Russia and the West, see Tang (2002: 37).

[11] It must be noted that the term "comprehensive security" used at the conference is NOT identical with the specific Japanese notion of 1981, but is a term to denote "security in all its aspects" (in Japanese, this is translated as *hookatsuteki anzen hoshoo*). Reference is made to factors such as politics, democracy, human rights, the economy and the environment. The important difference with the notion of comprehensive security as used in China and Southeast Asia is probably the explicit emphasis on democracy and human rights, keywords usually associated with US foreign policy.

[12] This is one of the reasons why the US is reticent when it comes to discussing the possibility of extending the presence of US forces in Central Asia, sent there in connection with the Afghan war.

position in the Gulf of Bengal. There have been numerous scenarios of the international system in Asia based on the analyses of strategic games among various "triangles," such as China, India, and the United States, including speculation about a rather unlikely triangular cooperation between India, Russia, and China to counter US influence on the Eurasian continent. Some Indian observers raised the specter of Sino-Russian cooperation:

> Sino-Russian relations have improved steadily since the May 1989 summit meeting at Beijing. The Moscow Declaration signed in mid-April 1997 is apparently aimed at working towards the creation of a new multipolar world order and ending US domination. The Moscow Declaration also signifies a major shift in Russian foreign policy in the Asian direction. Seeking to counter US clout and the North Atlantic Treaty Organization's (NATO's) eastward expansion, Russia is looking for new strategic partners in Asia, and China is foremost among them. Russia, China and three CAR nations (Kazakhstan, Tajikistan and Kyrgyzstan) have also signed an agreement to reduce troops along their common 7,000-km-long border, so as to reduce tensions and to make the border area more secure and tranquil. The new Russia-China friendship is also grounded in economic self-interest. Trade between the two countries increased sharply by approximately 25 percent in 1996-97. (Kanwal 1999)

The general consensus is that China and Russia benefit more from good relations with the United States than by allying themselves against the United States. Tang Jiaxuan rejected such proposals in a statement on February 2, 1999. It must be added that there is considerable uncertainty in China as to the direction of Russia's future development (Zhu 2002: 387ff.) The United States is also careful not to antagonize Russia in Central Asia (Blua 2002). Some Central Asian states likely welcome the US presence, trusting the US not to share the "imperialist" ambitions of either Russia or China in the region (Cohen). The willingness to engage in complex networks of cooperation is, however, not an expression of mutual trust, rather the outcome of a conscious strategy of diversifying linkages to reduce risks. Despite the conclusion of a friendship treaty with Russia in 2001, and extensive Russian supplies of arms to China, which is balanced by Russian supplies to India, it is fair to say that China remains cautious towards the possibility of a revival of Russian power. Zhu Tingchang (2002) has set out in detail favorable and unfavorable scenarios for future possible impact by Russia on China (P. 391ff). The Sino-Russian Friendship Treaty of July 16, 2001, is clearly aimed to neutralize the possibility of Russia cooperating with the United States against China (Kapila 2001).

It makes sense to analyze the dynamics of triangular relations between China, India, and the United States so long as it is understood that the "triangle" is not an equidistant one in terms of game theory, and that policymakers in these three countries will as a matter of course always include in their considerations factors outside the triangle.

Most Chinese observers agree that it is the United States and China that are the main competitors in the long run. China's strategy towards India is complicated by the fact that few in India believe seriously that China and the United States will become close partners, to the detriment of India: "The timing gives China a clear signal that the improvement after September 11 is a temporary improvement and not a permanent change" (Yan 2002). On the contrary, events since 1998 indicate a growing strategic closeness between the United States and India. The new Chinese Party leader Hu Jintao will have to move very carefully, lest he further encourages the United States and India from becoming allies in a move to constrain China. Some have argued that the Chinese military (i.e., the PLA) advocates a hard-line approach towards India (*Hindustan Times*, 16 November 2002). My personal exchange of opinion with specialists in Beijing and Shanghai, as well as extensive reading of Chinese language sources, confirms the existence of a wide variety of views on the matter. Similar to what research in other countries indicates, politicians are often more likely to exhibit a hard-line approach on issues of nuclear deterrence than military specialists who tend to be more cautious, yet Western specialists tend to identify the Chinese military with hard-liners (Sutter 1998; Robinson and Shambaugh 2003; Alagappa 2003). Chinese and Indian leaders frequently assert that they intend to avoid long-term strategic rivalry or even confrontation. Given the following statements that appeared in a recent editorial of the Hindu, these assertions are not mere verbal cosmetics:

> Yashwant Sinha, external affairs minister ... disappointed some Western and East Asian analysts who believed that the two Asian giants were bound to clash.... Zhu Rongji visited India early last year and PM Vajpayee will visit China this year.... India is keen on settling border issues.... China has recently settled some of its outstanding boundary disputes with Vietnam. Given that officialdom often passes off silence as policy towards China, Mr. Sinha has done well to lay down the contours of India's approach to China. (*The Hindu* 2003)

Caution is needed, however. Statements containing terms such as "rivalry" or "cooperation" are first of all a change in discourse, and do not necessarily indicate a genuine change of tactics or strategies. On November 8, 2002, Prime Minister Vajpayee referred to a "healthy competition between India and China, whereas India had previously argued that "its Look East Policy had nothing to do with any threat or competition from China" (*Indian Express* 2002). The issue is not so much the question of competition as such, but rather whether competition has or does not have an adversary character. As pointed out above, a key function of comprehensive security is to tone down the adversary character of the security discourse. The discussion below presents a succinct overview of the general notion of "comprehensive security" as it developed in Southeast and East Asia.

Differences in the Interpretation of Comprehensive Security in China and Japan

Security concepts are not created in a void. The occupation of Germany, Japan, and Korea demanded a comprehensive approach going far beyond military measures, and together the study of occupation policies provides an early example. As noted in the following statement: "NATO has had room for more than strictly military functions built into it from the beginning, including mechanisms for solving disputes, coordinating foreign and military policies, and consulting on political matters, and these have allowed it to serve nonmilitary functions" (McCalla 1996: 445-75).

Clearly, the study of comprehensive security should be conducted in a comparative fashion that avoids misperceptions of the term as a fuzzy or vague "Asian" notion. Recently, moves have been on the way in Japan, China, Korea, and India to modernize security decision-making by setting up National Security Councils that, as a rule, deal with a wide variety of issues affecting security, among which military security is an important but by far not the only factor (Report 2001). In addition to the discussion of "comprehensive security," there has recently been a spate of publications on alternative security concepts. [13] As noted above, there is no one single authoritative definition of the term "comprehensive security," not surprisingly for a term that was in the first instance not developed for political science analysis, but rather as a concept to guide policymaking. Therefore, I propose to focus on the various functions of the term, differing with circumstance and time. Non-Asian writers tend to emphasize conflict resolution, whereas the Japanese definition focuses on threat removal. Joe Camilleri's (1999) definition of the concept "comprehensive security" is an example of the first case:

> A particular practice or relationship may be deemed relevant to comprehensive security when it is likely to create new conflicts or exacerbate existing ones either between or within nations, especially to the extent that these are likely to involve the use of threat of force. As a corollary to this, a particular practice or relationship may be said to contribute to comprehensive security when it helps to resolve or obviate conflicts between and within nations, and especially armed conflicts. (P. 83)

A Japanese textbook definition, taken from a book edited by the Research group for comprehensive security of the Japanese Defense

[13] Below, please find a few English language titles on alternative security concepts particular relevant for the study of security in Asia: Amitav Acharya 1999; Buzan 1989; Mack 1993; Simon 1996; Alagappa 2003.

University, stresses the combination of military and nonmilitary aspects of security, as well as external and internal threats. It refers to the Japanese political context that demanded a restriction of the state's military power to the smallest level at which defense was feasible, while reserving a maximum scope for nonmilitary measures. Prime Minister Ohira brought up the notion in a report in 1980 designed to facilitate research on security (Booei 1999). [14] Subsequently, the meaning of the term was broadened by some, including Chinese authors, to cover specific areas of cooperation, such as international regime building. [15] Virtually unknown even to some Japanese specialists, the original Japanese notion of comprehensive security has found supporters in China. One recent widely sold Chinese book on strategy goes so far as to claim that current Chinese strategic thinking is derived from the Japanese concept, "Talking about China's security, the longer the more scholars identify with "comprehensive security" (zonghe anquanguan). One of the authors, Shen Weilie (2001) is professor for Strategic Studies at the National Defense University in Beijing. He relates the need for this concept to the open and dynamic (fluid) character of the international environment, where development and security are mutually dependent. According to him, the impact of the Asian economic crisis of 1997 and economic globalization has promoted the concept of "comprehensive security," and he regards this concept as being much wider in scope than traditional terms such as "common security" and "cooperative security." Not surprisingly, he advocates this concept since the end of the Cold War (in his opinion) signified the end of sharp ideological divisions and frontlines. It is ironic that in Japan, on the contrary, the concept was increasingly used to emphasize the community of values (an ideological element) between Japan and the United States, and that Japan increasingly uses this to emphasize the potential danger to Japan's security posed by China (Bao 2001: 243). In a recent research paper published in the United States, Chu Shulong (1999), then director of the North American Division of the China Institute of Contemporary International Relations in Beijing and a member of the CSCAP China National Committee (June 1999) refers to "comprehensive security" in the context of discussing China's New Security Concept (1997), also mentioning the "ASEAN Way," its use in CSCAP and the NEACD (Northeast Asia

[14] The famous Japanese scholar Eto Shinkichi was very much involved in committee work on comprehensive security. His voluminous book on the issue is a series of long essays on the historical background and issues related to security and Japanese-American relations over the whole post-war period, but hardly touches on conceptual issues (Eto 1991).

[15] Zhou Jihua lists a number of items that will contribute to the construction of a new international order in East Asia, among which a regional comprehensive security regime, but he refers mainly to general issues such as the environment (Zhou Jihua 2001).

Cooperation Dialogue) (pp. 8-11). He quotes with approval Chen Jian (assistant foreign minister of China): "Chen noted that the meeting of the twenty-one participants is "not to defuse a common threat, but rather to achieve a common goal: that is, regional peace and stability" (Chu 1999: 11).

Although the notion "comprehensive security" is usually considered to be a Japanese creation, similar notions have existed in Southeast Asia since the sixties, and continue to influence security thinking until the present. As Jose T. Almonte (1997), Presidential Security Advisor and Director-General, National Security Council, Republic of the Philippines, noted:

> Asians' long-term objective should be to replace security arrangements based on a military balance with mutual security based on economic cooperation—on mutually beneficial trade and investment. Under the balance-of-power concept, ethical principles have had no place in international relations. Maintaining peace through the workings of regional organizations—providing increasingly close rules, norms and procedures to bind states together cooperatively—offers a way of transcending a national-interest concept, which separate ethics from practicality. Through integration, the ethical course becomes the practical course by default. (P. 90)

Keeping in mind the wide variety of civilizations that coexist in Asia, Almonte plays down the role of cultural differences in security issues:

> The assumed opposition between Western and Asian values, however, is more of a political than cultural issue. There is no clash of civilizations of the kind imagined by Samuel Huntington in East Asia (although Lee Kuan Yew suggests there may be racist undertones in the US attitude towards a resurgent China). (P. 89) Indonesia has a long, and strong tradition of emphasizing overall "national resilience" (*Ketahanan nasional*). (Mack 1993; Chen 2001: 57ff., 74ff.,131)

It is in this spirit that Indonesia's foreign Minister Alatas argued that ASEAN's stability was best served by refraining to the largest extent possible from resisting large powers, but rather working towards establishing a new balance between the United States Japan, and China. He, too, eschews references to specific value orientations (Chen 2001: 78; Lee 1998). An interesting note is that Alan Larson (US Under Secretary of State for Economic, Business and Agricultural Affairs) referred to similar themes in an article entitled "Economic Strength, Resiliency Are Foundation of National Security" issued in December 2002 (Larsen 2002). Zhou Jihua has pointed out that Indonesia raised the concept of "national resilience" already during the sixties (Zhou 2001: 44ff). Although he puts this in the same category as Japan's "comprehensive security" (soogoo anzen hoshoo) he remains vague on a possible link. In Indonesia the concept was widened to cover "regional resilience" in the seventies by Indonesia's foreign Minister Malik, and supported by the Special Envoy

Keman from Thailand, and Malaysia's foreign Minister Shafi. At the ASEANs conference of high level officials held in Bangkok in June 1978, it was agreed that the term "national resilience" had laid the philosophical foundation for cooperation within ASEAN for the sake of national and regional resilience. The Manila Declaration of ASEAN issued in 1987 once more declared this to be the unified policy of ASEAN.

Comprehensive security, in any of its interpretations, assumes multilateral cooperation, and this also agrees with China's explicit quest for a multipolar world order (Yuan 2000). The interests of China and the United States do overlap on important issues, and there may be a commonality of views on the changing role of Islam in Southeast Asia (Yatai). The assumption here is that, in the Chinese view, multipolarity is not a value by itself, but a tool towards a strategic end, namely, the relative reduction of the relative superiority of the United States. Multilateralism in Europe, for instance, basically relies on "a common security model and their approach to cooperative security as a future security framework in the 21st century is based on the idea of multilateralism" (Lee 1998: 3).

By way of contrast, Japan has increasingly attached emphasis to the role of values in security discourse, and this is a tendency also conspicuous in the recent heavy emphasis in relations between the United States and India on shared values (see below). For the moment, it seems difficult to assess the direction in which Indian security perceptions will develop, all the more so because of the extremely wide range of opinions aired within the security community. It is true that comprehensive security as such has also been the topic of conferences held in India, such as a national seminar organized by the Delhi Policy Group in August 2001, titled "Comprehensive Security: Perspectives from Indian Regions." At times, security dialogues are clearly influenced by the notion even if the term is not mentioned (Rolfe 2002).

Conclusion

At the height of the conflict with the Soviet Union, Kissinger's realist approach deemphasized the role of ideological values in approaching security issues, and this was one of the factors facilitating the Sino-US rapprochement. More than a quarter of a century later, US policy has come full circle. US relations with post-Soviet Russia are getting closer, and India has become increasingly attractive to US policymakers in a global chess game that portrays China as a newly rising "challenger." This time, ideological values are given heavy emphasis in legitimizing the choice. As Blackwill (2002) states:

> In my view, close and cooperative relations between America and India will endure over the long run most importantly because of the convergence of their democratic values and vital national interests. Our democratic

> principles bind us—a common respect for individual freedom, the rule of law, the importance of civil society, and peaceful inter-state relations. With respect to overlapping vital national interests, let me now briefly share with you my "Big Three" for the next decade and beyond. They are to promote peace and freedom in Asia, combat international terrorism, and slow the spread of Weapons of Mass Destruction.

The title of my paper indicates that China and India will have to weigh the extent to which issues of ideological polarization will govern their relations. Both countries will inevitably face security dilemmas. Sharing concepts such as "comprehensive security" may provide greater leeway for policymakers who do not wish to become prisoners of man-made dilemmas. For policymakers to base themselves on nineteenth-century notions of European continental balance of power means asking the wrong questions. For China, the point is not to "redress the power imbalance" by seeking a counterweight to the United States; rather, China and India both need to reconsider what kind of global order they envisage and base their global strategies on these long-term concepts.

References

ALLISON, R.
2003 "Regional Threats and Prospects for Multilateral Defense Cooperation." 2002 Paper presented at the Conference on "Caspian Sea Basin Security," held at The University of Washington, Seattle, Washington on April 29 & 30, 2003.

ALMONTE, J.T.
1997-98 "Ensuring Security the 'ASEAN Way'." *Survival* 39(4): 80-92.

AMITAV, A.
1999 "Realism, Institutionalism and the Asian Economic Crisis." *Contemporary Southeast Asia* 21(1).

AYOOB, M.
2002 "Inequality and Theorizing in International Relations: The Case for Subaltern Realism." *International Studies*: 27-48.

BAO, X.
2001 "Lengzhan hou Riben de duobian anquan baozhang zhanlue." Pp. 242-253 in Xu, Y. (ed.), *Shiji zhiJjiao de Guoji Guanxi*. Shanghai: Shanghai Yuandong Chubanshe.

BARNETT, M.
2002 "Radical Chic? Subaltern Realism: A Rejoinder." *International Studies*: 49-113.

BLACKWILL, R.
2002 "Transcript: U.S. Ambassador Robert Blackwill Outlines the Growing Strength of U.S.-India." Addressing the Indian Chamber of Commerce in Kolkata on November 27, 2002.

BLUA, A. ET AL.
2002 "Russia: Shanghai Group Aims To Increase Economic Cooperation." *Radio Free Europe*.

BOOEI, D.
1999 Anzen hoshoogaku kenkyuukai comp. Anzen hoshoogaku nyuumon. Tokyo: Aki Shoboo.

BUZAN, B.
1989 *The Concept of National Security for Developing Countries*. Institute of SEA Studies.

CAMILLERI, J.
1997 "The Pacific House: The Emerging Architecture for Comprehensive Security,"

in Dickens, D. (ed.), *No Better Alternative: Towards Comprehensive and Co-operative Security in the Asia-Pacific*. Wellington: Centre For Strategic Studies.
CHEN, Q.
2001 *Lengzhan hou dongmeng guojia dui Hua zhengce yanjiu*. Beijing: Zhongguo Shehui Kexue Chubanshe.
CHU, S.
1997 *China and the U.S.-Japan and U.S.-Korea Alliances in a Changing Northeast Asia*. APARC Publication, Stanford University. Palo Alto: Stanford University APARC Publication.
DICKENS, D. (ED.)
1997 *No Better Alternative: Towards Comprehensive and Co-operative Security in the Asia-Pacific*. Wellington: Centre For Strategic Studies.
ESENOV, M.
2001 "Turkmenistan And Central Asian Regional Security." Pp. 244-254 in Gennady Chufrined, *The Security of the Caspian Sea Region*. Series: Stockholm International Peace Research Institute.
ETO, S. ET AL.
1991 *Soogoo anpo to mirai no sentaku*. Tokyo: Kodansha.
FANKONGBU
2001 *Zhongguo xiandai guoji guanxi yanjiusuo fankongbu yanjiu zhongxin comp. Guoji kongbuzhuyi yu fan kongbu douzheng*. Beijing: Shishi Chubanshe.
FU, X.
2003 "Yindu yuanhe hejunbei dongzuo pinfan." Xiandai guoji guanxi.
HANIFFA, A.
2001 "New Bush Aide Sees India as Top Regional Power." *rediff.com US edition*, May, 31.
JIN, Y. ET AL.
2001 *Yisilan yu Guoji Redian*. Beijing: Dongfang Chubanshe.
KANWAL, G.
1999 "China's Long March to World Power Status: Strategic Challenge for India." *Strategic Analysis* 22(11): 1713-28.
KAPILA, S.
2001 "Implications of the Sino-Russian Friendship Treaty," Article No: 559, September 4.
2001 "Bangladesh-China Defense Co-Operation Agreements Strategic Implications: An Analysis." *South Asia Analysis Group*, Paper No. 58214.
LARSON, A.
2002 "U.S. National Security Strategy: A New Era." Retrieved November 1, 2002 (*http://usinfo.state.gov/journals/itps/1202/ijpe/ijpe1202.htm*).
LEE, S.-G.
1998 "Multilateral Security in Europe and Northeast Asia: In Search of New Alternatives." *Seoul Journal of East Asia Affairs*.
LEIFER, M.
1996 "The ASEAN Regional Forum: Extending ASEAN's Model of Regional Security," Adelphi Paper 302 IISS: Oxford: Oxford University Press.
MA, J.
2002 *Guanzhu Yindu—jueqi zhong de daguo*. Tianjin: Tianjin renmin Chubanshe.
MACDONALD, J.A.
2003 "India, Pakistan: Impact on Caspian Region's Stability." Paper presented at the Conference on "Caspian Sea Basin Security," held at The University of Washington, Seattle, Washington on April 29 & 30, 2003.
MACK, A.
1993 "Concepts of Security in the Post-Cold War." Working Paper Research School of Pacific and Asian Studies, Australian National University.
MCCALLA, R.B.
1996 "NATO's Persistence After the Cold War." *International Organization* 50(3): 445-75.
MI, Q. ET AL.
2001 *Guoji guanxih yu dongya anqyuan*. Tianjin: Tianjin Renmin Chubanshe.

NAKANISHI, H.
1993 "Redefining Comprehensive Security in Japan." Kokubun Ryosei (ed.), *Challenges for China-Japan-U.S. Cooperation*. Tokyo: Japan Center for International Exchange.
QIAO, L. ET AL.
1999 *Chaoxianzhan, dui quanqiuhua shidai zhanzheng yu zhanfa de xiangding*. Beijing: Jiefangjun Wenyi Chupanshe.
RADTKE, K.W. AND R. FEDDEMA
2000 *Comprehensive Security*. Leiden: E.J. Brill.
RAMONET, I.
2003 "De la Guerre Perpétuelle." Le Monde Diplomatique, Internet edition.
RAPPAI, M.V.
1999 "India-China Relations and the Nuclear Realpolitik." *Strategic Analysis*, April (23): 15-26.
RASHID, A.
2003 "China Forced to Expand Role in Central Asia." *Central Asia-Caucasus Analyst* :International Eurasian Institute for Economic and Political Research.
REPORT
2001 "Report From The Conference On: Domestic Determinants Of Security: Security Institutions And Policy-Making Processes In The Asia-Pacific Region." Honolulu, Hawaii, January 10-11.
RICE, C.
2002 "A Balance Of Power That Favors Freedom." Retrieved November 1, 2002 (*http://usinfo.state.gov/journals/itps/1202/ijpe/ijpe1202.htm*).
ROLFE, J.
1997 *Unresolved Futures: Comprehensive Security in the Asia-Pacific*. Wellington: Centre For Strategic Studies.
2002 "Globalism, Regionalism and the War on Terror in South Asia." *Peace and Conflict Policy*, Article no. 897: Institute of Peace and Conflict Studies.
SHEN, W. ET AL.
2001 *Zhongguo guojia anquan dili*. Beijing: Shishi Chubanshe.
SIMON, S.W.
1996 "Alternative Visions of Security in the Asia Pacific." *Pacific Affairs*, 69(3).
SINGH, B.K.
2002 "Chinese Perspectives on the Kashmir Dispute" Article No: 696. February 17, 2002.
SOEYA, Y.
1998 "Japan's Dual Identity and the U.S.—Japan Alliance." Asia/Pacific Research Center.
SUBRAHMANYAM, K.
"Interview with K. Subrahmanyam" (Chairman—National Security Advisory Board (NSAB), India's Nuclear Doctrine, by Serge Berthier. Retrieved November 1, 2000 (*www.asian-affairs.com*).
SUN, P. ET AL.
2001 *Yindu guoqing yu zonghe guoli*. Beijing: Zhongguo Chengshi Chubanshe.
SUN, Z.
2001 *Zhongya xin geju yu diqu anquan*. Beijing: Zhongguo Shehui Kexue Chubanshe.
SUTTER, R.
1998 *Chinese Policy Priorities: Chinese Reaction to the South Asian Crisis*.
SWARAN, S.
1999 "The Kargil Conflict: Why and How of China's Neutrality." *Strategic Analysis* 23(7).
TANG, S.
2002 "2010-2015 nian de zhongguo zhoubian anquan huanjing: juedingxing yinsu he qushi zhanwang." *Zhanlue yu guanli* (5).
WANG, F. ET AL.
1997 *Heyinyun xia de nanya*. Beijing: Wenhua Yishu Chubanshe.
WANG, Z.
1997 *Bianyuan didai fazhan lun-shijie tixi yu dongnanya de fazhan*. Shanghai: Shanghai Renmin Chubanshe.

WOLF, C. ET AL.
2000 *Asian Economic Trends and Their Security Implications.* Santa Monica, CA: RAND.
XI, J. ET AL.
1996 *Shijie zhenzzhi xin geju yu guoji anquan.* Beijing: Junshi Kexue Chubanshe.
XIA, L.
2002 *Yatai diqu junbei kongzhi yeu anquan.* Shanghai: Shanghai Renmin Chubanshe.
XU, Y.
2001 *Shiji zhi jiao de guoji guanxi.* Shanghai: Shanghai Yuandong Chubanshe.
XUE, J. ET AL.
1999 *Zhongguo yu zhongya.* Beijing: Shehui Kexue Wenxian Chubanshe.
YAN, X.
2002 New York Times quoting Yan Xuetong (Director, Institute of International Studies at Qinghua University), July 26.
YAN, X. ET AL.
1991 *Zhongguo yu Yatai anquan. Lengzhan hou Yatai guojia de anquan zhanlue zouxiang.* Beijing: Shishi Chubanshe.
YANG, D. ET AL.
1999 *Shiju. Yi zhongguoren de shijiao kan shijie.* Beijing: Renmin Chubanshe.
YUAN, J.
2000 *China's Conditional Multilateralism and Great Power Entente.* U.S. Army War College, Strategic Studies Institute.
ZHAO, W.
2002 "YinZhong guanxi fengyunlu (1848-1999)." Zhongguo xiandai guoji guanxi yanjiusuo 2002 Yatai zhanluechang. Shijie zhuyao liliangde fazhan yu jiaozhu. Beijing: Shishi Chubanshe.
ZHOU, B.
2002 *Meiguo xin baquanzhuyi.* Tianjin: Tianjin Remin Chubanshe.
ZHOU, J.
2001 "Zonghe anquan baozhang yu dongya guoji xin zhixu," in Qingyu, M. (ed.), Guoji guanxi yu dongya anquan. Tianjin: Tianjin Renmin Chubanshe.
ZHOU, J. ET AL.
2002 *Meiguo anquan jiedu.* Beijing: Xinhua Chubanshe.
ZHU, T. ET AL.
2002 *Zhongguo zhoubian anquan huanjing yu anquan zhanliue.* Beijing: Shishi Chubanshe.
LUNCONG (EDITORIAL COMMITTEE)
2001 *Xiandai guoji guanxi luncong,* Vol. 1. Beijing: Shishi Chubanshe.

VIII. The US and the EU in CEA. Relations with Regional Powers*

MEHDI PARVIZI AMINEH AND HENK HOUWELING

ABSTRACT

The US and the EU are important external actors in the post-Soviet CEA region. One challenge confronting US policymakers is balancing commercial interests in the region with security interests and foreign policy goals. These include a desire to contain Iran, partly because of its support for radical Islamic forces in the Middle East, to prevent regional conflicts, assist NATO-member Turkey—a critical ally in an area that is of top US-security interest, and to normalize its relations with China, whose military potential and alliance with Russia is perceived as a threat to its own security interests. Commercially, the EU is not as involved in CEA as the US. The main powers in the EU—Britain, France and Germany—give priority to other regions over CEA. Britain puts emphasis on the Baltic States, France focuses on North Africa, and Germany has been more preoccupied with the development of Eastern Europe. As a group, the member countries of the EU act mostly in the context of economic assistance and diplomatic contacts. Military agreements have been signed on a bilateral basis mostly with Georgia.

Introduction

All US Presidents since Roosevelt agreed that the policy of domestic energy security of America extends beyond the legal borders of the country. Between the end of the 1960s and the oil price hike of 1973, energy security moved to the top of the American agenda. At that time, oil prices

* This article is partly based on article by Amineh (2003) in this journal.

had become volatile after decades of stable and low prices. American oil production from US legal territory reached its peak in 1970, with the number of drilling rigs down to one third of its 1955 level (Deffeyes 2001: 35).[1] Up to that time, US domestic high-cost producers had been able to get protection by quota against low-cost site producers operating abroad (Yergin 1991: 589). In the early 1970s, a buyer's market had turned into a supplier-dominated market. In April 1973, Nixon cancelled the quota system of imports: the United States consumer had made its entry into the world oil market. The US as a strategic actor did so, too, overruling domestic interest groups that, in the past, had succeeded in getting protection from political entrepreneurs against low-cost foreign producers. In all industrial economies, GDP growth and fossil energy consumption are strongly correlated. That is still true today. The global business revolution has not changed that. However, due to outsourcing parts of industrial manufacturing to late-industrializing countries, the contribution of energy use to GDP growth in high-income economies has substantially decreased.

US Domestic Consumption and Diversification: Linking-up with the CEA Region

In 2001, the United States consumed 19.6 million bbl/d of oil and 21.7 tcf of gas. The United States, thus, has the highest level of per capita consumption in the world. It has also the highest per capita emission of the world.[2] It is expected that by 2025, consumption in the United States will have risen to 28.3 million bbl/d of oil and 34.9 tcf of gas, respectively. In 2000, the United States imported 53% of its oil consumption, about one fourth of which comes from the Persian Gulf. It is expected that US oil imports will increase to 60% in 2020 (IEA 2002).

A major concern for the United States over the last ten years has been not only to secure adequate supplies of oil and gas for home consumption but also to diversify sources of supply. The US government considers the CEA to be a region of vital interest. It is the declared objective of the Bush government that such regions should not be dominated by anyone

[1] Geologist M. King Hubbert anticipated in 1956 on the basis of geological data and the fitting of consumption data to growth curves, the time of peak in US domestic oil output to the around 1972 (Deffeyes 2001: 135).

[2] The U.S. Senate represents about 5% of world population. In December 1997, it decided with 95 against 0 not to sign any treaty emerging from the climate change conference in Kyoto, December 1997, "unless the protocol or other agreements also mandates new specific scheduled commitments to limit or reduce greenhouse gas emissions for developing parties." In the 1990-1995 half decade, each American produced 5.3 tons of carbon, each Chinese 0.7 tons.

except the United States itself. Controlling the region helps to reduce US reliance on Persian Gulf oil. Diversification of supply is a stimulus for US involvement in the region and of its efforts to get access to its stock.

US Energy Policy Towards CEA: Overland Transport Routes

An important challenge for US policymakers involved in making energy policy is to strike a deal between private sector interests, to secure energy supply to the home market and to achieve foreign policy objectives transcending the security of domestic oil use. The interests of oil companies and government energy policy may follow different tracks. For example, pipeline diplomacy of the government is not dictated by efficiency. The case of Russia highlights the trade-off between private profit seeking and state policy.

One major obstacle to expanding the US role in the region is Russia's influence, in particular its control over oil and gas resources and transport routes. Also China and Iran are contenders for influence on decisions about location and construction of pipelines. Pipeline construction and management depends on vast amounts of money to be invested in projects, with depreciation times of decades and more. Technology and capital are US assets. However, pipeline diplomacy and pipeline commerce may, and do, collide. Pipeline diplomacy requires the US government to become involved as the driving force of the Baku-Tbilisi-Ceyhan (BTC) and the Trans-Caspian Gas Pipeline (TCGP), both of which circumvent Iran as well as Russia.

The BTC will have a length of 1,730 km with a section of 1,070 km in Turkey. The projected completion date is 2005. Its costs are estimated at US$3-4 billion. The BTC pipeline is by far the most expensive one of the Western options. Turkey has promised to cover the costs on its territory at US$1.4 billion. Based on a flow of 800,000 bbl/d, estimated transport costs from Baku to Italian ports are as high as US$2.80 per bbl. This is more than any other Western alternative pipeline option (Cohen 2002; AGOC, May 16, October 15, 2002; Soligo and Myers 1998).

According to Nana Janashia,[3] Director of the Caucasus Environmental NGO Network (CENN), the BTC is economically, politically, and environmentally not viable. Construction and transportation costs are very high, and Turkey's oil demand will probably not rise as much as expected in the near future. The BTC pipeline will only provide limited employment opportunities for the local population. Construction, maintenance, and protection will induce an overpricing of land and will damage agriculture, roads, and water supply. Corruption of Azerbaijan's and Georgia's

[3] Based on interview with Nana Janashia of CENN on March 3rd, 2003.

policymakers and noncompliance with European environmental legal standards pose a serious threat to the environment of the area. And last but not least, the pipeline is closely positioned to seven conflict areas:

(i) Nagorno-Karabakh, Armenia vs. Azerbaijan (15 km from the BTC pipeline)
(ii) Georgia vs. South Ossetia (55 km from the BTC pipeline)
(iii) North Ossetia vs. Ingushetia (220 km from the BTC pipeline)
(iv) Georgia vs. Abkhazia (130 km from the BTC pipeline)
(v) Russia vs. Chechnya (110 km from the BTC pipeline)
(vi) Russia vs. Dagestan (80 km from the BTC pipeline)
(vii) Turkey vs. PKK (the BTC pipeline will be running through Kurdish territory).

The BTC sister project is the TCGP from Turkmenbashy (Turkmenistan) via Baku and Tbilisi to Erzurum in Turkey. The pipeline will cost between US$2 to US$3 billion. Its initial throughput will be 565 billion cf, eventually rising to 1.1 tcf per year. Several TNOCs have appraised the feasibility of the project, one of which is the Transcaspian Gas Pipeline Project, a grouping of Bechtel, General Electric, and Royal Dutch Shell. The TCGP, however, encounters several challenges. It competes with the Russian Blue Stream pipeline.[4] It is also hampered by the lack of a legal regime for the Caspian Sea, and several Caspian littoral states are opposed to the pipeline on environmental grounds (Soligo and Jaffe 2002; EIA, June 2002). With a maximum of 2150 m below the Black Sea, the Blue Stream pipeline will be the deepest in underwater pipelines. Damage to the pipeline would not only involve expensive repair and cutting-of gas supplies, but would also have great ecological consequences. Interestingly enough, Moscow opposed the construction of the TCGP for ecological reasons but has no such concerns for the Blue Stream pipeline (Rasizade 2002).

Both the BTC and the TCGP pipelines are very costly projects. Two Washington-based independent research groups, the Carnegie Endowment

[4] The Blue Stream pipeline is an undersea route across the Black Sea from the Russian port of Tuapse to Samsun, Turkey, with which Russian gas will compete with Azeri and Turkmen gas. The costs are US$3.4 billion. The Turkish Company BOTAS and Russia's Gazprom built large parts of the on-shore pipeline. The Russian firm Stroitransgaz and the Japanese engineering firm Saipem built the undersea section. The companies involved in the project are Gazprom and Italy's ENI. Pipe-laying started in October 2001. The construction was finished on October 20th 2002. The Blue Stream has come under fire because of corruption. The United States has heavily opposed the Blue Stream pipeline, as it endangers the prospects for the TCGP. The pipeline also would increase Turkey's reliance on Russian gas from a current 66% to about 80%, and could hamper US efforts to reduce Russian influence in the South Caucasus (Amineh 2003: 198).

for International Peace and the Cato Institute (Carnegie Endowment for International Peace, March 19, 2001; Kober 2000) have criticized the BTC route as wasteful. Russian and Iranian alternatives are recommended, but the Bush administration appears determined to follow its own strategy. Economic efficiency and state strategy collide, with the latter winning over the first. Another major pipeline project of the US is the Central Asia Gas Pipeline (CentGas) from Dauletabad (Turkmenistan) via Herat (Afghanistan) to Multan (Pakistan) that could even be extended to India.

The pipeline will have a length of 1664 km. Its costs are estimated to be US$2.5 billion. It is expected that the pipeline will transport 1 MMbbbl/d. In the late 1990s, the Central Asia Gas Pipeline Ltd. and the CentGas Consortium began feasibility studies on the pipeline. This consortium was composed of Unocal with a share of 46.5%; the largest member Saudi Arabia's Delta Oil Co. Ltd. with 15%; Turkmenistan with 7%; Japan's Indonesia Petroleum Ltd. with 6.5%; Itochu with 6.5%; South Korea's Hyundai with 5%; and Pakistan's Crescent Group with 3.5% (AGOC, October 1, 2002). Gazprom agreed to take the remaining 10%. Construction plans of the CentGas Consortium were shelved in 1998. At that time, the Taleban had become too much of a destabilizing factor in the region.

During the 1980s, the United States armed, equipped, and trained Sunni Islamic forces to fight the Soviet army in Afghanistan. Amongst these was Osama bin Laden, now the United States' most wanted international terrorist. The United States welcomed[5] the Taleban's to power in Afghanistan in 1996. On the one hand, the United States hoped that the Taleban regime could contain the influence of the Shi'ite-based Islamic Republic of Iran in the region. On the other hand, it hoped that it would open the way for a centralized government in Afghanistan crucial to the CentGas consortium. The US-based Oil Company Unocal is said to have provided humanitarian aid to the Taleban, and the company's board expressed its support for the Taleban. A Unocal top executive stated, "If the Taleban lead to stability and international recognition, then it is positive. I understand Pakistan has already recognized the [Taleban] government. If the US follows, it will lead the way to international lending agencies coming" (*Pipeline News*, October 12-18, 1996).

After the war in Afghanistan and the expulsion of the Taleban regime at the end of 2001 the Bush administration discussed the composition of a new Afghan government. Oil diplomacy and company interests largely answer to whoever will govern the Afghans (*The New York Times*, December 15, 2001). President Bush appointed Zalmay Khalilzad, former

[5] For a detailed analysis of the Taleban period in Afghanistan, see Rashid (2000).

Unocal consultant, as his special envoy to Afghanistan. The nomination underscores the strategic and private financial interests at stake in the US military campaign in Afghanistan. Khalilzad is intimately involved in the long-running US efforts to directly access oil and gas resources of the region. As an adviser for Unocal, Khalilzad made a risk analysis of a proposed gas pipeline from Turkmenistan across Afghanistan and Pakistan to the Indian Ocean. Also current Afghan President Karzai, before he was made President of Afghanistan, was a consultant to Unocal. The fact that Karzai and Khalilzad are experts in the oil business make them ideal partners for oil companies and the United States to revitalize this project.

Bush has a long and personal relationship with Enron's former CEO Kenneth Lay, who gave a generous contribution to Bush's election campaign. Only one month after Bush had become President, Cheney met with Kenneth Lay and other Enron executives. At that time, Enron stood to benefit from the CentGas pipeline. The conclusion therefore is that the Liberal State in America wears a business hat even in matters of state-strategy (see also Introduction on Democracy in Outward-Oriented Capitalism). As has been noted, "The US government has become an instrument of a segment of American society: corporate business. It has become, as others than myself have already recognized, 'America Incorporated... If a candidate today is not acceptable to the corporate mainstream, he is unelectable. Corporate money determines national policy" (Callari, May 22, 2002).

While this all does not add up to a conspiracy theory, it does indicate a significant money subtext to Operation Enduring Freedom (Almaty, *Financial Times*, December 25, 2001). On May 30, 2002, Afghanistan's then interim President Karzai traveled to Islamabad to meet with Turkmenistan's President Niyazov and Pakistan's President Musharraf to discuss the pipeline construction. Later in 2002, ministers of the three countries met in Kabul to discuss a feasibility study financed by the Asian Development Bank (ADB). On December 27, 2002, Karzai, Pakistan's Prime Minister Mir Zafarullah Jamali, and Turkmen President Saparmurat Niyazov signed an agreement to build the CentGas pipeline.

Until now, no TNOCs have been named in the context of the CentGas pipeline. Afghan Reconstruction Minister, Amin Fahrhang, explained that Afghanistan is negotiating with TotalFinaElf and ExxonMobil. TNOCs will officially be invited when the ADB feasibility study of US$1.5 million is completed in 2003 (AGOC, October 1, 2002). The Afghan Mines Minister Juma Mohammadi underlined the importance of the pipeline for Afghanistan and the region: "This will be beneficial for the economic development of Afghanistan, by creating jobs and providing gas for our needs... The project will make a substantial contribution to the economic development of Afghanistan, Pakistan and other countries" (AGOC, October 1, 2002).

The objective of US policymakers is not only to obtain oil and gas from CEA but also to control its flow to oil and gas markets in the West and in Southeast Asia. US economic interests are combined with strategic interests to weaken Russian and Iranian influence in the region and also to ensure better control over both resources and the shipping lanes in the Persian Gulf. Rivalries that are played out here will have a great impact on the shaping of post-Soviet CEA, resulting also in worldwide consequences.

The Military Connection

The belief that economic dynamism comes from technological innovation in a healthy private economy and a balanced government budget underlies US armament acquisition policy of Truman, Eisenhower, and Nixon. Kennedy, Reagan, and Bush Jr., however, resorted to deficit spending to build military capacity; the latter two Presidents increased deficit spending also for financing tax relief, from which high-income earners profit most (which translates into the ability of these professionals to raise record amounts of money for financing election campaigns).

The Clinton-Presidency, following Eisenhower, emphasized again a healthy economy as the core of United States security. For the Clinton Presidency, a healthy economy implied productive superiority of US enterprises over foreign competitors. Clinton elevated competitiveness of US enterprises into a pillar of the national security of the US. Officials of his administration based trade policy explicitly on a twofold track. One track is based on Ricardo and is suitable for raw material producers and subcontractors. The other one derives from Schumpeter and is fit for high-tech producing America: "A nation's comparative advantage is less a function of its national factor endowments and more a function of strategic interactions between its firms and governments and the firms and governments in other nations. In such industries, comparative advantage is created, not endowed by nature" (Laura Tyson, Chairperson of the Council of Economic Advisors of President Clinton, as cited in King 1995, 137).

In Clinton's world of man-made comparative advantage, some of his trade officials believed that state agents had to provide a helping hand as salesmen: "The Clinton Administration and its successors will inevitably continue to play hardball in helping American firms to lock-up contracts abroad. Foreign governments will learn that the United States will not roll-over when confronted with their aggressive tactics and at the same time the cost of intervention will rise for them" (Garten, November-December, 1995: 58).

In external commercial policy, such as emphasis implied the demand on Japan and South Korea "to open-up" markets for US investors and products and to adapt domestic institutions that were believed to obstruct access but fell beyond the scope of international trade law. Bill Clinton articulated this perspective in a speech at the Georgetown University during his 1992 electoral campaign: "Our economic strength must become a central defining element of our national security policy. ... We must organize to compete and win in the global economy" (*Harvard International Review*, Summer 1992: 26-27). The great discovery in commercial policy of the Clinton administration is that Ricardo's free trade theory, which underlies neoclassical free trade theory, is good for others but does not apply to US technology intensive manufactures. "Schumpeter" therefore has come alongside "Ricardo"[6] in America's external commercial policy.

In the external policy of President Bush, the commercial element gets a less prominent role, whereas military policy is more visible than in the Clinton Administration. The military cannot create trade flows itself, nor is that institution able to contribute to financial stability. However, implementing regime change in sovereign states does create opportunities for US businessmen, such as weapons suppliers, foreign investors, sellers of candy bars, and distributors of the Holy Bible. The eagerness of the Bush government to expand its military forces in the Persian Gulf as well as in the Caspian Sea is a mixture of private commercial interest, domestic energy security and global security objectives. NATO expansion is good for weapons producers in the United States, though not for Russia's armaments exporting sector. One may expect that the long-term stationing of US military forces in the CEA states will contribute to US armaments sales to host governments, in particular to its oil and gas exporters.

US Military in CEA

Since the September 11th attacks, the Bush administration has made a virtue out of necessity, using its need for bases in the Afghanistan campaign to establish a military foothold throughout CEA. Currently the US military is involved in the following Eurasian countries: Afghanistan (combat role), Pakistan (bases), Uzbekistan (base), Tajikistan (base), Kyrgyzstan (base), and Georgia (military advisers and base). The aim is to create a platform that can take action against any group perceived as a danger (MacAskill, March 14-20, 2002). As part of the antiterrorism campaign, for the first time in the CEA history, US military forces have been stationed in Uzbekistan, Kyrgyzstan, and Tajikistan.

[6] "Ricardo" seems to be in particular good for African exporters of primary commodities. Indeed, if their productivity increases, importers see a price fall instead of higher incomes for Africans.

In January 2002, 3000 US troops arrived in Kyrgyzstan as a supplement to the already stationed 1500 military personnel in Uzbekistan. Uzbekistan has agreed to the deployment of 1500 US troops at its Hanabad air base. In return, the United States promised to provide its host US$160 million in aid in 2002, an increase of US$100 million over earlier figures. These funds got allocated despite an undemocratic government and international critique on its respect for human rights. Only two days before US Secretary of State Colin Powell visited Uzbekistan in December 2001, the Uzbek parliament had granted a lifetime term of office for President Islam Karimov. The United States is also carrying out a more extensive military build-up in Kyrgyzstan. A 37-acre air-force base for 3000 personnel at the Manas airport near Bishkek is currently under construction. The site has been granted by the Kyrgyz government to the United States for 12 months, with a possible extension. The US has also planned to relocate fighter jets from Pakistan to Uzbekistan and Kyrgyzstan. The impact of US military personnel in Kyrgyzstan on the stability of CEA is hard to predict. More important, will it be successful in development? The Kyrgyz President Akaev hopes that the US will increase its political and economic support for his country.

The most important military relationship that the United States has established in the region is with Georgia. On February 27, 2002, the US government declared it would provide Georgia with military support worth US$64 million.[7] Cooperation with Azerbaijan also has significantly increased. Both countries provided the US with over-flight rights for air strikes on Afghanistan. Washington reacted in January by lifting an eight-year ban on aid to the Azeri government. This policy was aimed at ending the Azeri blockade of Nagorno-Karabakh. On March 28, 2002, the US government promised Azerbaijan US$4.4 million, the same amount it has promised to Armenia. The US force established in CEA could help to gain support from the countries in CEA. At the same time extended presence of US military in CEA could lead to rivalry with Russia, Iran, and China.

Russian Military in CEA

There have been no significant changes in the number of Russian troops in CEA since September 11, 2001. In Georgia, there are still about 2000 troops of the Russian peacemaking contingent under the aegis of the Newly Independent States (NIS). According to the information agency PRIMA, in October 2002, the size of the forces of the 12th military base in Adzhariya was reduced by 300. At the same time, 130 Russian recruits, who should be

[7] Based on interviews with local officials who prefer to remain anonymous.

based in Batumi, have not received permission from the Georgian Ministry of Defense.

About 10-12 thousand Russian troops are currently defending the Tajik border in the Pamirs and Tien Shan. The 201st motor-infantry division is located in Tajikistan. After September 11, 2001, and the subsequent US military expansion in the region, Russia seriously considered possibilities for strengthening its military presence in CEA. Whether or not planned reinforcements are a response to terrorist threats or reaction to US military presence is hard to say.

President Putin advanced the idea of forming a collective force from the elite troops of the NIS armies for antiterrorism operations. To consider this initiative, the defense ministers of the NIS held a summit in June 2002. The antiterrorism centre of NIS, under the leadership of Boris Mel'nikov, worked out a model operation in case of terrorist penetration into any of the countries. Putin's initiative resulted in establishing the Collective Forces of Quick Response (CFQR) with its staff in Bishkek. At present, the number of the military personnel is 1300 (one battalion each from Russia, Kazakhstan, Kyrgyzstan, and Tajikistan). Operations would start from the military airfield in the town of Kant near Bishkek. By June 2002, the Russian minister of Defense Sergei Ivanov had signed documents with Kyrgyz authorities on bilateral planning of military training and technical cooperation. Within the last year, Russia has supplied Kyrgyzstan with various technical equipment, and created a few joint ventures on developing special intercommunication and other military techniques, and the Russian Ministry of Atomic Manufacturing has provided Kyrgyz frontier troops with the apparatus to protect its borders. Kyrgyz authorities are also considering the problem of land lending to Russian troops. Russia will provide Kyrgyzstan with an antiaircraft rocket complex S-300. Russian military academies receive students from Kyrgyzstan and Kazakhstan on favorable terms (about 600 persons every year). New treaties of military cooperation have been signed between Russia and Armenia. Military advisers and specialists have been sent from Moscow to Armenia. Military training of Russian and Armenian troops took place in August 2002 near Giumri. The 102nd Russian military base functions on the territory of Armenia. Military training of the NIS countries, signatory states of the Treaty of Collective Security, were held in Kazakhstan in June 2002.[8] Russian security officials claim that Russia has top-secret military facilities in CEA and that NATO and the United States are keen to get information on these installations. In Kazakhstan, there is the Sary-Shagan antimissile launching site and a radar station. Also in Kyrgyzstan, the Russian navy

[8] This information is based on archival material.

has a long-distance communications centre and on Lake Issyk-Kul a testing site for nuclear submarines' rockets and space surveillance station at Nurek in Tajikistan (Rasizade 2002: 267).

Current US-Russian Relations in CEA

Putin's decision to support the US-led antiterrorism campaign after the September 11th attacks had few critics among the Russian political elite but also few enthusiasts. According to Grigory Yavlinsky, leader of the centrist political party Yabloko, Putin called shortly after the attacks for a meeting of all leaders from both houses of the Russian parliament, 21 in total. The objective of the meeting was to discuss how to respond. One of the attendants advocated that Russia should support the Taleban, and two were in favor of an unconditional support for the antiterrorism campaign. The other participants were of the opinion that Russia should remain neutral. Thus, Putin's decision to support the United States did not have a broad basis among the Russian political elite. This is not just due to the fact that thousands of military members, politicians, and civil servants have grown up in suspicion of the West and therefore view attempts towards policy convergence with the West, and in particular with the United States, with caution. Putin is a respected president, both among the Russian population and the political elite. But he cannot afford to shift too far to the United States even though the latter is now silent about the war in Chechnya. In any way, hard-liners among the political elite hold their tongues because of respect for the president (Cottrell, May 22, 2002). More important for the future is that the US exploits Russia's weakness wherever it sees a chance to do so while at the same time it tries to be very friendly to Putin. The refusal by the Russian leadership to endorse America's war in Iraq has not changed that. Unlike France, US policymakers seem to "forgive" Putin for what is seen as "defiance" of America. Why? Our speculation is that Russia's position of being in the middle between China/dynamic East Asia and a divided Western Europe gives it policy alternatives and that the US is aware of that.

Have the September 11, 2001, attacks on the United States and the war in Afghanistan resulted in silent coordination between the United Sates and Russia in dividing the area into mutually accepted zones of influence, and if so, are the implicit rules that such cooperation requires well understood on both sides? Or is cooperation ad hoc and easily reversible? Evidence is hard to find. As long as the United States does Russia's work in Afghanistan, and thus helps to protect Russia's southern border, the two views are hard to distinguish.

However, the further penetration of NATO onto the doorstep of the Russian Federation in Eastern Europe, in the Baltic States, in the

Caucasus, and in the Black Sea, is likely to provoke second thoughts in the Kremlin, despite the warm words of the Bush administration for Putin. The deployment of a missile shield combined with the creation of NATO rapid response force with a global mandate will not convince Russia of America's respect for Russian interests. It should be noted that the stability of the territorial East-West divide in Europe during the cold war was based on such respect. If countries in Eastern Europe follow the Polish example of equipping themselves with advanced US warplanes, Russian leaders could become interested in buying some Chinese medicine. That could provoke an open struggle in Russia between "Westernizers" and those in support of Russia concentrating its energies in Asia. NATO's ambition is to equip a Response Force of 21000 and give it a worldwide mission:

> "Effective military forces, able to deploy to wherever the Alliance decides, are essential to the Alliance ability to achieve its wider security objectives as well as its core function of collective defense." [9]

The accomplishment of this mission will not be welcome news in Russia *and* China. Shifting NATO bases from Western Europe to Central and Eastern Europe, in particular to Rumania's Black Sea coast will bring US troop deployments up to Russia's western borders. Despite Russia's limited cooperation with NATO, it will not get a voice in the use or threat of use of weapons of mass destruction by the alliance (Sammon, May 29, 2002). The loss of Iraq as weapons contractor and the more than probable loss of Russia's oil contracts are directly hurting Russia' economic interest. America's threat of war against Iran cuts into Russia's nuclear sector. Putin may find it difficult to explain to his critics at home why the double standard policy, which is so obviously practiced by the United States and its supporters in NATO is also good for Russia. The US influence over oil price will pit the interests of the largest consumer nation against the interest of Russia as an oil exporter (Blagov, May 30, 2002).

For Russia, cooperation with the United States may be a temporary improvisation coming out of a domestic compromise between the short-term objectives of mastering the threat of terror and the longer-term objective of reestablishing Russian dominance in CEA. As has been stated by Mikhail Margelov, chairman of the International Affairs Committee of the Federation Council, which is the upper house of the Russian parliament, and one of the President's closest advisers:

> In Afghanistan, the Americans are doing our job for us. ... I hate to say this, but fortunately for us the Americans got involved. ... In Georgia, before last

[9] NATO Headquarters, Press Release, Ministerial Meeting of the Defense Planning Committee and the Nuclear Planning Group, Brussels, June 12, 2003.

> week it was only our question. Now it is an American question. We share the responsibility. ... We opened the first front against international terrorism in 1999 in Dagestan. Now the Americans have opened the second front. September 11th showed that we have a common enemy. Either we fight that together, or we will be killed. (Peer, March 18, 2002)

In the Introduction, we emphasized that governments by acting in a system, cannot do only one thing at a time. Russia is reported to have shared intelligence and cooperated with the United States to stabilize a post-Taleban regime in Afghanistan (Blank 2002: 9). However, not all interests have such areas of overlap. On March 26, 2003, the Russian foreign minister Igor Ivanov accused the US of waging a propaganda campaign, by claiming that Russian companies had sold banned military equipment to Iraq: "We are seriously concerned by the attempts of certain circles in the United States to drag Russia into an information war over Iraq by making unfounded accusations that Russian companies supplied some military equipment to Iraq, bypassing the sanctions" (Isachenkov 2003). He called the war illegitimate and stated that only international inspectors could determine whether Iraq possessed weapons of mass destruction.

To counter-balance concerted efforts by the United States to achieve both dominance in CEA, as well as some kind of world leadership, Russia has entered into a number of strategic alliances with regional powers, especially Iran and China. Russia delivers weapons, military technology and also assists in the building of Iran's nuclear field in Bushier. In Iran, Russia meets Japan as a potential partner. The Japanese government continues negotiating with the Iranian government, despite US pressure to pull out, on a contract that would give a Japanese consortium rights to develop the Azadegan oil field in the north of the country. Russia cooperates with China not only in different bilateral political, economic and security terms, but also in the context of the Shanghai Co-operation Organization (SCO). As all of the Central Asian states (except for Turkmenistan) Russia and China belong to the SCO, this could be viewed as a direct attempt to reduce the rationale for a Western security presence in the region (Amineh 2003: Chapter 4).

On May 28, 2003, China and Russia signed an agreement for the construction of a 2,400 km pipeline from Russia's Siberian oilfield to the Chinese City of Daqing. The cost is estimated at US$2.5 billion. The Russian firm Yukos and the China National Petroleum Corporation (CNPC) back this pipeline. It is agreed that CNPC will purchase 5.13 billion bbl of Russian oil, worth some US$150 billion, between 2005 and 2030.[10] Another pipeline that is under discussion would pump up

[10] "China and Russia Ink Oil Pipeline Agreement," May 29, 2003 (see *http://www.china.org.cn*).

to 1.6 MM bbl/d from oil fields near the shores of Lake Baikal, north of Mongolia to Nakhodka and then to Japan. Japan is competing here with China in the Russian Far East to get access to Russian oil. Japan has offered help to Russia if Japan's preferred pipeline from Angarsk, north of Mongolia, to Nakhodka is build first. It would cost US$5.2 and have a length of 3,800 km. The Russian state-run oil firm Rosneft backs this pipeline. The projects would give Russian oil companies a major boost in export capacity and strengthen Russia's position as oil deliverer to the Asia-Pacific region (AGOC, April 17, 2003).

Current US-Chinese Relations

Revolutionary China under Mao played a crucial role in bringing the Cold War into East and Southeast Asia (Chen 2001). China compelled the United States and the Soviet Union to turn attention away from each other in Europe and to divide attention and energy between both sides of the Eurasian landmass. China's decision to fight the United States in the Korean War diverted US resources to East Asia. That war brought Japan back onto the world's industrial landscape. China's initial support to Hanoi in the Vietnam War had the same effect on South Korea; however, unlike the Japanese in the Korean War, the Koreans paid the price in battle deaths in the Vietnamese War. In terms of attention and resources, the Cold War was not that bipolar. In terms of lives and ammunition spent, the bipolar world was particularly unipolar: East Asians were on the receiving side of the great Cold War divide. That insight could help in achieving greater insight into the post-Cold War diplomacy of the United States, Russia, and China.

The Cultural Revolution, which followed at the heels of the Great Leap Forward disaster and the inner-party struggle these upheavals induced, prepared the way for the strategic reversal in US-Chinese relations of the 1970s. The connection between foreign policy changing course and domestic order change, which is a theme in critical geopolitics and rejected by neorealists, is highlighted by China's simultaneous transition towards a market economy. In the realist school, there is no reason why international realignments should coincide with domestic order change. In critical geopolitics, both levels are coupled (see Introduction).

China and India are entering into competition for oil at a time in which experts expect a speedy decline of the world stock after the peak in world oil production is reached. Each of these countries has hundreds of millions of people whose oil consumption is zero or close to zero.

The scope for future global economic expansion on the basis of fossil energy is determined by the proportion of the stock not yet exhausted and by the share of economic growth accounted for by the use of fossil-based energy. Under the assumption that no new major oil fields will be discovered, and there are good grounds for making it (Deffeyes 2001: Chapters 7 and 8) the time in which supply-induced scarcity will appear at the horizon can be calculated by current and expected annual extraction rates and the size of new discoveries. Some experts expect that world oil production will reach its maximum value somewhere in this decade (Deffeyes, 2001: 158); others are more optimistic and shift that date to about 2050. Supply-induced scarcity, or its fear among major users, is a powerful force of competition (Andrews-Speed, Xuanli Liao, and Dannreuther 2002: Chapter 3).

China lost energy self-sufficiency in 1995, when it turned into a net importer of oil. In 2010, China is expected to be the largest oil importer of the world. Its foreign policy too is becoming a hostage of domestic society needs whose satisfaction is required for international survival of state and society. It is no surprise, therefore, that China's government sees in the Caspian Sea oil a source of future supply. In its "Develop the West" policy, the Chinese are reaching out from Xinjiang to Kazakhstan and Kyrgyzstan. In the Far West, China meets the largest energy consumer in the world, the United States. Both countries differ in several respects. China is in the midst of a transition towards industrial capitalism. The United States has turned into a less fuel-intensive service economy. Both transitions are coupled, as the United States is China's best external customer of manufactured products. For some time to come, that coupling process will continue. No one knows how long it will take for China to become less export dependent. However, it seems extremely unlikely that it will be possible to mobilize another 300 million or more people for work when about 20% of their production goes to overseas customers. A second no less important difference between both countries is that United States has fortified itself militarily on China's front door and on the country's backdoor, whereas for China, the United States is on the other side of the Pacific. The United States accesses more easily than China the oil fields in Latin America and Western Africa.

President Bush redefined the relations between China and the United States from Clinton's endorsed strategic Partnership to that of strategic competitor. Probably few would disagree with the report in the *South China Morning Post* of June 4, 2003, about the forces at work in the location where both meet:

> President Hu Jintao yesterday agreed to boost energy co-operation and vastly increase trade with neighboring Kazakhstan, reasserting China's interests in a region where the US has dramatically increased its presence and brought its troops nearer to the Chinese border. Mr. Hu and Nursultan Nazarbayev, the Kazakh president, signed agreements to revitalize work on an oil pipeline from Kazakhstan to China and increase China's participation in the Kazakh

> energy sector, establishing a program for co-operation, which will run to 2008. Mr. Hu said Kazakhstan could be involved in implementing his government's program to develop its northwestern Xinjiang autonomous region, which borders Kazakhstan. The central government faces strong separatist sentiments in Xinjiang, whose ethnic Uygur are close to Kazakhs. China sees Xinjiang's economic development as a way to ease tension. Mr. Hu was in Moscow last week to take part in a summit of the Shanghai Co-operation Organization, a security grouping of Russia, China and four Central Asian nations in which China plays a central role. With a crucial member, Uzbekistan, actively seeking closer relations with the US, the group's task of countering American influence in the region will be difficult.

However, a belt of untapped energy wells beyond its borders surrounds China, unlike the United States. It extends from the Russian Far East, CEA, and the Middle East to the south. It is here where our geopolitical hypothesis may help to anticipate what lies ahead (see Introduction).

During the last decade, China's military modernization attracted international attention. China has one of the largest militaries in the world. Currently, defense expenditures are at the level of US$52 billion, out of a GDP of US$1315 billion. Defense outlays buy, among other military equipment, 370 strategic and tactical nuclear weapons. A number of available warheads are placed on top of 20 intercontinental ballistic missiles; 230 nuclear weapons are believed to be deployed on aircraft, missiles, and submarines, with regional capabilities. The remaining warheads, 150 in number, are believed to be held in reserve (*Bulletin of the Atomic Scientists* 1999; SIPRI 1999). In 2000, the total strength of its military personnel was estimated at 2.5 million, 1.8 million of which are ground forces of the People's Liberation Army (International Institute for Strategic Studies 2000: 186). The People's Liberation Army Air Force possesses about 4,350 aircraft, and is planning to produce Su-27s, with the Chinese designation J-11, under license from Russia (*Washington Post,* October 2, 2000: A17). Interestingly, China is also planning to buy one-to-four AWAC aircraft from Israel.[11]

This force will be too large to be removed by preventive strikes: a portion of the force is likely to survive and thus available for retaliation. Experts believe that it will be cheaper to exhaust the Missile Shield by expanding the size of the missile force and/or the number of warheads than to make the defensive system full proof. Advances in warhead technology and guidance of the 1960s and 1970s, such as placing more warheads on top of one missile and getting them independently targeted and even maneuverable after release, brought the superpowers in the Cold War to the momentous decision to largely abstain from missile defense. The

[11] The sale of the plane to China is opposed by the United States on the grounds that its radar contains restricted US components.

conclusion is that China will have a retaliatory capacity, irrespective of what the United States does. The US Air Force will therefore not fly with impunity over Chinese territory. It is even more unlikely for early proliferator Britain to require late-proliferating Iran to "open-up" the sites for inspectors, as the Blair government did in Iran as a follow-up to America's threat of pre-emptive strikes against that country. However, the United States may wish to have its spy planes flying as close to the Kyrgyz-Kazakh-Chinese border as to the mainland's coast in the Strait of Taiwan. The unfriendly, though sportsmanlike, encounter between the US spy-plane and a Chinese jet fighter in spring 2001 highlights the new risk of China's regained status in US security policy of being a "strategic adversary." Not all US leaders believe that China has already entered into the zone of the civilized. However, the option of the redeemer nation to fight a new Boxer War has moved beyond the horizon of the plausible. But the risk of external strangulation remains. Chinese analysts consider the United States as a major threat to China's energy security (Feng 1999: 1-5). China's vulnerability to energy interdiction by the United States has increased due to the spread of US military force onto its Western border.

At present times, US-Chinese relations hang in a precarious balance between crisis and cooperation. The September 11th attacks increased the US need for Chinese support in the war against terrorism. China is in need of US recognition that its separatists in the Far West are terrorists instead of freedom fighters that deserve US support. In North Korea, the United States seeks to get China in the co-driver seat of the US non-proliferation policy. China has a stake in getting nuclear weapons out of North Korean hands peacefully. A US attack on North Korea is likely to bring war to the whole Korean peninsula, which will endanger Chinese economic development. Even worse, it could bring US troops to the Yalu. US passivity in the face of North Korea's claim to be a nuclear power may induce Japan to pass the nuclear threshold, which the Chinese are trying to avoid.

At the Asia-Pacific Economic Cooperation (APEC) meeting of October 2001, China declared that it would offer assistance to the United States to fight international terrorism (Shuja 2002). China joined that war, and used it to its own advantage, in the western border region. Changes in the vocabulary in Chinese-American relations on who is a terrorist to whom reflect ad hoc policy changes. China's continued upward mobility, however, would bring positional change at the dyadic, regional, and world level. In the light of past experience, it is unlikely that those times will the best for peace (Midlarsky 2000: Parts II and III). Is the US military move into West Asia and CEA anticipating those times, and if so, will anticipation have the effect of shifting forward the timing of confrontational behavior? When the Iraq war was underway, China made its opposition clear. On March 25, 2003, newly elected Chinese Prime Minister Wen Jiabao declared that China stands for a peaceful settlement of the Iraq issue within the UN framework. Every effort should be made to avoid a war.[12] Once government documentation has become available, it will be possible to figure out who was waiting for whom in publicly announcing resistance to the Anglo-Saxon attack. Such knowledge would help to get a better insight into how China perceives itself in relation to the current distribution of power in the world.[13]

Current US-Iranian Relations

The US seems to have persuaded China that missile and nuclear weapon sales to countries such as Iran[14] or Syria would be counterproductive to Chinese-US relations. It is reported that China no longer supplies Iran with Silkworm missiles, it broke its commitment to provide Syria with M-9 missiles, and has suspended its nuclear energy cooperation agreement with Iran (Shuja 2002; Xuetong 2000: 8-10; Ahrari 2001).

China and Iran are raising the symbolism of both being ancient civilizations that were brutalized by the industrialized West. Effects of worsening US-Iranian relations may have a dyad crossing effect and spread to US-Chinese and US-Russian relations. US client states among EU members will then have to decide which new subcontraction missions serve their interests and which are too risky to undertake.[15] The tensions between the US and Iran may pose an obstacle to intensive cooperation between Iran and CEA as these countries may be wary of incurring US disapproval. In the 1990s, the US put severe restrictions on economic cooperation with Iran. In 1992, the Iran Non-Proliferation Act was passed by Congress, followed by the Iran Libya Sanctions Act (ILSA) on August 5, 1996. These acts prohibit investment in Iran's energy sector. The ILSA was extended in 2001 until 2006, designed to punish foreign countries or companies spending US$20 million or more in Iran's or Libya's oil business (Lorenzetti 2002; Lapidus 2001).

[12] See *http://www.chinaembassy.org*

[13] Our guess is that Germany came first, then France, that jumped on the German shoulder, for its own reasons, to be followed by outspoken resistance by Russia and China.

[14] In November 2004 Iran came to an agreement with the EU to stop its uranium enrichment programme.

[15] To avoid surprise, decision-makers could do worse than spend some of their precious time by reading Jervis (1997: 253-295); see also our discussion in the introduction on assumptions of endogeneity in social science research designs.

However, following the attacks of September 11, 2001, in New York and Washington, President Khatami, in the UN General Assembly meeting in November, expressed sympathy with the United States. He strongly condemned the terrorist attacks against US citizens. It should be hypothesized that Khatami also spoke for a domestic audience opposed to clerical rule. Iran today has a split state-personality and society. The Iranian leader also criticized US bombing operations in Afghanistan as destabilizing and cruel. The already tense atmosphere between the two countries worsened with Bush's January 2002, *State of the Union* address, in which he characterized Iran as belonging to the "axis of evil" alongside the US former ally Iraq, and North Korea. The September 2002 US *National Security Strategy* document announced the US natural right as sovereign state to wage "pre-emptive" war against its self-declared brutish regimes. That document will not have assured the Iran leadership of a US benign intention. However, all political factions of the Iranian political elite, the left, the center and the right, refused to be put into the box of evil powers.

The US accuses Iran of continuing to arm warlords, including Ismail Khan and his group *Jamiat-e Islami* in Herat, and Rashid Dostum and his group *Junbish-e Milli-e Islami* in Mazar-e-Sharif in Afghanistan. Until recently, these groups were united in the Afghan Northern Alliance, a complex of various non-Pashtun rebel movements. Iran and Russia previously supported the Northern Alliance in its fight against the Pashtun dominated Taleban (Symon, September 19, 2001). On August 8, 1998, Taleban guards in the city of Mazar-e Sharif in Afghanistan killed eleven Iranians (ten diplomats and one journalist) (*Amnesty International*, September 3, 1998). Iran and the Taleban, thus, were not cooperating.

According to official US sources, however, members of the Iranian regime (the right faction and Hizbollah factions) did have contacts with the Taleban and its supporters long before the September 11th attacks. For example, the Egyptian based *Al-Jihad* is alleged to be in contact with segments of the Iranian government. According to an *Insight Magazine* report during the 1990s, al-Zawahiri, head of *al-Jihad*, was believed to be the second person behind Osama bin Laden and repeatedly traveled to Iran on the invitation of the Minister of Intelligence and Security Ali Fallahian. Ahmad Vahidi, according to reports, is the commander of the Qods force, a special-operations unit conducting foreign terrorist operations. According to US and European intelligence reports, in the months following the September 11th attacks, members of *al-Jihad* traveled through the Iranian city of Mashad to join Osama bin Laden's forces in Afghanistan (Timmerman 2001).

Another reason for Bush's "axis of evil" speech was Israel's seizure of the *Karine-A*, a ship loaded with arms, on January 3, 2002. Israel stated that these arms originally came from Iran and were intended for the Palestinian Authority. Officials involved in the Afghan peace process asserted that Iran's foreign policy has changed from an ideology of revolutionary Islam to a more pragmatic approach that is motivated by national interests (Dinmore and Khalat 2002).

Current US-Turkish Relations

There is much US support for its ally Turkey and for Turkey's policies in CEA. However, Turkish initial refusal to allow the United States to open a second front in Iraq through Turkey will give both allies food for thought about the future of their relationship. When the United States acquires new bases close to the Black Sea, its military operations will be able to bypass Turkey. The United States has urged Turkey to fill the vacuum in the region since the disintegration of the Soviet Union and may still think that Turkey could serve as a counterweight to Russian and Iranian influence there. The United States and Turkey share the following objectives in the region:

(i) To prevent Russia from regaining dominance over CEA. However, in case Turkey is punished by the US for its "defiance" and the US fails in Iraq to prevent Kurds from carving out a proto-state, US-Turkish relations will deteriorate. In those circumstances, Turkey may move closer to Russia.
(ii) To bring under control ethnic conflicts in Georgia, Chechnya, and Nagorno-Karabakh.
(iii) To support Turkey's desire to join the EU.
(iv) To support the BTC oil pipeline running through Turkey instead of Russia or Iran (Kirisci, November, 1998: 3).

Turkey has suggested that the United States include political and military issues into the preliminary agreement between Turkey, the US, and Georgia, which was concluded in early 2001 (Safrastyan, Winter 2001). Nonetheless, Turkey's reasserting itself as a major actor in the region could become problematic for the United States. The Turkish refusal to provide transit rights for US military forces deployed to the second front in Iraq highlights the potential for diverging interest. Turkey and Iran, fueled by dangers of Azeri separatism, are competitors. The conflict between Azerbaijan and Armenia and the clash of rival Islamic and Turkish secularist ideologies could result in proxy wars between Turkey and Iran in the Transcaucasus.

The European Union and CEA

EU authorities fear fast-track depletion of North Sea oil deposits. That fear seems to be correct (Deffeyes 2001: 155ff.). Consequently, diversifying to secure supplies is on their agenda, reflecting the need for a long-term EU common energy policy. The EU imports about 70% of its total oil consumption, and 40% of gas consumption. Up to 40% of the EU's gas imports currently comes and will continue to come from Russia. The EU candidate states have an oil dependence of 90-94% and a gas dependence of 60-90%. OPEC represents 45% of current EU oil imports. Both the launching of the EU-Russia strategic energy partnership on November 30, 2000, in Paris, as well as the vast energy potential of CEA have refocused the EU's attention. Although the Caspian region could not substitute OPEC imports, it surely could provide an alternative. As we see it, the EU is in a favorable position for CEA regional involvement. It is not under the suspicion of harboring superpower aspirations, as is United States, and is an attractive and geographically close importer of CEA oil and gas resources. The EU follows a different policy from that of the United States with regard to Iran and does not incur Russian suspicion with regard to its eastward expansion.

The main instruments of European strategy in CEA since December 1991 have been the Agreements on Partnership and Cooperation signed with all CEA countries (except for Tajikistan) and with Belarus, Moldova, Mongolia, Russia, and Ukraine. These do not promise admission to the EU, but are aimed at affecting European interests in co-operation with the CIS on a bilateral basis. The agreements can be voided in the case of human rights violations in CEA, with the intention that the EU can put more political pressure on the CEA governments. Other important instruments of the EU to pursue its interests in CEA are the Technical Assistance to the Commonwealth of Independent States (TACIS) program, Transport Corridor Europe-Caucasus-Central Asia (TRACECA) initiated at a conference in Brussels in 1993, as well as the Black Sea Regional Energy Centre (BSREC), the Black Sea Environmental Program (BSEP), and Interstate Oil and Gas Transport to Europe (INOGATE). The organization for Security and Co-operation in Europe (OSCE) is another international institution that is directly linked to European interests and connects Europe to CEA. Since the end of the Cold War, OSCE documents, especially the Paris Charter of 1990, make CEA an integral part of the European security system.

EU regional ambitions will not materialize unless the European Common Foreign and Security Policy (ECFSP) project gets beyond intergovernmental cooperation and an effective military force is being created, one capable of operating in EU border regions.

Discussions on a Common Foreign and Security Policy were initiated at the Hague European Council Meeting of 1987. At that meeting, European leaders did not go beyond the observation that without common foreign and defense policies, the project of integration would remain incomplete. The collapse of the Soviet Union, however, made the creation of such a policy more, not less, urgent, in particular for relations between France and Germany. At the special European Council meeting in Dublin of June 1990, President Mitterrand and Chancellor Kohl proposed in a joint initiative to develop a common European foreign and security policy. The Maastricht Treaty does institute a Common Foreign and Security Policy; however, that policy is largely organized along the model of inter-governmental cooperation.

It is important, at this point, to refer to two developments in these policy areas. The first is the struggle for primacy: who will control that policy? States, as members of the EU, in particular those among the big three, or the EU as the organization of a democratically governed political community of peoples of Europe? The second is the outcome of that struggle in terms of the creation of a military force at the EU-level capable of projecting military power into the CEA region, and the political control of such a force.

Due to the strong linkage between both issues, we reflect here on both at the same time. As a follow-up to the Treaty on European Union, mentioned above, members of the Western European Union, which can operate outside Europe, decided in 1992 to implement peace-keeping and humanitarian operations. At the 1996 NATO Summit in Berlin, agreement was reached that the United States allowed WEU to use in such operations NATO's military capacity and planning, which the United States could veto. The European Deputy of NATO's SACEUR would be responsible for the conduct of military operations, whereas the WEU would be responsible for political strategy.

The European Council meetings of Cologne and Helsinki decided in 1999 to strengthen the Common Foreign and Security Policy of the EU by creating a European Security and Defense Policy. To that effect, a Rapid Reaction Force of 60,000 military personnel, including navy and air forces, for peace-keeping operations (so-called "Petersburg tasks") was instituted as a permanent body of the Union. This force should be able to operate outside NATO. However, the Rapid Reaction Force is not a standing European Army under EU-command; but it may become one. EU members have to assign military forces to the Rapid Reaction Force. The model of inter-governmental decision-making ("2nd pillar") has been maintained. Crisis-management has been added to its mission. The Laeken European Council of December 2001 declared, two years earlier than planned, the structure of European Security and Defense Policy to be "operational," without precisely saying what that meant.[16] To highlight the crisis in Atlantic's defense, we make passing reference to NATO's setting up its own Rapid Reaction Force and to the Brussels meeting of Belgium, France, Germany, and Luxemburg of April 29, 2003, on how to improve the defense capacity of Europe. Interrelated with this issue is the relationship between Europe and the United States in NATO. The United States has a strong preference in favor of a military-capable Europe that is prepared to fight in and outside the NATO area as part of US-led NATO force, the deployment and use of which are planned in the Pentagon. Will America ever find a united Europe prepared to join it in its global redeemer mission? The "coalition of the willing" is more likely outcome.

Though the European Union may soon have a ready force of its own for peace-keeping and crisis-management, the EU as a whole does not have a comprehensive policy agenda for safeguarding its interests in the CEA-region. The EU apparatus has to incorporate in its development policy differences between member states in regional priorities. France is more oriented towards Northern Africa, Germany towards Eastern Europe, and Britain towards the Baltic States. Apart from some significant investments by French TNCs, France has not been a major player in CEA. French naval forces participated in the PfP's Co-operative Partner 99, and in the feasibility plan to raise sunken ships in the harbor of Poti in October 1999. France has also provided Georgia with equipment and supplies for a military hospital worth US$2 million. It has offered French language courses at the Georgian military training center Khodjori, as well as the training of Georgian pilots to integrate them into the French peace-keeping force in Kosovo. Germany has been more preoccupied with transitions in Eastern Europe. Its activities in CEA are mainly based on education and advice. Ten Georgian servicemen trained in Germany in 1999. Henning von Ondarza, a Bundeswehr four-star General, is a member of the three-man International Security Advisory Board to the Georgian government. In 1990, Germany exported many weapons belonging to the former East German army to Turkey. Turkey has probably reexported some of the weapons and used them in its support for Azerbaijan in the Nagorno-Karabakh conflict with Armenia. During the period between 1991-94, Germany exported 9,000 former East German army trucks to Russia, Ukraine, and Kazakhstan, and a coastal patrol boat to Georgia.

---

[16] For the purpose of this article, we have to omit reference to the competitive expansion of NATO and the EU. It should be noted, however, that no professor of international relations could have provided the new members of these organizations with a better introduction to the realities of intra-core conflict than they got from practice between the last months of 2002 and the Anglo-Saxon war against Iraq, and being an object in that reality.

Britain has focused special attention on the Baltic States. Commercially, Britain has had an active presence, not least through two of its highest-profile enterprises, BP-Amoco, especially in the AIOC consortium and the British Bank of the Middle East in Armenia. Like France, it provides little military assistance to CEA. Assistance takes place mainly through NATO-sponsored educational programs. For example, the War Studies Program at King's College in London has run a special one-year Master's program for civilians working in the ministries of defense in one of the former Soviet republics, and a British general is head of the International Security Advisory Board in Georgia (Bhatty and Bronson, Autumn, 2000: 136-137).

PROSPECT

The increasing involvement of the US and the EU, the conflicts within NATO, the uncertainty about the future direction of Russian foreign policy, the involvement of China, Iran, and Turkey in the region, and the activities of TNC's, all underscore the following:

(i) the significance of the oil and gas resources in CEA,
(ii) the potential for conflict for the control of these resources, and
(iii) the absence of an agreed-upon policy framework for managing that competition.

What we are witnessing now is a recomposition of the geostrategic map not only for CEA but also for the whole world. Tensions could be further aggravated by disparities in military power if conflicts were to escalate. The Eurasian region includes states with the largest armed forces in Europe and Asia: Russia, Turkey, Ukraine, Iran, Pakistan, China, India, and Uzbekistan. The region also has four nuclear-armed countries—Russia, China, Pakistan and India—making it a dangerous potential flash point of global significance.

Further, security risks concern the US/NATO involvement in numerous political and economic crises in post-Soviet CEA, the war on terrorism in Afghanistan, and the crisis in Iraq. All these conflicts have been closely connected to the activities of the single military world power with global reach capacity, the United States, and to the responses its activities provoke, in particular to its support for the territorial expansion by settlers into Palestinian lands enforced by nuclear armed Israel. Deepened US presence in CEA puts the Russian leadership in the awkward position of choosing to lean towards a divided and paralyzed Europe, to remain active at the world level by aligning itself with the United States that seems intent on destroying Russia as a world power by other means, or to unify strategically with rising China.

This set-up could also easily lead to tensions among CEA rulers wanting to perpetuate authoritarian regimes and to gain outside support for themselves and their regional ambitions. There is no simple way to resolve all of these tensions peacefully and amicably. It is, hence, unlikely that we can expect true stability in CEA anytime soon. Instability is all the more likely if economic development fails. Political stability has no easy ride in new countries with large amounts of natural resources with strong demand on the world market. When it comes to the extraction of Caspian oil and gas resources, it would be ideal if it were so that no one will win the game unless everyone wins, and that everyone is fully aware of this and acts accordingly. Without concerted efforts in this regard, Central Eurasia and also the Middle East could become even less stable in the near future than these regions already are now.

An alternative to the new Great Game would be for all outsiders to invest in the preservation of local stability, beginning with greater respect for local cultures, assistance in solving social problems, and overcoming the politics of stagnation without imposing a redeemer model from the outside. Both the US and EU have potential to strengthen economic prosperity, security and political stability amongst the post-Soviet CEA countries. If conflict prevails over stability in CEA, this will have a great effect not only on the region's security but also on global security as a whole.[17] It seems, for the medium-term, inevitable that energy politics in CEA and the Caspian region will embrace elements of both competition and cooperation.

References

AHRARI, E.M.
2001 "Iran, China, and Russia: The Emerging Anti-US Nexus?" *Security Dialogue* 32(4).
ALEXANDER'S GAS AND OIL CONNECTION
2003 "Moscow Plans to Develop New Markets."
2002 *"The Energy Wars of the Caucasus."*
2002 "Construction of Turkmenistan-Afghanistan-Pakistan Pipeline to Start in 2004."
2002 *"LUKoil Decides Not to Join Baku-Ceyhan Pipeline Project."*
AMINEH, M.P.
2003 *Globalization, Geopolitics, and Energy Security in Central Eurasia and the Caspian Region.* Den Haag: Clingendael International Energy Programme.
AMNESTY INTERNATIONAL
1998 "Afghanistan-Thousands of Civilians Killed Following Taleban Takeover of Mazar-e Sharif." *ASA.*
BHATTY, R. AND R. BRONSON
2002 "NATO's Mixed Signals in the Caucasus and Central Asia." *Survival* 42(3): 136-137.
BLANK, S.
2002 "Is Russia's War on Terrorism the Same as America's?" *Central Asia-Caucasus Analyst.*

[17] See Amineh (2003: Conclusion).

CALLARI, R.
2002 "Energy Interests, the US Government, and the Post-Taleban Trans-Afghan Pipeline." *Central Asia-Caucasus Analyst.*
CARNEGIE ENDOWMENT FOR INTERNATIONAL PEACE
2001 *"Caspian Oil: How Soon and at What Cost?"* Meeting Report 3(8).
COHEN, A.
2002 "Caspian Pipeline Construction Efforts Move Forward." *Eurasia Insight.*
DEFFEYES, K.S.
2001 *Hubbert's Peak: The Impending World Oil Shortage.* Princeton: Princeton University Press.
DINMORE, G. AND R. KHALAT
2002 "America's New Enemy." *Financial Times.*
EDUCATIONAL FOUNDATION FOR NUCLEAR SCIENCE
1999 *Bulletin of the Atomic Scientists.*
EIA
2002 "Central Asian Region Analysis Brief." *EIA.*
FENG, X.
1999 "Views on Some Hot-Spot Issues in International Situation." Xiandai Guoji Guanxi (Contemporary International Relations) (12): 1-5.
Financial Times
2001 "West Plans Oil Pipeline via Afghanistan." *Financial Times.*
GARTEN, J.E.
1995 "Is America Abandoning Multilateral Trade?" *Foreign Affairs* 76(6): 58.
Harvard International Review
1992 "A New Covenant for American Security." *Harvard International Review*, p. 26-27.
INTERNATIONAL INSTITUTE FOR STRATEGIC STUDIES
2002 *Military Balance 1999-2000.* London.
ISACHENKOV, V.
2003 "Russia: US Trying to 'Destroy' Iraq." Cnews.
KING, P. (ED.)
1995 *International Economics and International Economic Policy: A Reader.* New York: McGaw-Hill.
KOBER, S.
2002 "The Great Game, Round 2: Washington's Misguided Support for the Baku-Ceyhan Oil Pipeline." *Cato Foreign Policy Briefing* no. 63.
LAPIDUS, G.W.
2001 "Central Asia in Russian and American Foreign Policy after September 11, 2001." Presentation from 'Central Asia and Russia: Responses to the War on Terrorism,' a panel discussion held at the University of California, Berkeley, October 29, 2001.
LORENZETTI, M.
2002 "Oil Industry Sees Mixed Signals on US Iran Policy." *Oil and Gas Journal.*
MIDLARSKY, M.I. (ED.)
2000 *Handbook of War Studies II.* Ann Arbor: Michigan University Press.
PEER, Q.
2002 "Putin Plays a Weak Hand Well." *Financial Times*: 7, March 18.
Pipeline News
1996 *Pipeline News* no. 33.
RASIZADE, A.
2002 "Washington and the 'Great Game' in Central Asia." *Contemporary Review* 280(1636).
SAFRASTYAN, R.
2001 "Turkey and Eurasia in the Aftermath of the September 11 Tragedy: Some Observations on Geopolitics and Foreign Policy." *CCAsP Newsletter.*
SHUJA, S.M.
2002 *"North-East Asia and US policy."* Contemporary Review. August.
SOLIGO, R. AND JAFFE, A.M.
2002 "The Economics of Pipeline Routes: The Conundrum of Oil Exports from the Caspian Basin," in Kalyuzhnona, Y. et al. (eds.), *Energy in the Caspian Region-Present and Future.* New York: Palgrave.

SOLIGO, R. AND J. MYERS
1998 The Economics of Pipeline Routes. *Baker Institute* Working Paper.
South China Morning Post
2000 "Journey for lost treasure. One million Chinese Cultural Relics Believed to be Kept in Overseas Museums." *South China Morning Post*, June 21.
SYMON, F.
2001 "Afghanistan Northern Alliance." *BBC News* September 19.
TIMMERMAN, K.R.
2001 "Iran Cosponsers al-Qaeda terrorism." *Insight Magazine*, December 3.
WASHINGTON POST
2000 "Russians help China modernize its arsenal." *Washington Post*, A17.
XUETONG, Y.
2000 "China's Strategic Security Environment." *Shijie Zhishi* (World Affairs) (3): 8-10.
YERGIN, D.
1991 *The Prize: The Epic Quest for Oil, Money and Power*. New York: Simon and Schuster.

IX. Paradigms of Iranian Policy in Central Eurasia and Beyond

EVA RAKEL

ABSTRACT

Iran and CEA have historically close links going back as far as the sixth century BC when the Persian Achaemenid Empire conquered the region. For a long time, Persian was the main language of the elite in CEA. Since the disintegration of the Soviet Union, Iran has been determined to re-strengthen its position in CEA, particularly in economic and security terms. Iran is an active player in the Economic Co-operation Organization (ECO). It promotes the construction of southern pipelines from CEA to export the region's oil and gas resources as it hopes to profit from it for its own oil and gas export. However, it has to be noted that Iran in no way is a dominant player in the region. The rivalry between the various political factions of the Iranian political elite—the Conservative Traditional Right (*Rast-e Sonati*), Traditionalist left (*Chap-e Sonati*), Revolutionary or *New* Left or Hizbollah, Conservative Modern Right *Rast-e Modern*—leads to incoherence in Iran's foreign policy and makes Iran an unreliable actor to cooperate with not only for the countries of CEA but also for other countries interested in the region (i.e., the United States, European Union, Turkey, Russia, China, Saudi Arabia). Additionally, the great national economic problems in Iran are an obstacle for Iran to become more active economically in CEA.

Introduction

Iran and Central Eurasia (CEA) have historically close links, which go back as far as the sixth century BC when the Persian Achaemenid Empire conquered the region. For a long time, Persian was the main language of the elite in CEA and remained so even after the Safavid Empire

(1501-1772) had established Shi' Islam as the state religion, while the major part of CEA remained Sunni. However, since the emergence of the Safavid Empire and Shi'ism as state religion, Iran had no real opportunity to influence CEA. This changed when the Soviet Union disintegrated.

In the nineteenth century, Iran found itself involved in the "Great Game" in the region between Tsarist Russia and the British Empire. At stake for Britain was retention of its authority over the Indian continent. Russia tried to counter British influence and secure access to the warm seas. The first local uprisings in Iran against the penetration of the country by European powers occurred in 1903, two years before the first Russian Revolution spread into Iran. Iran's policy was based on finding a possibility for equilibrium between the two powers and remain independent itself, but the policy failed to achieve that objective. During the Second World War, Britain and the Soviet Union occupied the country, which served as a United States (US) supply link to the beleaguered Soviets and thus brought domestic transport in Iran partially under US control. The US got involved in training Iranian army and police forces. In 1943, the US even managed to get a national appointed as director-general of Iranian finance. His trips into the Soviet sphere soon brought conflict among the Russians as well as within the Iranian government. America's objective has not changed since Secretary of State Hull wrote in 1943 to President Roosevelt that no great power other than the United States should be established at the Persian Gulf opposite the American petroleum development in Saudi Arabia. After the Second World War, the Iranian government formed a strategic alliance with the US, both to strengthen its role in global politics and to assure its domestic socioeconomic development and stability (Amineh 1999: 213). The first serious Cold War confrontation between the Soviet Union and the US involved Iran and oil. The "first Iranian oil crisis" lasted from November 1945 to June 1946. The first crisis would not be the last for Iran. In a sense, it continues until today. After the independence of the Indian subcontinent in 1947 and Britain's withdrawal from the Persian Gulf, the battle for dominance between the US and the Soviet Union was less for the conquest of territories but more for the United States to gain influence on the region's energy resources. For the Soviet Union, military security against the US was an important factor. Iran feared expansion of its northern neighbor. These early conflict episodes at the beginning of the Cold War, therefore, had a great influence on Iran's foreign and domestic policy. In foreign policy, Iran did not, unlike during the nineteenth century, search for a balance of power (Nahavandi 1996). The politics of balancing involves conflicting domestic society actors and their interests (see Introduction of this book). Since the disintegration of the Soviet Union, Iran finds itself involved in a new Great Game played out between state as well as non-state actors, for the control of the energy resources of CEA. Several observers in Iran saw the death of Ayatollah Khomeini in 1989 and the demise of the Soviet Union, as an opportunity for others to gain more influence in the region. Because of its natural resources, its strategic location as a historical, geographical, and economic link between East and West, the Persian Gulf, and the Caspian region, Iran inevitably plays an important role in the region, even globaly, regarding peace and security, both as an object as well as an acting subject. Because of its geographic position, Iran would be an attractive partner for cooperation with the CEA countries and could be considered as a bridge in the Eurasian corridor. It is a natural transit link and the shortest route between the Caspian Sea and the Persian Gulf. Iran has major oil terminals as well as the necessary infrastructure in terms of export outlets for the region's oil and gas resources. In addition, Iran is an active player in the Economic Co-operation Organization (ECO). It also promotes the construction of southward pipelines from CEA that could be connected to its own pipeline infrastructure.

Despite this potential for cooperation, it has to be noted that Iran in no way is a dominant player in the region. Iran's own internal economic and political problems, which are made worse by the US effort to isolate Iran, hamper investments in the region. The rivalry between the various political factions of the Iranian political elite, which include the Conservative Traditional Right (*Rast-e Sonati*), Traditionalist left (*Chap-e Sonati*), Revolutionary or *New* Left or Hizbollah, and the Conservative Modern Right (*Rast-e Modern*), leads to incoherence in its foreign policy. Iranians do not agree on a single strategic concept for the country's development, which is nothing new in modern Iranian history.

Iranian foreign policymaking today does not only revolve around President Khatami (Traditionalist Left faction); the Supreme Leader, Ayatollah Khamenei (Conservative Traditional Right) carries out his own foreign policy through institutions manned by his favorites: the "cultural bureaus" of the Iranian embassies. These cultural bureaus act independently from the embassies, which have to implement government policy. While Khatami strives for a "dialogue between Civilizations," or a "Détente" in foreign politics, Supreme Leader Khamenei undermines these attempts, particularly by continuing his support of Islamist radical groups in other Muslim countries, such as Hizbollah (Party of God) in Lebanon and Hamas (The Islamic Resistance Movement *Harakat al-Muqawamah al-Islamiyyah*) in Gaza/West Bank. Iran's conflictual relation with the United States is an obstacle for cooperation with the countries of CEA. These countries of the region do not want to get drawn into these conflicts.

Threat perception in Iran is focused on the United States and its cooperation with Turkey, Israel, and some Arab states. To prevent an American/Turkish coalition from gaining more power in CEA, and because of the US policy to isolate Iran, the Iranian government has entered into a strategic alliance with Russia. Russia has become one of the main arms suppliers to Iran since the Iranian Islamic Revolution. At the same time, Iran hopes to improve trade relations with the European Union (EU), which it sees as a means to strengthen its national economy and free the country from isolation in international politics.

Historic rivalry in CEA between Iran and Turkey has not come to an end. It is visible in the involvement of both countries in conflicts in the region. For example, during the war between Armenia and Azerbaijan on Nagorno-Karabakh, Iran sided with Armenia, being afraid that a winning Azerbaijan in this conflict could stimulate the Azeri population in Iran to start a social upheaval and demand reunification with Azerbaijan. Turkey sided with Azerbaijan in the war. However, Turkey and Iran also cooperate in the context of the exploitation and export of the oil and gas resources in CEA. There exists also a historic rivalry with Saudi Arabia, particularly regarding the ownership of the Abu Musa and the Greater and Lesser Tunb Islands. However, since President Rafsanjani came to power in 1989, a gradual improvement in the relations between the two countries can be noted. At the same time, Iran sees its own security threatened by already existing and possibly further conflicts in CEA, such as the war in Tajikistan, the war in Nagorno-Karabakh, and the drug business.

This article will be organized as follows: part one discusses the conflict between the various factions of the Iranian political elite and its influence on foreign policy; part two analyzes Iran's relations with other countries interested in CEA (i.e., US, EU, Turkey, Russia, Saudi Arabia) and Iran's role in CEA since 1991. It ends with the prospects for relations between Iran and CEA.

1. The Different Factions in the Iranian Islamic Political Elite

One important obstacle to the development of a coherent foreign policy in Iran is the rivalry between the different political factions. Generally, we can distinguish between four factions in Iran: the Conservative Traditionalist Right, the Conservative Modern Right, the Traditionalist left, and Revolutionary Left or Hizbollah. This distinction refers to their respective position on social and economic issues in the Islamic context of contemporary Iran and their foreign policies. What these groups have in common is their opposition to Shah Mohammad Reza Pahlavi's regime as well as their loyalty to the person and political teachings of Ayatollah Khomeini.

Conservative Traditionalist Right: This faction emerged out of the dissolution of the first ruling Party after the Iranian Islamic Revolution, the Islamic Republic Party, in 1987. The strongest organized group within this faction is the Militant Clergy Association (*Jame'-e Ruhaniyat-e Mobarez*), headed by Ayatollah Mohammad Reza Mahdavi-Kani, former minister of the Interior. Its supporters come mainly from the traditional middle class and the bazaar economic sector. The most prominent members of the Militant Clergy Association are Parliament Speaker Ali Akbar Nateq-Nuri; Ali Larijani, head of state radio and television; and Morteza Nabavi, editor of the newspaper *Resalat*. Domestically, the Conservative Traditionalist Right faction believes that the *velayat-e faqih* (rule of Supreme Islamic Jurist/Supreme Leader) belongs to the clergy and that the people should not play any role in the election of the Supreme Leader. The Supreme Leader stands above the constitution and the clergy controls the politics of the country without being responsible to the people. In its foreign policy, the Militant Clergy Association supports the export of the revolution and is heavily opposed to the United States. Economically, it strives for privatization and opposes state economy (Mortadji Hodjat 1999).

Conservative Modern Right: When the legislative elections took place in 1996, the supporters of President Rafsanjani founded the Servants of Reconstruction (*Kargozaran-e Sazandegi*). The Conservative Modern Right faction is mainly supported by techno-bureaucrats: modern professional associations, employer organizations, as well as the modern business-oriented urban-middle class and industrial groups (Buchta 1996: 52-55). Its leading figures are Gholam-Hussein Karbashi, the former mayor of Tehran; Ayatollah Mahajerani, Khatami's minister of culture and Islamic guidance; Mohsen Norbakhsh, Central-Bank governor; Mohammad Hashemi, Rafsanjani's brother; and Faezeh Hashemi, Rafsanjani's daughter. Domestically, the main goal of the Conservative Modern Right faction is the transformation of Iran into a modern state with a theocratic ideology. Economically, it supports privatization of the economy, foreign investment in Iran, and industrialization. A precondition for investment is the breaking of Iran's economic isolation. Therefore, during Rafsanjani's presidency, there has been a shift in its foreign political orientation from the export of the revolution towards détente in relations with the Middle East, especially Saudi Arabia, and Western countries (Mortadji Hodjat 1999: 190, 195). The main organs of this faction are the daily newspapers *Ettela'at* and *Hamshari*.

Traditionalist Left: This faction emerged out of a split within the Militant Clergy Association shortly after its foundation because of ideological differences and personal ambitions of its members. The middle class and lower middle class mainly support this group. Its official organization, the

Combatant Clerics Society (*Majma-e Ruhaniyun-e Mobarez*), was founded by Ayatollah Khoinhia, responsible for the American embassy hostage taking in 1979 and founder of the critical newspaper *Salaam;* Mehdi Karrubi, former director of the Martyrs Foundation and Parliamentary Speaker; and Mohtashami, former minister of the Interior Department and ambassador to Syria. The Combatant Clerics Society consists solely of clerics. Other members of this faction are loosely connected in the *Khat-e Emami* (Line of the Imam) that supported Ayatollah Khomeini in the early years of the revolution. Some of them were members of a student organization of the same name that was responsible for the assault on the US embassy in 1979. However, more recently, they have moderated their position. The Traditionalists Left faction supported Khatami's candidacy for president. A third important group within the Left faction is the Mojahedin of the Islamic Revolution (*Sazeman-e mojahedin-e enqelab-e eslami*), founded by Behzad Nabavi. He is editor of the biweekly newspaper *Asr-e Ma.* It is composed exclusively of laypersons, [1] militant Islamists as well as a group of ex-Maoists of the party of *Ranjbaran* (proletarian). A student group close to the Mojahedin of the Islamic Revolution is the Organization pro-student Office for Consolidating Unity (*Daftar-e Tahkim-e Vahdat*). Another student group that has developed in the last years is led by Heshmatollah Tabarzadi, a radical turned democrat. He is president of the Union of Islamic University Associations and editor of *Payman-e Daneshju.* This group is close to the writer and Islamic thinker Abdolkarim Soroush, who argues that Islam and democracy are compatible and calls to end the monopoly of the clerics on state political power (see Amineh 2003). Other prominent members of the Traditionalist Left faction are Ali Abdi, journalist and former hostage-taker of the American Embassy; former Prime Minister Mohandes Mir-Hosein Musavi; and Mohammad Reza Khatami, Khatami's brother and head of the Islamic Iran Participation Front (*Jebhe-ye Mosharekat-e Iran-e Islami*). In general, the Traditionalist Left faction promotes the election of the Supreme Leader by the people and demands the limitation of his power to those who are envisaged by the constitution. Economically, it supports the intervention of the state and nationalization policies. Since the election of Khatami as president, the more moderate wing within the Left faction has become dominant. In 1998, a new group of clerics, religious laypersons, Islam-oriented workers, and Islamic women's activists was formed within the Islamic left, the Islamic Participation Party of Iran (*hezb-e mosharakat-e Iran-e eslami*) (*Aban*, December 12, 1998; *Ettela'at,* December 23, 1998). It supports President Khatami.

[1] On the history of the origins of the Organization of Mojahedin, see Rahnema (1990).

Revolutionary Left: In the mid-1990s, the Hizbollah factional group emerged. Its first organization was created in early 1996, called the Union for the Defense of the Values of the Islamic Revolution (*Jame'-e defa'-e az Arzeshha-ye Enqelab-e Islami*). Its leader is the former Intelligence Minister Hojjatoleslam Mohammad Mohammadi Rayshahri. The policy of the organization is not yet clear. On the one hand, it supports the Islamic Left position on rigid state control and radical sociopolitical egalitarianism (*Ettela'at,* April 8, 1997). On the other hand, it supports a totalitarian theocracy and repression of dissidents.[2] Another group within the Hizbollah is the *Ansar-e Hizbollah,* led by Masoud Dehnamaki,[3] whose political methods can be compared to those of Mussolini's fascists. In the last years, it has carried out various physical assaults on liberal intellectuals, political figures, and newspaper offices. Another important figure that belongs to the Hizbollah is Ayatollah Ahmad Jannati, head of the Council of Guardian (*shura-ye negahban*) which is mainly supported by the low and middle classes (mostly peasants). The Hizbollah's major organs are the newspapers *Sobh, Kayhan,* and *Jomhuri Islami.* However, the Revolutionary Left, until now, has played a minor role in the decision-making process of the Iranian Islamic regime. Since the election of Khatami as president in 1997 and again in 2001 no significant change in political life in Iran can be noted, which is still dominated by the Conservative Traditionalist Right faction.

The political regime in contemporary Iran is based on the principle of the *velayat-e* faqih. As long as this principle exists, it is the religious leader who takes the ultimate decisions on important external and internal matters. Only when the principle of the *velayat-e* faqih is abolished will there be a possibility of transition towards more democratic state and society relations in Iran (Amineh 1999).

2. Iran's Foreign Policy Since the Election of Khatami as President in 1997

Under the rule of Khomeini, the foreign policy of Iran was directed towards confrontation with the West. During Rafsanjani's period, this attitude changed to a more pragmatic orientation, not least because of his attempt to improve the devastating economic situation of his country and attract Foreign Direct Investment (FDI). Khatami's presidency inaugurated important changes in the Iranian foreign policy. It can be

[2] For a general overview on the political program of the New Left see *'Asr-e ma*, March 7, 1995, pp. 2-7.

[3] See *al-Mujaz 'an Iran,* bo. 87, December 1998, pp. 18-19 on the origins of the Ansar-e Hizbollah.

said that his election as President led to a shift in Iran's foreign policy from confrontation to conciliation. However, Khatami does not have total control over the Iranian foreign policy. Supreme leader Khamenei has his own network of institutions. Ayatollah Khamenei integrates his representatives abroad in cultural organizations and Islamic centers. He also conducts his foreign policy through organizations like the Islamic Propagation Organization (*sazeman-e tablighat-e eslami*), the Hajj and Welfare Organization led by Mohammad Mohammadi Rayshahri, and the Society for Reconciliation Among Islamic Sects (*majma'-e jahani-ye baraye taqrib-e baine mazaheb-e eslami*). The most important pillar of the supreme leader's foreign policy is the "cultural bureaus" of the Iranian embassies that actually act independently of the embassies. Their most important goals are to give financial support to radical Islamist groups abroad (Buchta 2000: 50). While Khatami strives for a "Dialogue between Civilizations" or policy of "Détente," Supreme Leader Khamenei undermines these attempts, particularly by continuing the support of Islamist radical groups in other Muslim countries such as Hizbollah in Lebanon and Hamas in Gaza/West Bank. It has also been said that the Traditional Right faction of the Iranian political elite has supported the Taleban and Osama bin Laden's *al-Qaeda* forces (Timmerman, December 3, 2001).

To reconstruct its poor economy, Iran has left the confrontational approach behind it and seeks cooperation within Eurasia and beyond. In this context, Iran follows five main objectives, which are to:

- Maintain stability on its northern and southern borders;
- Establish friendly relations with the Arab states;
- Increase its influence and activity in the independent states of CEA;
- Safeguard the existence and territorial status quo of the multiethnic Iranian nation state;
- Make best use of its geographical position as energy transit country between the Caspian Sea and the Persian Gulf.

Generally, the Conservative Traditionalist Right faction of the Iranian political elite has also accepted this policy of Khatami. As has been stated by Javad Larijani, member of the Committee for Foreign Policy of Parliament, who is believed to be a member of the Conservative Traditionalist Right faction: "The motto 'Détente' is very interesting, the motto 'Dialogue between Civilizations' a pertinent view. The fact, that we have a better image in the world and acknowledge the world, is very encouraging. However, we are concerned about the inefficiency of the diplomatic establishment" (interview with Neshat of June 6, 1999, cited in SWB ME/3555 MED/6, June 8, 1999).

The most important success of the first four years of Khatami's presidency was that he was able to improve Iran's position in the international scene. Even his internal enemies had to recognize his successful foreign policy, not least because of the necessity to secure Iran's oil income, which is central to the development of the country's economy. The improvement of the international climate particularly became apparent in Khatami's interview with the American television channel CNN on January 7, 1998. Here he made clear his aim to improve the relations with the United States, though he was very cautious in this matter:

> [N]othing should prevent dialogue and understanding between two nations, especially between their scholars and thinkers. Right now, I recommend the exchange of professors, writers, scholars, artists, journalists, and tourists. A large number of educated and noble Iranians now reside in the US as representatives of the Iranian nation. This shows that there is no hostility between the two nations. But the dialogue between civilizations and nations is different from political relations. In regard to political relations, we have to consider the factors, which lead to the severance of relations. If some day another situation is to emerge, we must definitely consider the roots and relevant factors and try to eliminate them. (cited in Fürtig 1998: 307)

The General Assembly of the UN on November 4, 1998, proclaimed the year 2001 as the "United Nations Year of Dialogue among Civilizations." This resolution originally had been Khatami's proposal after his election as president and his chairmanship of the Organisation of Islamic Conferences (OIC). Khatami's slogan of a "Dialogue between Civilizations" is interesting for at least three reasons:

- The discussion of the term civilization is rather a modern phenomenon;
- Dialogue between civilizations' indicates that Khatami recognizes civilizations other than the Islamic one;
- While Khomeini reduced Islam to power, Khatami adds civilization to power.

However, there are serious differences between the Conservative Traditionalist Right and the Conservative Left factions among the Iranian political elite with regard to Iran's relations with the United States. While Khatami seeks a dialogue with the West, Khamenei considers a "dialogue with America [...] even more harmful than establishing ties with that country" (Barraclough 1999: 12).

At the same time, the 1990s witnessed the tightening of US policy towards Iran. In 1992 the Iran Non-Proliferation Act was passed extending the export sanctions on Iraq and Iran, followed by the Iran Libya Sanctions Act (ILSA) that prohibited all investment in Iran's energy sector. President Bush extended the ILSA in 2001 until the year 2006.

Iranian-US Relations

The attacks of September 11, 2001, in New York and Washington created a good opportunity for Iran to improve relations with the United States, after a period of confrontation for as long as 23 years. President Mohammad Khatami showed his sympathy with the victims of the terrorist attacks. In an address to the UN General Assembly in New York in November 2001, he declared that Iran strongly condemned these attacks against US citizens. In Tehran, masses were on the street to express their sympathy.

However, Iran is not happy with how the United States fights its war against terrorism, to say the least. Iran criticized US attacks in Afghanistan, saying that the bombing killed civilians and risked destabilizing the region. The Iranian government also proposed that the UN and not the United States should head any effort of a global war against terrorism. At the same time, Iran accused the United States of only following its selfish interests after it had called several Islamic Lebanese and Palestinian groups terrorists, who were, according to Iran, national liberation organizations (Recknagel 2001). According to official American sources, members of the Iranian regime (the Conservative Traditionalist Right and the Hizbollah factions) established contacts with the Taleban long before the September 11 attacks. For example, the Egyptian *Al-Jihad* organization is supposed to be in contact with segments of the Iranian government. According to a report of the *Insight* magazine (Timmerman, December 3, 2001), during the 1990s al-Zawahiri, head of *al-Jihad* and the second person behind Osama bin Laden, repeatedly traveled to Iran on the invitation of the Minister of Intelligence and Security Ali Fallahian, and Ahmad Vahidi, who, according to reports, is the commander of the *Qods* force, a special-operations unit conducting foreign terrorist operations. According to American and European intelligence reports during the months following the attacks of September 11 in New York and Washington, members of *al-Jihad* traveled through the Iranian city of Mashad to join Osama bin Laden's forces in Afghanistan (Timmerman, December 3, 2001).

At the same time, however, European and American officials believed that Iran has acted responsibly in Afghanistan. Officials involved in the Afghan peace process say that Iran's foreign policy has changed from an ideology of revolutionary Islam to a more pragmatic approach that is motivated by national interests (Dinmore and Khalaf 2002). The already tense atmosphere between the two countries changed for the worse when President Bush in his State of Union speech in January 2002 referred to Iran as being part of the "axis of evil," alongside Iraq and North Korea. All political factions in Iran rejected this statement.

A determining factor in Iran's regional foreign policy, as well as its relations with the United States, is Israel. Iran does not recognize the legal existence of that state. Tehran critiques the Arabic-Israeli peace process. This has severe consequences for its relations with Europe and the United States but also for its relations with Arab countries, Turkey and the countries of CEA where Israel is developing into an influential actor. The Conservative Traditionalist Right faction of the Iranian political elite would probably resist a change of Iran's general foreign policy towards Israel by the Khatami government. It can be said that the Khatami government seems to give the Israeli State a de facto recognition, while following the official policy of nonrecognition (Reissner 1999).

Another reason for Bush's State of the Union speech was probably the seizure of the *Karine-A*, a ship loaded with arms, by Israel on January 3, 2002. Israel stated that these arms originally came from Iran and were destined for the Palestinian Authority. As noted above, America and Iran disagree on who is a terrorist and who is a freedom fighter in the conflict between colonizing Israel and Palestine. This disagreement suggested that Iran would support radical groups, respectively, national liberation movements, in the Middle East (Dinmore and Khalaf 2002). Pragmatism is evident in Iran's attitude towards Salman Rushdie. Iran's Foreign Minister Kamal Kharrazi declared that "the Government of the Islamic Republic ... has no intention, nor is it going to take any action whatsoever to threaten the life of the author of the Satanic Verses or anybody associated with his work, nor will it encourage or assist anybody to do so." One day earlier Khatami had declared, "we should consider the Rushdie affair as completely closed." The conservatives however, reaffirmed Khomeini's *fatwa* and even raised the reward on the head of Salman Rushdie (Rieck 2000: 138-41).

Another major concern for the United States is Iran's nuclear program. Both US intelligence and the International Atomic Energy Agency agree that unless stopped from the inside or outside, Iran will have produced one or more nuclear weapons within five years.[4] The nuclear problem in Iran might solve itself if the refomers of the Iranian political élite will triumph. However, in the short term this is rather unlikely. The United States has to be aware that a military intervention in Iran could easily backfire on Iranian domestic policies, undermining or forestalling the prospects for a "velvet revolution" in Iran. Thus, dealing with the worst-case scenario of

[4] In October 2004 the Iranian parliament passed a bill that demands from the Iranian government to continue the uranium enrichment programme. All 247 members of parliament supported the bill (BBC, 31 October 2004). Only one month later it came to an agreement with the EU to stop uranium enrichment for the time being.

Iranian concervatives possessing nuclear weapons will make this scenario even more likely (Pollack 2003: 5-7).

The tension between the United States and Iran poses an obstacle for intensive cooperation between Iran and CEA. The countries anticipate US disapproval. The tension also hampers US TNCs to invest in Iran. How the relations between the United States and Iran will develop in the future is not yet clear. In the beginning of March 2002, Deputy Secretary of State Richard Armitage and White House officials warned Iran not to put pressure on its Caspian neighbors to follow a policy that would interfere with US political and commercial interest in the region. Additionally, Iran's position in the Caspian Sea worsened when, on July 23, 2001, two Iranian Air Force planes flew over a BP/Amoco survey ship that was exploiting the Araz-Alov—Shargh (Azerbaijani name)/Alborz (Iranian name) block licensed by Azerbaijan. On the same day, an Iranian gunboat entered Azerbaijan's territorial waters and threatened to fire on the survey ship. Iran aims to enforce its own claims on this part of the Caspian Sea (Olson 2002). BP/Amoco even suspended work in the Araz-Alov—Shargh (Azerbaijani name) Alborz (Iranian name) block (Amineh 2003).

The European Union (EU) follows a different policy towards Iran than the United States. European leaders tend to consider engagement with Iran as necessary for bolstering Khatami's reformist government.

Iranian-EU relations

The EU foreign policy Chief Javier Solana demanded that the United States act multilaterally and not unilaterally (Black 2002). Secretary of State Colin Powell explained that Bush's remarks about Iran is belonging to the axis of evil did not mean that the United States sought to invade these countries but that "action is going to be required" (BBC, February, 2002). In June 2003, the EU demanded that Iran accepts greater inspections of its nuclear program if it has to convince those people who believed to be developing nuclear weapons. With the US government putting greater pressure on the Iranian government and President Bush backing recent student protests in Iran, the EU foreign ministers signaled that they would not further develop trade and political relations with Iran if it did not comply with the demands (Black and Steele 2003: 1).

Iran's relations with European countries are improving. Since 1999, Khatami visited several European countries: France, Germany, Greece, Italy, Vatican City, and Spain. Europe is becoming increasingly more oil and gas dependent (see articles by Amineh and Houweling). Although the Caspian region could not be a substitute for imports from the OPEC, it surely could be an alternative source of supply. On February 23, 1998, the EU foreign ministers decided to resume ministerial contacts with Iran.

On June 17, 2002, the EU gave the green light to launch formal trade relations with Iran, despite heavy pressure from the United States. The major aim is to give backing to the Traditionalist Left faction of President Mohammad Khatami. This could increase European advantage over their Amerian counterparts. The EU is Iran's main trading partner. In 2000, EU's imports from Iran totaled 8 billion Euros, more than 80% of which consisted of oil products. Exports to Iran amounted to 5.2 billion Euros. American-based TNCs have argued that US unilateral sanctions gave their European rivals an unfair advantage (Dempsey 2002: 1). Iran profits from the division among the major powers and also cooperates with Russia.

Iranian-Russian Relations

Despite their historic rivalry after the disintegration of the Soviet Union, Iran and Russia realized that they had similar interests in CEA. As has been observed by Amineh (2002: 293), the Russo-Iranian alliance may turn into an important geopolitical fact in the post-Cold War era of the region. Iran and Russia have several similar objectives in their policy towards the region: both countries support Armenia against Azerbaijan in the Nagorno Karabakh conflict; both oppose any Turkish influence in CEA; Iran nor Russia has friendly relations with Uzbekistan and Georgia (Roy 1998); both have supported the Northern Alliance in Afghanistan against the Taleban; and both oppose the construction of the Main Export Pipeline (MEP) from Baku via Tbilisi to Ceyhan and the Trans Caspian Gas Pipeline (TCGP) from Turkmenbashy via Baku to Erzurum in Turkey.

However, the Russo-Iranian relations are fragile. While the Russians see themselves as the last bulwark against Islamic fundamentalism, the Iranians consider Russia as a "newcomer" in the region. Russia does not want Iran to create an area of influence in its own backyard as could be seen in the civil war in Tajikistan. Some Russians feared that the civil war in Tajikistan would be vulnerable to Iranian-style "fundamentalist" political Islam. Therefore, Russia saw the necessity to keep its border guards at the Afghan-Tajik border (Amineh 2000: 587). In the Nagorno-Karabakh conflict, when being threatened by an Armenian attack, Iranian military crossed the river Araxes and got an immediate response from Russia. Russia made it clear that its good relations with Iran would depend on Iran's acceptance of Russia's supremacy in the South Caucasus (Ramezanzadeh 1996).

In 1995, Russia and Iran embarked on what had been called by the Russian ambassador in Iran a "strategic relationship." Russia saw the alliance with Iran as a counterbalance to NATO's expansion towards the East and the South, to Western efforts to get control of the energy resources of the region, and to activities of Turkey in the CEA. Iran needs Russia as an ally to deal with the various social upheavals in the region as well as an arms supplier. Currently, there are only a small number of countries that sell weapons to Iran (Amineh 2002: 293). Russia is an important arms supplier; it has planned to sell military and other equipment to Iran worth $4 billion between 1997 and 2007. However, Iran has to meet its financial obligations.[5] On October 3, 2001, Russia signed a new military accord with Iran that could lead to US$300 million in annual sales of jets, missiles, and other weapons, despite US fears that arms sales to Iran could further destabilize the Middle East. According to Sergei Yastrzhembsky, a Kremlin spokesman, rejected American claims that Iran supported anti-Israel groups such as Hizbullah. Yastrzhembsky knows of no evidence for that accusation. He also claimed that "Iran is much closer to us than to the US and we are not going to act to the detriment of our national interest and our national security" (cited in LaFranière 2001) The Iranian Defense Minister, Ali Shamkhani, referring to Israel, explained that the agreement with Russia "is not aimed against any country" (cited in LaFranière 2001). Russia's defense industry depends for its survival heavily on export. Iran is interested in Russia's mid-range air defense systems (the S-300 missile) as well as Sukhoi fighter jets, MiG-29 fighters, and anti-ship missiles (LaFranière 2001). In Bushire, an Iranian port on the Persian Gulf, Russia also is building a 1,000-megawatt power station that will include 6 nuclear reactors (Myers 2002). While US President Bush sees Iran as part of the axis of evil, Dmitri Trenin, analyst at the Moscow Center of the Carnegie Endowment for International Peace, stated that Russia views Iran as "a good citizen of the region, or not much worse than the others" (cited in Baker 2002: A15). He noted that Iran did not become involved in the Chechnya conflict but remained neutral in the battle between Islamic separatists and Russian troops there. However, Russia is reported to have pressured Iran to accept the "additional protocol" of the International Atomic Energy Agency. In 2000, trade volume between Iran and Russia reached US$900 million compared to US$600 million in 1999. The total value of goods transacted by Iran and Russia in 2000 amounted to US$600 million, with Russia having a share of US$550 million and Iran of US$50 million. According to these statistics, more than 50% of Russian exports to Iran were composed of machinery and industrial equipment. Additionally Russia exported ironware, chemical fertilizers, paper, wood, and timber while importing from Iran fruits, vegetables, fruit juice, and tomato paste (Iran Commerce 2001).

---

[5] For a Russian view of the Russian-Iranian relations, see Vishniakov, Vikotr, 1999 "Russian-Iranian Relations and regional Stability," International Affairs (Moscow), vol. 45(1): 143-53.

Iran and Russia oppose the construction of the MEP route and the TCGP. Both are worried by Georgia's and Azerbaijan's willingness to cooperate with NATO (*Ettela'at*, February 15, 1999). To slow down Turkmenistan's activity in the construction of the TCGP, Russia signed in December 1999 an agreement with Turkmenistan to buy Turkmen gas at US$36 per 1,000 cm^3 (Novoprudsky 1999; Dubnov 2000). In an attempt to persuade TNCs not to proceed with the construction of the MEP route, Iran reduced the costs of its oil swaps with Turkmenistan, Kazakhstan, and Azerbaijan by 30%, starting in 2000 (*Ettela'at*, November 23, 1999). While the future of the TCGP remains uncertain, the MEP route was inaugurated on September 18, 2002. Although Russia and Iran have a similar policy regarding the MEP route and the TCGP, the long-term interests of both countries in Caspian energy differ. Both promote the construction of pipelines through their own territory and oppose pipelines through each other's country. When the Soviet Union collapsed, Russia and Iran initially followed a similar policy regarding the division of the Caspian Sea among the Caspian littoral states. Until recently, both countries pledged for a continuation of the treaties of 1921 and 1940. However, more recently, Russia has started to conclude bilateral agreements with Kazakhstan and Azerbaijan to set their mutual national borders and to attract FDI. These agreements have been heavily opposed by Iran (Amineh 2003).

Iranian-Turkish Relations

Traditionally, Persia/Iran and Ottoman/Turkey are rivals for dominance of the trade and transit routes across CEA. With the disintegration of the Soviet Union, Iran and Turkey have yet again entered into competition for influence in the region, but in a much more moderate way. There is also cooperation between the two countries, for example within the Economic Co-operation Organization (ECO). Additionally, the two countries aim to limit Russia's influence in the region and, thus, hope to prevent Russia's economic dominance there. More important, Turkey and Iran signed in 1996 a US$23 billion deal on gas delivery to Turkey for 23 years. The pipeline from Tabriz in western Iran to the Turkish capital Ankara opened on December 10, 2001. It has a length of 2,577 km and is expected have a 4 billion m^3 flow in 2002. Export is expected to increase to 10 billion m^3 in 2007. However, Turkey suspended imports on June 24, 2002, because of price cuts of Russian gas. So far in this year, Turkey has met only a fraction of its purchase pledges. There have been warnings of broader tensions. In September 2002 Tehran Radio said the decision to stop the gas flow was "not a friendly act" (Lelyveld 2002). The United States objects the deal between Iran and Turkey because it is a rival to the TCGP from Turkmenbashy (Turkmenistan) via Baku and Tbilisi to Erzurum in

Turkey. Turkey argues that it needs both pipelines to meet its rising energy demands (AGOC, November 13, 2002). Momentarily, Turkey gets about 70% of its gas from Russia. The delivery is expected to increase when the controversial Blue Stream pipeline running under the Black Sea has been completed in 2005. For the United States, the Blue Stream is another competitor to its favored TCGP pipeline.

Obstacles to friendly relations between Turkey and Iran are not difficult to locate. Revival of a pan-Turkish movement, that disappeared together with the Ottoman Empire in the beginning of the twentieth century, is a possibility. It would be a threat to Iran and its cultural ally Tajikistan. With the establishment of the Black Sea Economic Co-operation Organization, Turkey has created its own economic cooperation organization, which excludes Iran. Turkey has also entered into cooperation with Israel against Syria and Iraq, and its good relationship with Azerbaijan forces Iran to turn even more towards CEA and Russia to compensate for its forced isolation (Amineh 2003).

Iranian-Saudi Arabian Relations

Relations between Iran and Saudi Arabia are improving since President Rafsanjani re-oriented Iran's regional policy. However, in 1992, the relations between both countries experienced a backlash when Iran declared ownership rights over the Abu Musa and the Greater and Lesser Tunb Islands, whose sovereignty is disputed. Saudi Arabia, for its part, recognized Iran's importance as a neighboring country. Saudi Arabia supported the US's containment policy but rejected isolating Iran. Additionally, both countries have realized that the oil business requires cooperation. Unless they act united, Western governments and oil companies will be able to divide and rule. Mecca and Medina, the most sacred cities of Islam, get large numbers of believers from Iran. Both have a stake in managing the stream of visitors. Without the support of Saudi Arabia, during the meeting of the OIC at the beginning of December 1997 in Tehran, Iran would never have been able to obtain the presidency of the organization, which it obtained for five years. The OIC covers 56 Muslim States. The participation of Crown Prince Abdullah from Saudi Arabia in the meeting in Tehran was considered to be a success for rapprochement between both countries. At the beginning of 1998, former President Rafsanjani visited Saudi Arabia. His visit had already been planned during his presidency. The policy of closer relations between the two countries reached its peak in May 1999, when Khatami visited Saudi Arabia (Reissner 1999: 47-49).

The shift of Iran towards a more moderate policy towards the Persian Gulf States and the Middle East has left it with limited room for maneuver in CEA. The countries of the CEA are eager to find partners outside the region to cooperate with. To profit from their demand to break out of their isolation, Iran will have to develop a more creative policy towards the region.

Iranian-Central Eurasian Relations

Despite leaders' frequent reference to the "common cultural heritage" of Islam and the Persian language, Iran and the CEA countries do not have that much in common. The Shi'itization of Iran in the 16th century under the Safavid Empire created a cultural boundary between Iran and the Persian-speaking countries, whose religion was Sunni Islam (Afghanistan, Northern India, and Central Asia). The integration of CEA into the Soviet Union in the beginning of the twentieth century prevented economic relations between Iran and the region. The disintegration of the Soviet Union in 1991 and the disappearance of the 1,700 km long border between Iran and the Soviet Union had an important geopolitical impact on Iran and is of crucial importance for its future. While the road network to CEA and Europe was totally blocked for Iran during the Soviet era, since 1991, the door towards Europe has been reopened (Nahavandi 1996: 2). For obvious reasons, Iran has not played the "ethnic card" in CEA. For example, it would never show solidarity with Azerbaijanis, as it fears the rise of nationalist feelings among its own Azeri population. Despite Azerbaijan's linguistic identification with the Turks, Azerbaijan and Iran do have strong cultural and religious ties. Azerbaijan is the only former Soviet republic in which Shi'a Islam is the dominant religion. Playing the "ethnic card" in Tajikistan could antagonize the other ethnic groups in the region (Uzbeks, Kyrgyz, and Kazakhs). The Tajiks are culturally closer to Iran than the other ethnic groups in the region (Efegil and Stone 2001: 362).

Iran recognized the independence of CEA countries in 1991, hoping it could profit economically from the reestablishment of good relations with these countries. Former Iranian President Hashemi Rafsanjani repeatedly declared that, with the independence of the CEA states, a new "economic trade center" had emerged. It is also the shortest link between the Caspian Sea and the open seas.

Iran aims to accomplish the following objectives:

- Expand its infrastructure, especially its railway network;
- Gain political and economic influence in CEA through ECO;
- Acquire shares in a number of Caspian oil and gas development and export ventures.

For the countries in CEA, Iran is their main link to international markets. The states of CEA have all requested Iran to connect its railway network to rail in Turkmenistan, through which the other countries in the region could gain access to Iran's railways, linking CEA, Russia, and the Persian Gulf. In March 1995, the Iranian and the Central Asian presidents opened the 700 km railroad connecting the Iranian city of Bafq to the Iranian Persian Gulf port of Bandar Abbas. The construction of this line completed the rail link between the Iranian city of Mashad and the Persian Gulf. The line that connects Iran with Turkmenistan (the Tejen-Sarakhs-Mashad line) was completed in March 1996. It is 140 km long and enables CEA and Russia to access Europe via Turkey and to reach the Persian Gulf, Pakistan, and India by a short and time-saving route. With the inauguration of this transport link, the countries of CEA now have access to the Persian Gulf; it provides an alternative rail link to the Russian railway system. A major project under way is the Trans-Asian Railway (TAR), which will connect Singapore with Istanbul. The United Nations Economic and Social Commission for Asia and the Pacific initiated the TAR in the 1960s. The main direct route will have a length of 14,000 km. Currently, the total length of missing part is 1950 km, of which 1400 km extends between Bangladesh and Thailand, and 450 km in Iran between Kerman and Zahidan in southeastern Iran. Construction of the Iranian link started in February 1995 (Peimani 1998: 109; Kasuga 1997).

In addition to bilateral and multilateral transport agreements between Iran and the countries of CEA, the ECO is a forum for region-wide cooperation. The ECO was first established in 1977 between Iran, Turkey, and Pakistan in the context of Regional Co-operation and Development. The organization survived until the Islamic Revolution in Iran in 1979. In 1985, the organization was reestablished as the ECO. In 1990, the Treaty of Izmir, from 1977, was modified to create a proper legal basis for ECO and to open the treaty for accession by new member-states. The breakthrough of the ECO took place in 1992 at the Tehran Summit, which paved the way for the expansion of the organization from three to ten members, which now also include Azerbaijan, Kazakhstan, Kyrgyzstan, Tajikistan, Turkmenistan, Uzbekistan, and Afghanistan.

The ECO is a large economic cooperation organization. Its member-states together have a population of 300 million and cover an area of 7 million km^2. Since 1993, members of ECO have concluded agreements on cooperation in transport, transit trade, the simplification of visa procedures, cooperation against smuggling and customs fraud. During the ECO Tehran Summit in June 2000, the member-countries focused on energy cooperation and the development and implementation of trade agreements. Trade, transport, energy and industrial/agricultural cooperation constitute the core priority areas of the ECO. Despite these many agreements, the ECO record in promoting regional trade is not very impressive. To promote trade integration, ECO member-countries have to overcome a variety of problems, the most important of which are the absence of a dense network of transportation links and limited financial resources (Afrasiabi 2000; *Ettela'at*, April 1, 1996: 1). The oil and gas wealth is important for economic prosperity in the region. One problem is to get to the market of high-income importing countries.

Iran plays an important role in the exploitation and export of these resources. It has the world's second largest proven natural gas reserves (estimated at 942.2 trillion cubic feet [tcf]) after Russia and also ranks second in proven oil resources (11.4%, estimated at more than 130 billion barrels [bbl]). In 2003, Iran produced 3,852 thousand barrels per day [Ubbl/d] (BP 2004). It also has an extensive pipeline network to which possible pipelines from CEA could be connected.

Iran's chief foreign policy aim has been to prevent the United States and its allies, Turkey and Saudi Arabia, to fill the vacuum that has been left in CEA after 1991. Iran knows that it is not able to fill this vacuum by itself and therefore has played what Roy (1998) has called the Russian card on a North-South strategic axis (Moscow-Yerevan-Tehran) in opposition to the East-West axis (Washington-Ankara-Baku-Tashkent). This strategic double axis becomes obvious in the competition between the various existing and proposed pipelines: East-West pipelines for the United States (Trans-Caspian, Baku, Georgia, Turkey); North-East pipelines for Russian and Iran (Baku-Novorossiysk-CPC connections with Iranian networks to the Persian Gulf). Iran supports the following already existing and possible pipelines:

- Tabriz-Ankara pipeline from Tabriz (Iran) to Ankara (Turkey);
- the Korpezhe-Kurt-Kui pipeline from Korpezhe (Turkmenistan) to Kurt-Kui (Iran);
- Iran oil swap pipeline from Neka (Iran) to Tehran (Iran);
- the Baku-Tabriz pipeline from Baku (Azerbaijan) to Tabriz (Iran);
- Tehran-Kharg Island pipeline from Tehran (Iran) to Kharg Island (Iran).

Iran's closest relations are with Turkmenistan. This is mainly due to geographic proximity and mutual interests in the exploitation and export of Turkmenistan's oil and especially gas resources. Relations with Kazakhstan are also developing well, particularly with regard to the oil business. Uzbekistan is the only country in CEA that supports the US sanctions against Iran and openly accuses Iran of exporting Islamic fundamentalism. Uzbekistan tries to distance itself from Iran also in terms of culture. It is undeniable that the region's most important cultural centers, Samarkand and Bukhara, which are the cradle of Iranian civilization, are in Uzbekistan

(Efegil and Stone 2001: 356-7). Additionally Uzbekistan has a substantial Tajik population, which culturally is the closest to Iran. In 1999, the Tajik population in Uzbekistan amounted to 5.5% of the total population (World Bank 2002: Table 2.1). Iranian relations with Azerbaijan fluctuate, depending on the Azerbaijan government's foreign policy orientation. The remaining countries believe to have good relations with Iran or at least have no hostile relations with it (Efegil and Stone 2001: 357).

America's opposition to a more active involvement of Iran in CEA has hampered the strengthening of ties between Iran and the region. For example, Georgia tried to establish close ties with Iran. Former president Eduard Shevardnadze even made an attempt to help improve relations between Iran and the West but finally had to give up due to US opposition. Georgia then turned towards Turkey as a possible actor to cooperate with. On a per capita basis, Georgia is the largest recipient of US aid in the former Soviet Union (Salukvadze 2002).

When Haydar Aliyev became president in Azerbaijan in 1993, he aimed at establishing friendly relations with Iran. However, the US and its regional allies, Turkey and Israel, put him under pressure. Iran has to compete with Western countries for influence in the region. In matters of financial assistance and technology, Western countries have more to offer than Iran. International financial institutions, such as the World Bank, the International Monetary Fund, the European Development Bank, and the European Bank for Reconstruction and Development are active in the region (Hunter 1996: 62). This puts them in a much more advantageous position than Iran to have influence in CEA.

Prospects for Relations Between Iran and CEA

Until now, Iran has failed to develop a coherent long-term foreign policy towards CEA. Instead, it has limited itself to responding to developments beyond borders. Despite its reference to the "common cultural heritage" of Islam and the Persian language, Iran has neither the expertise nor the leverage to become a serious actor in the region. Since 1991, Iran has attempted to establish economic relations with the countries of CEA, especially in trade, transport, and the construction of pipelines. It also has tried to strengthen cultural and scientific links with the region, emphasizing the historical "Persian background" of the culture of CEA. Since 1991, it has been sending preachers and religious agents to the region to promote Islamist policy. Most of these preachers come from Iran's own Turkic-speaking population. Despite this common background, Iran's economic involvement in the region is still limited with the exception of its increasing relations with Turkmenistan. Iran's problematic relations with the West, especially with the US, pose an obstacle to Iran's ambitions in CEA. Most of all, however, the rivalry between the different political factions in Iran's leadership prevents the country from developing a coherent foreign policy. Factionalism, economic distress, and social turmoil in Iran even pose a threat to the survival of the Islamic Republic of Iran. The discontent of the so-called *Third Force* (those born after the Revolution of 1979) and the questioning of the legitimacy of the concept of the velayat-e faqih by the public and even by the press, seem to prepare the ground for a possible political change in Iran. In any case, Iran has never been as politically and ideologically polarized as it is today. Whatever changes might take place, they will also affect Iran's relations with the countries of CEA.

References

ALEXANDER'S GAS AND OIL CONNECTION (AGOC)
2002 "Iran Approves Turkish Gas Deal."
AFRASIABI, K.L.
2000 "The Economic Cooperation Organization Aims to Bolster Regional Trade Opportunities." Part I September 28.
AMINEH, M.P.
2003 *Globalization, Geopolitics and Energy Security in the Central Eurasia and the Caspian Region.* Den Haag: Clingendael.
2002 *"Sicherheit und Entwicklung in Eurasien—neue Gedanken zur Geopolitik im Zeltalter der Globalisierung."* Pp. 267-301 in Reiter, E. (ed.), *Jahrbuch fürInternationale Sicherheitspolitik.* Volume I. Hamburg: MITTLER.
2000 *"Konflikt und Zusammenarbeit um die Kontrolle der Erdölressourcen im post-sowjetischen Eurasien: Rußland, Türkei und Iran."* Pp. 575-620 in Reiter, E. (ed.), *Jahrbuch für Internationale Sicherheitspolitik.* Hamburg: Mittler.
1999 *Die Globale Kapitalistische Expansion und Iran-Eine Studie der Iranischen Politischen Ökonomie 1500-1980.* Hamburg, Münster: Lit Verlag.
AMNESTY INTERNATIONAL
1998 "Afghanistan-Thousands of Civilians Killed Following Taleban Takeover of Mazar-e Sharif." ASA.
BAKER, P.
2002 "Russia Plans 5 More Nuclear Plants in Iran." *Washington Post Foreign Services,* July 27, A15.
BARRACLOUGH, S.
1999 "Khatami and Consensual Politics of the Islamic Republic." *Journal of South Asia and Middle East Studies* 22(2).
BBC
2002 "US Softens Line on Eevil Axis."
2004 "Iranian MPs approve uranium bill."
BLACK, I. AND J. STEELE
2003 "Tehran Faces Pressure to Allow Nuclear Inspections." *The Guardian Weekly,* June 19-25.
BLACK, I. ET AL.
2002 "Bush's Axis of Evil' Widens Split with Allies." *The Guardian Weekly,* February 7-13.
BP
2004 *Statistical Review of World Energy.*
BUCHTA, W.
2000 *Who Rules Iran?* Washington: Washington Institute of Near East Policy.
1996 *"Irans fraktionierte Führungselite und die fünften Parlamentswahlen." KAS/Auslandsinformationen*: 52-55. August.

CALDWELL, D.
1996 "Flashpoints in the Gulf: Abu Musa and the Tunb Islands." *Middle East Policy*: 50-57, March.
DEMPSEY, J.
"EU Backs Formal Launch of Trade Ties with Iran." Financial Times, June 18.
DINMORE, G. AND R. KHALAF
2002 "America's New Enemy." *Financial Times,* February 1.
DUBNOV, A.
2000 "The Big Transcaspian Pipeline Project Flops." *Moscow News*, January 26-February 1.
EFEGIL, E. AND L.A. STONE
2001 "Iran's Interests in Central Asia: A Contemporary Assessment." *Central Asian Survey* 20(3): 353-365.
ENERGY INFORMATION ADMINISTRATION
2002 "Caspian Sea Region: Reserves and Pipelines." July.
IRAN COMMERCE
2001 "Iran-Russia Trade Volume Hits 900 Million Dollars." No. 2.
LAFRANIÈRE, S.
2001 "Russia Signs Arms Deal with Iran." *The Guardian*, October 3.
FÜRTIG, H.
1998 *Islamische Weltauffassung und außenpolitische Konzeptionen der Iranischen Staatsführung seit dem Tod Ajatollah Khomeini.* Berlin: Das Arabische Buch, Zentrum Moderner Orient, Geisteswissenschaftliche Zentren Berlin e.V., Studien 8.
HODJAT, M.
1999 *Jenaha-ye siassi da Iran Emrouz (Les Tendances Politiques dans l'iran d'aujourd'hui).* Tehran: Maghso-Negar 1378.
HUNTER, S.
1996 "Iran, Russia and the Southern Republics of ex-Russia." Goftego.
KASUGA, K.
1997 "Trans-Asian Railway." *Japan Railway & Transport Review*, June.
LELYVELD, M.
2002 "Turkey: Ankara cuts Gas Prices after Russian Concessions." *Eurasia Insight,* October 6.
MYERS, S.L.
2002 "Russia May Reconsider its Nuclear Projects with Iran." *International Herald Tribune*, August 3-4.
NAHAVANDI, F.
1996 "Russia, Iran and Azerbaijan-The Historic Origins of Iranian Foreign Policy," in Coppetiers, Bruno (ed.), *Contested Borders in the Caucasus.* Brussels: VUB Press.
NOVOPRUDSKY, S.
1999 "Patriotic Feat: Russia Buys Turkmen Gas Out from Under Turkey and Ukraine." *Izvestia*, December 21.
OLSON, R.
2002 "Turkey-Iran Relations, 2000-2001: the Caspian, Azerbaijan and the Kurds." *Middle East Policy*: 111-129, June 1.
PEIMANI, H.
1998 *Regional Security and the Future of Central Asia-The Competition of Iran, Turkey, and Russia.* Westport, Connecticut & London: Praeger.
POLLACK, K.M.
2003 "Securing the Gulf." *Foreign Affairs* 82(4): 2-16.
RAHNEMA, A.
1990 *The Secular Miracle: Religion, Politics, and Economic Policy in Iran.* London: Zed Books.
RAMEZANZADEH, A.
1996 "Iran's Role as Mediator in the Nagorno-Karabakh Crisis," in Coppetiers, Bruno (ed.), *Contested Borders in the Caucasus.* Brussels: VUB Press.
RECKNAGEL, C.
"US and Iran Chart Different Courses over War on Terror in 2001." *Eurasia Insight*, December 14.

REISSNER, J.
1999 "Iran unter Khatami-Grenzen der Reformierbarketi des politischen Systems der Islamischen Republik." *Stiftung Wissenschaft und Politik.*
RIECK, A.
2000 "Iran: Towards an End of Anti-Western Isolationism?" in Hafez. K. (ed.), *The Islamic World and the West: An Introduction in Political Cultures and International Relations.* Leiden, Boston, Köln: Brill.
ROY, O.
1998 "The Iranian Foreign Policy Toward Central Asia." *Eurasia Insight.*
SALUKVADZE, K.
2002 "The US on the 'Silk Road' of Expansion to Eurasia?" *Central Asia-Caucasus Analyst,* March 13.
SYMON, F.
2001 "Afghanistan's Northern Alliance." *BBC News,* September 19.
TIMMERMAN, K.R.
2001 "Iran Cosponsors al-Qaeda Terrorism." *Insight Magazine,* December 3.
VISHNIAKOV, V.
1999 "Russian-Iranian Relations and Regional Stability." *International Affairs* 45(1): 143-53.
WORLD BANK
2002 *World Development Indicators.* Washington, DC: World Bank.

PART FOUR

LOCAL CONFLICTS

X. Growing Tension and the Threat of War in the Southern Caspian Sea: The Unsettled Division Dispute and Regional Rivalry

HOOMAN PEIMANI

ABSTRACT

The absence of an acceptable legal regime for the division of the Caspian Sea among its five littoral states has created grounds for conflicts, crises, and wars in the Caspian region, a situation worsened since 2001 when Iran, Azerbaijan, and Turkmenistan found each other on a collision course over the ownership of certain offshore oilfields. The region has since been heading towards militarization, while the persistence of conflicts over the Caspian Sea's division has prepared the ground for military conflicts. Fear of lagging behind in an arms and the manipulation of conflicts by the United States and Turkey have further encouraged militarization. Against this background, certain factors, including Turkey's efforts to deny Iran political and economic gains in the Caspian region, the growing American military presence in Eurasia, and the expanding American-Azeri military ties since 11 September 2001 will likely contribute to the creation of a suitable ground for a military conflict in the Caspian region.

Introduction

The absence of an agreed-upon legal regime for the division of the Caspian Sea among its five littoral states (Iran, Russia, Azerbaijan, Kazakhstan, and Turkmenistan) has created grounds for conflicts, crises, and even wars in the Caspian region since the fall of the Soviet Union in 1991. Due to the high political, economic, and military/security stakes of the Caspian littoral states in its division, the situation has become especially tense since 2001. In that year, fear of losing their desired share of the Caspian Sea pitted the southern Caspian countries (i.e., Iran, Azerbaijan and Turkmenistan) against each other. Having a military dimension, their confrontation over certain oil fields had the potential to escalate into a full-scale military conflict, a possibility still well in place after approximately two years. The failure of efforts to find a consensus among the five littoral states, such as the unsuccessful Ashgabat summit meeting in April 2002, has since worsened the already tense situation. There are enough grounds for the belief that future conflicts have the potential to escalate into military confrontation. To obtain their desired share of the Caspian Sea and/or to preserve their national interests threatened by their Caspian neighbors, the littoral countries, in effort to boost their military forces, have initiated an undeclared arms race and the gradual militarization of the Caspian region. Fear of lagging behind, on the one hand, and the manipulation of conflicts by countries such as the United States and Turkey, with long-term interests in the Caspian region, on the other, have further encouraged the militarization process. In this context, Iran's concern about Turkey's role in the region and, in particular, its growing military and economic cooperation with Azerbaijan, in concert with the American government, have created anxiety in Iran, a country fearful of its complete encirclement by enemy states. Such concern will likely encourage Iran to strengthen its military forces, in general, and those along its long border with the Caucasus, the Caspian Sea, and Central Asia, in particular. Against this background, certain factors will have a major negative impact on the situation and will likely contribute to the creation of a suitable ground for conflicts, including military ones, in the Caspian region. Of those, the major ones are the growing American military presence in Eurasia and the expanding American-Azerbaijani military cooperation since September 11, 2001.

The collapse of the Soviet Union opened a new era in the Caspian Sea region. Among other factors, the Caspian oil and gas resources have since elevated the region's international status, while pitting its littoral states against one another over their division. Having the world's fifth largest oil reserves and its second largest natural gas deposits, the offshore Caspian fossil-energy resources are not of vital importance to Iran. Nor are they for Russia, a country with the world's largest gas reserves and significant oil deposits. Nevertheless, for a variety of political, economic, and security reasons, including its energy resources, the two regional powers have vested interests in the Caspian Sea. During the Soviet era, they agreed to use jointly the resources of the largest closed lake on Earth. However, the

Caspian oil and gas resources are of great importance to the other littoral states. The latter have considered them as their main source of revenue and a means to preserve their sovereignty since their independence in 1991, when the Soviet Union fell. In short, all the five littoral states have strong reasons to insist on a formula for the division of the Caspian Sea, which will leave them with the largest possible share.

Apart from political conflicts among the littoral states and their manipulation by non-littoral powers, this reality has created a major obstacle to their acceptance of a legal regime for the Caspian Sea binding on all of them. The littoral states' inability to agree on such a legal regime has created uncertainty about the ownership of many offshore oil fields and prevented their development, while making the Caspian states hostile to each other for their developing disputed oilfields. Unsurprisingly, the situation is therefore ripe for tension between and among them. This fragile situation could easily escalate into military confrontations, given the existence of other sources of conflict in the bilateral relations of the littoral states arising from their political concerns and military/security considerations. Among them, the major one is a growing concern in both Tehran and Moscow about the expanding political, economic, and military presence of the United States in other littoral countries. Such concern has damaged, to a varying extent, ties between and among the Caspian countries. In the southern Caspian region, various tensions and grievances in Azeri-Iranian bilateral relations have created grounds for major conflicts with significant impacts on all the littoral countries. Such impacts will be the result of Iran's troubled relations with the United States and Turkey, the two major backers of the Azeri regime, which could conceivably drag the latter into any future conflict between Iran and Azerbaijan only to turn it into a major regional conflict. For the Azeris, those concerns and grievances include Iran's friendly ties with Armenia and its alleged providing military assistance to the Armenians. For the Iranians, they include the official and unofficial Azeri claims to Iran's Azerbaijan in different forms, the harassment by Azeri police and border guards of Iranian truck drivers passing through Azerbaijan, and the latter's growing ties with Israel. Additionally, the two sides have accused each other of supporting the other side's illegal opposition groups.

The regional political fragility became evident in July 2001, when the southern Caspian region found itself on the verge of a military confrontation. At that time, disputes over the ownership of certain Caspian offshore oilfields between Iran and Azerbaijan and between Azerbaijan and Turkmenistan reached an unprecedented stage of hostility. Iran and Turkmenistan accused Azerbaijan of illegal development and operation of certain disputed oil fields to which all the three states have ownership claims. They also accused Azerbaijan of efforts to develop other disputed oilfields with the assistance of foreign oil companies. Furthermore, they accused Azerbaijan of violating their territorial waters with its military and nonmilitary vessels, while Azerbaijan accused them of the same violations. In particular, one incident in that month had the potential to develop into a full-scale military confrontation, although Iran evaluated it as a minor issue in Iranian-Azeri relations. Thus, Azerbaijan harshly reacted to Iran's use of its navy to force a BP ship to leave the disputed Alov or Alborz oil field, as called by the two countries, respectively, and to stop its unilateral oil exploration there on Azerbaijan's behalf. The extent and nature of Iran's bloodless use of force are a matter of disagreement between the two countries. However, certain developments created suitable grounds for rapid deterioration of Azeri-Iranian relations and a sudden escalation of the incident to war. One was the Azeri government's description of the incident as a major violation of its territorial integrity and its threat to use military force in any future incident over disputed oil fields. Another was Turkey's dispatch of a small number of fighter jets to Azerbaijan under the pretext of participating in a previously-arranged air show (CAN, August 17, 2001). Azerbaijan's official and unofficial references to Turkey's move as a clear sign of its determination to defend the Azeris in any future confrontation with Iran offset Turkey's official statements downgrading the move's significance. Of course, the Turks made sure that the "air show" had left no doubt in anyone's mind about their taking sides with the Azeris. Yet another development was the simultaneous official visit to Azerbaijan of Turkish Chairman of the Joint Chiefs of Staff General Hussein Kivrigoglou, which was treated in the same manner by the Turks and the Azeris (CAN, September 21, 2001). Not only did the two developments create tension in Iran's ties with Azerbaijan, they provoked the disapproval of other littoral states and particularly of Russia and Turkmenistan. Finally, the sale of two American military boats to Azerbaijan added fuel to its conflicts with Iran and Turkmenistan (CAN, October 15, 2001). The latter expressed deep concern about the transaction, which they portrayed as a threat to their national security and a provocative act leading to an arms race, while the American government downgraded the boats' military significance. In particular, Turkmenistan's reaction was very strong and included its revelation of its purchase of Ukrainian military boats, which, in turn, provoked a harsh Azeri reaction (Ibid.). Briefly, in their reaction to the August incident, the Caspian littoral states, excluding Azerbaijan, warned against foreign involvement in their regional affairs, against the militarization of the Caspian Sea, against the use of force for settling territorial disputes, and against the threat of escalation of such disputes into war.

Added to such a history of tension in the Caspian region, certain turns in the American foreign policy towards the Caspian region since late 2001 have created concern about an emerging arms race there. Such an arms race could contribute to a military confrontation there, particularly between Iran and Azerbaijan, with a dire impact on the stability of the Caucasus, a region prone to war and instability. These turns have included the sudden and rapid expansion in the post-September 11 era of the American military forces in Central Asia and the Caucasus and in the countries in their close proximity. Justified as a necessity for its "war on terrorism" in Afghanistan, the American government has deployed its forces in Afghanistan, Pakistan, and several other countries in West Asia, which has created security concerns not only in Iran, but also in other regional countries, including China and Russia. In particular, the pattern of American military deployment in Central Asia and the Persian Gulf cannot be justified for the requirements of a limited warfare in Afghanistan against the remnants of the Taleban and al-Qaeda. Having secured an airbase in Uzbekistan, neighboring Afghanistan, the Americans have also obtained the right to use another airbase, this time in Kyrgyzstan, without any apparent necessity or usefulness, as the country does not share a border with Afghanistan. The Kazakh government has turned down the American request for a third airbase in Kazakhstan, a country far from Afghanistan. However, it has granted them over-flight and emergency landing rights, while they have also received over-flight rights from Turkmenistan. Despite their build up in Central Asia and Afghanistan, the Americans have also deployed their troops in Georgia, who are referred to as "military advisers." Moreover, they have expanded their naval and air forces in the Persian Gulf and the Arabian Sea with no obvious direct relevance to their type of operation in landlocked Afghanistan, which is separated from the Arabian Sea by Pakistan. In his August 23, 2002, statements, General Tommy L. Franks, commander of the American forces in Afghanistan, Central Asia, and the Persian Gulf, confirmed the suspicions of the regional powers (Iran, China, India, and Russia) about the American plan to take advantage of the opportunity to stay in West Asia for a long time. Meeting with Uzbek President Islam Karimov in Tashkent, he stated that the American military presence in Central Asia and Afghanistan would increase, the Americans would expand their military relations with the Central Asian countries, and the American forces would stay longer than expected in Afghanistan (REF/RL, August 24, 2002). On the same day, an American Congressional delegation visiting Tashkent stressed the American government's determination to stay in the region (Ibid.).

Another major turn in the American foreign policy towards the Caspian region has been the inclusion of military cooperation in the American bilateral relations with the Caucasian countries. Thus, the American government lifted a ban on selling arms to Azerbaijan and Armenia and concluded military cooperation agreements with Azerbaijan in March and April 2002 (Peimani, April 19, 2002). Added to that, the US government made clear, through the various statements of visiting American State Department officials to Azerbaijan, its commitment to Azerbaijan's defense and security and to the improvement of Azerbaijan's military capability to meet any future Iranian military challenge. They also stressed the American support of Azerbaijan in any future confrontation with Iran over the disputed oilfields.

As the mentioned developments have contributed to the expansion of a sentiment of mistrust and animosity in the Caspian region, unwanted escalation of conflicts has made both Baku and Tehran interested in easing tensions in their bilateral relations. In this regard, the most significant development has been the Azeri president's official visit to Iran. After about three years of repeated rescheduling, President Haidar Aliev paid a three-day visit to Iran on May 18, 2002, which was a major positive move in Azeri-Iranian relations. The visit clearly demonstrated Azerbaijan's interest in improving ties with its large neighboring country without regard to American disapproval. The visit was not very important in terms of its economic achievement, although Iran and Azerbaijan signed ten agreements on economic cooperation (*Ettellat*, May 21, 2002: 2). Of these, only two are of significance for both sides and especially for Azerbaijan, which has long sought access to open seas and the international markets via Iran to reduce its over reliance on Georgia and Russia in that regard. The latter requires connecting the Azeri railroad network to the Iranian one to make Iran's Persian Gulf port of Bandar Abbas available to the Azeris. To that end, Iran agreed to finance and build two railroad lines to connect Azerbaijan to Iran's Caspian Sea port of Astara, and to connect the latter to Iran's north-south railway via the city of Gazvin (Ibid.). Iran also agreed to construct a highway between Baku and Astara to facilitate Azerbaijan's land link to the international markets via Iranian highway networks (Ibid.). However, it is not certain whether those projects will be implemented at all, as their respective agreements lack a clear timetable for their implementation.

Lacking a strong economic significance, the importance of President Aliev's visit lay in its political dimension. In particular, it reflected the Azeri government's effort to distance itself from the United States in its hostile approach to Iran, a country demonized by the Americans as a member of the "axis of evil." This is notwithstanding Azerbaijan's close and growing ties with the United States, a major source of tension in its relations with Iran since the mid-1990s. For economic, political, and military/security reasons, the Azeris have sought extensive and multidimensional relations with the Americans. Yet, the Azeris also realize those relations could not solve all their problems and particularly secure their peace, stability and territorial integrity. As regional powers with certain influence in Armenia, Iran and Russia are indispensable allies for Azerbaijan, which has suffered from its prolonged conflict with Armenia over its breakaway region of Nagorno Karabakh, which is now under Armenian control. In part, President Aliev's visit to Iran was a step towards securing its assistance in that regard.

In spite of the importance of President Aliev's visit, Azeri-Iranian relations are still far from being on an irreversible peaceful and constructive path. During his visit, President Aliev had friendly talks with Iran's President Mohammad Khatami, its Foreign Affairs Minister Kamal Kharrazi and its Spiritual Leader Ayatollah Ali Khamenei. The Iranian leaders emphasized their interest in expanding friendly ties with Azerbaijan and avoided dwelling on the well-known sources of conflicts and disagreements in their bilateral relations. Nevertheless, in a polite form, they clearly stated their concern over Baku's expanding ties with Israel and the United States, in particular its expanding military relations with Washington. Along the same line, they also reiterated their stance on the division of the Caspian Sea among its littoral countries, another major source of conflict and tension between Iran and Azerbaijan. Iran advocates its division into equal shares among the five littoral countries and opposes the development of its oilfields in the absence of an agreement to that effect among all the littoral states. Azerbaijan on the other hand, supports its division into five unequal shares based on the length of the littoral countries' coastlines, while developing certain oil fields to which it claims ownership in the absence of such an agreement. During their talks with President Aliev, the Iranians clearly expressed their opposition to such deals between the littoral countries and the development of the Caspian offshore oilfields in the absence of a consensus among all the littoral states (CAN, May 30, 2002).

President Aliev's visit at least helped the two sides agree on negotiations as a means of settling their disputes, although it did not remove their conflicting views on them, which had created major tensions in Iranian-Azeri relations. The Azeris hinted at their possible flexibility in settling their disputes with the Iranians over certain offshore oil fields to which both sides claimed ownership. As a result, the two sides agreed on their legal experts' work on the issue and on their meeting in Baku in June. While there is no sign of Baku's acceptance of Tehran's formula for the Caspian Sea's division, the prevailing positive mood during President Aliev's visit and his extended invitation to President Khatami for an official visit to Azerbaijan created grounds to hope for improving ties between the two countries.

However, certain factors will contribute to tension and hostility in Azeri-Iranian relations despite an apparent effort by the two sides to forge better ties. In this regard, Azerbaijan's relations with the United States and Turkey and the latter's troubled relations with Iran are the major contributing factors. As discussed earlier, in the post-September 11 era, the American political and military presence in the Caucasus and Central Asia has significantly increased. Added to its deployment of troops in Afghanistan and Pakistan, and the expansion of its military forces in the Persian Gulf, the American government has also increased its military presence in Turkey, a neighbor of Iran where the United States has permanent military bases. These recent developments have taken place within the context of a growing American political and economic influence in the Caucasus and Central Asia, in general, and in Azerbaijan, in particular, which began in the mid-1990s. Iran considers all these developments as a threat to its national security and also to its interests in the Caucasus. The latter will likely make Azeri-Iranian relations susceptible to tension and crises even if the two sides can settle their Caspian Sea-related disputes on Iran's terms. Yet, such a scenario will be unlikely in the foreseeable future, given the clear American promise of political and military support to Azerbaijan in its dealing with Iran over those disputes.

Thus, another source of conflict in the relations of Azerbaijan and Iran is the unresolved issue of dividing the Caspian Sea. Of course, this is not just a cause of tension in Iranian-Azeri relations as the issue is of importance to all the five Caspian states. Nevertheless, it will likely become a major concern for the two Caspian neighbors because of a certain development in 2002. In that year, Russia settled its disagreements with Kazakhstan and Azerbaijan over the division of the Caspian Sea, not as a result of an agreement among all the Caspian littoral states, but through concluding bilateral agreements. Leaving Iran in an unfavorable situation to claim its desired share when three out of five countries had agreed on their sea boundaries, such an approach to dealing with the Caspian Sea legal regime would certainly contribute to major conflicts between Iran and its neighboring Azerbaijan over the ownership of many offshore oilfields.

Finding a legal regime for dividing the Caspian Sea has been a source of conflict among its five littoral countries since the Soviet Union's break-up in 1991. Until 1999, Iran and Russia opposed the division of the Caspian Sea into national zones with exclusive rights to all its resources in favor of dividing it based on an agreement among all the littoral states for its common use as provided by the two Iranian-Soviet agreements of 1921 and 1940. They based their claims on the nonapplicability of an international law regarding the high seas providing for exclusive national zones to the Caspian Sea, which is the world's largest landlocked lake.

Eager to put their hands on its offshore oil and gas fields to address their serious financial problems, the other three states were insisting on its division into such zones. Russia gave up that policy in 1999, when it found large offshore oil reserves close to its Caspian coastline. While Iran has insisted on nondevelopment policy of offshore fossil energy resources in the absence of a legal regime, the other Caspian states have taken steps to develop those oilfields, which they consider to fall within their territorial waters. Unsurprisingly, this policy has created conflict between and among the littoral states as there are many double and triple claims to different parts of the Caspian Sea, which are rich in fossil energy. Several meetings among the littoral states' officials to agree on a legal regime, including the most recent one held in Ashgabat in April 2002, have all failed to achieve that objective, because of their conflicting interests.

In the absence of a binding legal regime, Russia has sought to address the division issue through bilateral agreements. The result has been the conclusion of two separate agreements between Russia and its two neighbors: Kazakhstan in May 2002 and Azerbaijan in September 2002. The details of the one with Azerbaijan have yet to be released. However, if Russia's agreement with Kazakhstan is of any indication, it probably only settled territorial issues, including the ownership of fossil energy resources, to the extent agreeable to both sides when they signed it. The settlement of harder issues, such as the ownership disputes of oil fields close to both countries, was left for future agreements. Russia's conclusion of agreements with its immediate neighbors can only create a false perception of addressing the unresolved issue of dividing the Caspian Sea, while practically leaving most of the sources of conflicts intact. At best, these agreements could temporarily settle territorial disputes in the northern Caspian Sea for as long as the development of the oil fields in the disputed zones is not feasible for any one, for one reason or another. Any future unilateral attempt for their development will surely lead to conflicts.

Even though this arrangement may be suitable for its signatories for a while, it has not resolved the legal issue forever. In addition to the unresolved territorial disputes between Kazakhstan and its neighboring Turkmenistan, the other three neighbors, Iran, Azerbaijan and Turkmenistan, are yet to settle major disputes, which have pitted them against each other since the mid-1990s. As mentioned before, their intensity was about to push them to a military confrontation in 2001, which could have escalated into a dangerous level because of Turkey's intervention. For example, the three countries have claims to three major offshore oil fields (Azeri, Cheraq and Guneshli) now being developed by Azerbaijan. Relations between Turkmenistan and Azerbaijan have deteriorated over their disputed ownership and that of other oil fields such as Serdar, or Kapaz as they call them, respectively. Iran and neighboring Azerbaijan claim the ownership of what they refer to respectively as the Alborz or the Alov oilfield. Azerbaijan's effort to begin its development unilaterally provoked an Iranian show of force in July 2001. Turkmenistan's frustration at settling its disputes with Azerbaijan makes it close to Iran's view of demanding a multilateral agreement on the legal regime. However, as the latter does not seem to be feasible in the near future, Russia's bilateral agreements have inclined Turkmenistan to hint at its flexibility. Thus, on September 26, 2002, Turkmen President Safarmurad Niyazov mentioned to visiting Russian Energy Minister Igor Yousefov that the Caspian states could sign an agreement to divide the Caspian Sea without Iran's participation, thus, leaving Iran as the only littoral state insisting on a multilateral agreement (REF/RL, September 27, 2002). Given the depth of conflict between Turkmenistan and Azerbaijan as reflected in their beefing up their naval forces in 2001 and also in their closing down their embassies in each other's capital for "technical" reasons, bilateral agreements do not seem to be a practical solution to Turkmenistan's problem. However, they can surely damage its peaceful and extensive ties with neighboring Iran, with which it shares a 920-kilometer border. Due to the importance of their multidimensional relations and their common concern about the American presence in the Caspian region, major disagreements between Iran and Russia over the Caspian Sea will not put them on a collision course, at least in the short run. However, such is not the case when it comes to Azerbaijan's disagreements with Iran and Turkmenistan. As the latter have many reasons for grievances with Azerbaijan, their unresolved territorial disputes with their Caspian neighbor could well burst into a major crisis, and possibly a military conflict.

Yet another contributing source of conflict in Iranian-Azeri relations is the extremely irritating issue of the construction of the Baku-Tbilisi-Ceyhan oil pipeline. Its raison d'être is to exclude Iran from the lucrative business of exporting Caspian oil and particularly from exporting Azeri oil. Through Georgia, the envisaged pipeline connects Azerbaijan to Turkey's Mediterranean port of Ceyhan where oil tankers can deliver Azeri oil, as well as that of Kazakhstan and Turkmenistan, to the international markets. The pipeline is a controversial project for exporting Caspian oil since many factors, including its length, direction, cost and nonsecure nature, make it an inefficient means of oil export for many oil companies operating in the Caspian region. Clearly, political motives have inspired its ion.

Being neighboring countries with extensive oil export infrastructure and access to open seas, Iran and Russia offer "natural" export routes to their Caspian neighbors. Yet, only a fraction of the existing oil exports are conducted via their routes, thanks to the American government's opposition to their use as the long-term export routes. As a result of its troubled relations with the United States, Iran's share is very insignificant and much smaller than that of Russia, although it offers the shortest, the cheapest, and the most secure route. For political and strategic considerations, the United States, whose companies dominate the Caspian oil industry, has sought to bypass Iran and Russia for Caspian oil exports. Being partly used by Kazakhstan, the existing route via Georgia connecting Azerbaijan to the Black Sea port of Supsa is only a short-term solution.

Having secured the American support, Turkey has sought to establish itself as the main Caspian export route. Beside economic gains, that status would increase Turkey's political influence in its neighboring Caucasus and in Central Asia, the two regions of great importance for the Turks, while significantly increasing their international political and economic influence. If it becomes fully operational and enjoys the commitment of major Caspian oil exporters, the Baku-Tbilisi-Ceyhan pipeline will help Turkey achieve its regional and international political and economic objectives. When operational, the pipeline will connect the three landlocked Caspian oil exporters (Azerbaijan, Kazakhstan and Turkmenistan) to the international markets via Turkey, while bypassing two regional powers (Iran and Russia) with strong interests in the Caspian region. Being mainly a political project, the pipeline is meant to ensure the latter's minimum involvement in the Caspian energy industry. For its lack of economic sensibility, most major oil developers (American and non-American alike), especially those operating in Kazakhstan and Turkmenistan, have not shown any strong interest in the project.

As the "heart" of the Turkish export route, the Baku-Tbilisi-Ceyhan pipeline suffers from major deficiencies. It is long (about 1,700 kilometers) and costly. Its cost estimated at $2.4 billion to $4 billion (Peimani 2001: 78-79). It also lacks security, as it has to pass through highly unstable Azerbaijan and Georgia. Apart from the threat of their armed independence movements, there are doubts about the long-term stability of their fragile political systems. The Turkish part of the pipeline will pass through Turkey's rebellious Kurdish region, which could burst into another round of civil war. Moreover, the Turkish route's westward direction would make it suitable to supply the European market whose fuel imports will only grow by a million barrels per day in the next decade whereas that of the Asian market will increase by 10 million barrels per day. There is not even certainty about the availability of the European market since the existing unused OPEC and Russian export capacities could meet most, if not all, of the increased European demand. The shortcomings of the Baku-Tbilisi-Ceyhan pipeline have dissuaded many Caspian oil developers from committing themselves to its construction and/or its use. Its main proponents have been Azerbaijan, Georgia and Turkey, which have obvious interests in the project. The United States has also promoted it mainly for political considerations. For about a decade, its government has tried unsuccessfully to convince the majority of the American and non-American oil companies to opt for the pipeline. The major oil developers in Kazakhstan, whose commitment to use the pipeline is a necessity for its economic viability, have refused to do so. Instead, they have reduced their reliance on the Georgian route via Azerbaijan (Baku-Supsa pipeline) and have increased their export via the Russian oil pipeline network, while exploring the possibility of export through a pipeline via Iran. Currently, they export most of their oil via a Russian pipeline connecting Kazakhstan to the Black Sea port of Novorossiysk.

Despite its shortcomings, after over eight years of delay, the construction of the Baku-Tbilisi-Ceyhan oil pipeline began in September 2002 to create yet another source of conflict in Iran's relations with Azerbaijan and Turkey. It is a provocative development, as the pipeline is meant to deny Iran and Russia a significant amount of annual revenues in transit fees and deprive them of an additional source of power and influence in the Caspian region, while uplifting the regional political and economic power of their rival, Turkey, in a strategically neighboring region where they seek to limit its power and influence. Serving clear anti-Iranian and anti-Russian objectives, there is no doubt that its construction will worsen the existing sentiment of mistrust and suspicion towards the United States, the main promoter of the pipeline, in both Russia and Iran. The American oil companies will be the pipeline's major beneficiaries, if its construction continues as planned and if it becomes fully operational in 2004. Beside its impact on Iranian-American relations, the development will surely contribute to worsening Iranian ties with Turkey, the major regional beneficiary of the pipeline.

Apart from the negative impact of the Iranian-Azeri territorial disputes on Iranian-Turkish relations, the issue of exporting Iranian natural gas to Turkey has also become a source of conflict between Iran and Turkey. On October 7, 2002, Turkish Minister of Energy Zaki Chekan arrived in Tehran to negotiate Turkey's resumption of imports of Iranian natural gas, which the Turks stopped in June 2002 for its "poor quality" (*Hamsahrye*, October 10, 2002). As the quality has remained the same as agreed upon when the export began in December 2001, evidence suggests that two factors were the main reasons for that behavior: Turkey's poor economy and American's efforts to isolate Iran. According to Mr. Chekan, there were three major issues to be resolved between the two countries: the lower than agreed upon quality of the exported gas, the annual amount of exports and their price. Being raised after months of imports, the quality issue seemed to be nothing more than a ploy to create grounds for renegotiating the gas agreement. The Iranian oil ministry's readiness to let foreign experts test the gas quality confirmed this point. Thus, apart from a political factor, two economic reasons seem to be the major causes of the dispute. According to the Turkish-Iranian gas agreement, Turkey was obliged to import from Iran 4 billion cubic feet of gas in 2002. The annual import will increase to 10 billion cubic feet by 2010. In reality, Turkey does not need and cannot afford such volume of imports. In determining their future annual gas requirements in the 1990s, the Turks based their calculations on unrealistic economic growth rates. They therefore signed agreements for large annual imports from Iran, Russia, Azerbaijan, Turkmenistan, Egypt, and Nigeria. Not only have those economic growth rates not been achieved, Turkey has experienced a severe economic recession since 2001, while its foreign debt has soared to over $117 billion (Eia, July 22). Unsurprisingly, its gas consumption was estimated to decrease by 14% in 2002 compared to that of 2001 (RFE/RL, 7 October 2002).

In such a situation, in October 2002, the Turks had to convince the Iranians to agree on a lower volume of annual exports and on a price reduction since the Turks had stated that the price agreed on was now very high for them. As the Russians, who, in late 2002, began to export gas to Turkey via the sea-based Blue Stream Pipeline, were forced to give a 9% discount on their gas price, the Iranians found it difficult to insist on their previously-agreed price. Thanks to the American pressure, the discount is estimated to help Turkey save about $280 million on its gas imports from Russia over the next three years (RFE/RL, October, 7 2002). Legally speaking, Turkey, which suddenly stopped its gas imports from Iran, was in violation of its agreement signed with Iran in 1996. At that time, the two sides agreed on a schedule for annual gas exports from Iran to Turkey at a certain price. Iran has invested in a pipeline for this purpose based on a specified amount of exports at a sensible price justifying its investment. However, from its beginning, Turkey has failed to meet its obligations under the agreement. Its failure to finish its part of the pipeline on time delayed the exports by several months. After that costly delay for Iran, Turkey unilaterally stopped importing only after about seven months, creating both financial losses and technical difficulties for Iran, which had to divert the ready-to-export gas to other projects, such as injecting its oil wells.

Since the Iranian-Turkish gas agreement is based on the concept of take or pay, Turkey is compelled to pay for the amount of gas agreed to be exported from Iran, if it refuses to import it. However, even in the best circumstances, the agreement would require a lengthy legal battle, which, in 2002, Iran did not seem to be interested in, at least for political and security reasons. The deployment of American troops in Afghanistan, Pakistan, Central Asia, the Caucasus, and the southern Persian Gulf countries has almost completed Iran's encirclement by the hostile Americans. For that matter, the Iranian government did not wish then, nor does it wish now, to deteriorate its ties with Turkey, a NATO country hosting American military forces, to turn the Turks into an active hostile nation as well. It is wonder that Iranian Minister of Oil Bijan Namdar Zangeneh did not reject Iran's willingness to accommodate Turkey's requests when he commented on October 7, 2002 in his talks with his Turkish counterpart and the possibility of Iran's flexibility. Thus, he stated, "In trade negotiations, everything is possible" (*Hamsahrye*, October 10, 2002: 2).

The mentioned economic difficulties were part of the reason for Turkey's sudden cut in gas imports. Yet, it is not a secret that the American pressure on the Turks also played a significant role in their behavior. Being contrary to the American policy of isolating Iran and weakening its economy, the Americans have opposed the Iranian-Turkish gas agreement since the Turkish government under the Welfare Party (Refah), a party with religious tendencies and a positive attitude towards Iran, signed it in 1996. For that matter, the Americans considered to apply to Turkey the D'Amato Act, providing economic sanctions on countries or companies investing in the fossil energy industries of Iran and Libya. Apart from technical difficulty of applying the Act, as the gas agreement did not require Turkey's investment in Iran, the Americans decided not to impose any sanction on Turkey mainly to avoid a conflict with their regional ally. This was notwithstanding the fact that the agreement violated the spirit of the Act, seeking to deny Iran economic gains and political influence. Nevertheless, the Americans did not hide their disappointment by the export of Iranian gas to Turkey via an Iranian pipeline. The latter weakened the American case to push for the controversial Baku-Tbilisi-Ceyhan oil pipeline partly justified on the alleged nonreliability of the Iranian route for any type of oil and gas export. Turkey's cutting its imports about two months before the pipeline's construction began suggests political considerations as a factor in Turkey's decision. Regardless of the Americans' hope, Turkey cannot abrogate its gas agreement with Iran without paying a price in cash and in deteriorating its ties with a large neighbor. However, it certainly took advantage of a guaranteed American backing in its dealing with Iran to squeeze the Iranians for concessions when they were in a tight economic and political spot, a function of their international isolation and extensive mismanagement of their economy by their ruling elite. As Tehran was becoming increasingly concerned about the American policy towards Iran and its surrounding regions, the Turks eventually got their desired gas deal from a fearful regime eager to secure Turkey's friendship. Seeking to avoid antagonizing its relations with its neighbor and to resume its exports to Turkey for economic reasons, Iran finally agreed to revise its gas agreement with Turkey to decrease the gas price and to exempt the Turks from their commitment to buy 9 billion cubic meters of gas annually (*Hamsharye*, October 10, 2002: 2). As both sides adhered to a policy of not disclosing the details of their October 2002 agreement, including the new agreed gas price, evidence suggests that the Iranian had to accept at least a 9% discount to match the Russian discount.

Despite various reasons for conflicts and periodic tensions and their regional rivalry, both Iran and Turkey have sought to improve their relations. Friendly ties are equally important for the two neighbors, requiring a long period of peace and stability for them to address their numerous economic problems. Permanent tensions and conflicts and their possible escalation to major political and military confrontations will surely not serve their long-term national interests. Despite fluctuations, the two sides have made efforts with some success to reduce tension and expand their bilateral relations. The official visit to Tehran of Turkish president Ahmed Necdet Sezer on June 17, 2002, should be seen as an effort to improve relations. Given the ceremonial status of the Turkish presidency, the visit could not, and was not meant to, help expand ties drastically, a fact reflected in the conclusion of only two minor economic and cultural agreements. However, it was surely important for its political significance, as it indicated the value of friendly ties with Iran for the Turks and their reluctance to follow—on just about every issue of the American policy towards Iran, a member of the "axis of evil"—a policy aimed at its total regional and international isolation. President Sezer stressed this point as he expressed Turkey's "deep interests" in expanding bilateral relations with Iran in political, economic and cultural fields, this is notwithstanding the importance of ties with the United States for the Turks and their NATO membership. However, one should not exaggerate the importance of political moves such as the mentioned visit for the improvement of Iranian-Turkish relations, given the two sides' conflicting interests in Central Asia and the Caucasus.

The absence of a legal regime for the division of the Caspian Sea has not only prevented the full development of its rich oil and gas resources, but also created grounds for hostility and confrontation among its littoral states. Bilateral agreements such as the ones between Russia and its neighboring Azerbaijan and Kazakhstan cannot address the unresolved issue of a legal regime for the Caspian Sea forever. Peace and stability in the Caspian region require a legal regime acceptable to all parties in order to remove a major source of tension and conflict and to create mechanisms for the peaceful settlement of future disputes. Unless such arrangement is reached, the Caspian Sea's rich resources will likely help deteriorate ties between and among its littoral countries, all of which have other reasons for unhappiness with their neighbors. This has clearly reflected in Iranian-Azeri relations since 2001.

Within the context of their military ties with the Americans since March 2001, any major efforts on the part of the Azeris to boost their military capabilities, especially their naval power, will surely provoke a reaction in kind by all other Caspian states, of course, to a varying extent. Given the history of conflict and mistrust between Azerbaijan and its two Caspian neighbors, Iran and Turkmenistan, an arms race will likely put all these countries on a dangerous path. Since there are various sources of grievances and conflicts between and among the three Caspian neighbors, such an arms race could even lead to military confrontation in the southern Caspian Sea. Yet, any arms race will inevitably drag in not only other Caspian states such as Russia, but also certain other countries, namely, the United States and Turkey. As reflected in its new policy towards Azerbaijan in the post-September 11 era, the United States' arming the Azeris and taking sides with them in their disputes with Iran will likely widen the scope of any future regional arms race and military confrontation to include Russia. Having lost its preeminent superpower status, the latter is a dissatisfied regional power sharing Iran's concern about a growing American presence along its borders.

Moreover, Turkey's behavior in Central Asia and the Caucasus, including its expanding ties with Azerbaijan and its efforts to deny Iran any economic and political gain in that country, has made the Iranian regime concerned about Turkey's role in the pace of events in those for Iranians important regions. Given Iran's estranged relations with the United States, whose forces have permanent bases in Turkey, the latter's receiving full American backing in pursuing its national interests in the Caucasus and Central Asia, and, in particular, in the Caspian region, has created a major security concern for Iran. Its leadership fears complete encirclement by enemy states. The expansion of the American military presence along its eastern and southern borders in the post-September 11 era and the prospect of a future pro-American government in Iraq, its western neighbor along with Turkey, have aggravated that fear. Such fear will likely encourage it to strengthen its military forces, in general, and those along its long border with the Caucasus, the Caspian Sea, and Turkey, in particular.

In conclusion, at least three major factors have negatively affected the pace of events in the Caspian region since the initiation of the

American "war on terrorism" in West Asia and its proximity. These are the growing American military presence in Eurasia, the expanding American-Azerbaijani military cooperation and the growing Turkish rivalry with Iran in that region. The resulting deteriorating situation will likely contribute to the creation of a suitable ground for conflicts, including military ones, in the Caspian region. For such a scenario to be avoided, a thaw in Iranian-American relations will be an absolute necessity. Geared to a fundamental change in the nature of the Iranian political system and a major shift in the American foreign policy to West Asia, that development will help ease tension in that region by eliminating sources of friction and conflict in Iran's relations with Azerbaijan and Turkey.

References

Tehran Ettellaat
2002 "Iran and Azerbaijan Expand Their Bilateral Relations." *Tehran Ettellaat*, May 21, p. 2.
"IRANIAN-AZERBAIJANI TENSION PERSISTS"
2001 *Central Asia News (CAN)*, Digest Number 416. August 17, 2001.
PEIMANI, H.
2002 "Caspian Sea Divide No Closer to Closure." *Asia Times* online. Hong Kong. April 19.
2001 *The Caspian Pipeline Dilemma: Political Games and Economic Losses*. Westport, CT: Praeger.
RADIO FREE EUROPE/RADIO LIBERTY (REF/RL)
2002 October 7.
2002 September 27.
2002 August 24.
CENTRAL ASIA NEWS
2001a "Turkey Has Sent a Note to Iran." *Central Asia News (CAN)*, Digest Number 528. September 21.
2001b "Turkmenistan and Ukraine Expand Military Cupertino." *Central Asia News (CAN)*, Digest Number 611. October 15, 2001.
2002 "Tehran Gains from Aliev's Visit." *Central Asia News (CAN)*, Digest Number 1126, May 30.
2002 "Turkey." *Eia Country Analysis Brief*. Retrieved November 20, 2002 (*http://www.eia.doe.gov*).
Hamsahrye
2002 "Zangene: We resolved Our Export Problem." *Hamsahrye* (Tehran), October 10, 2002.

XI. The 'Power of Water' in a Divided Central Asia

MAX SPOOR AND ANATOLY KRUTOV

ABSTRACT

In the not-too-distant future, the former Soviet Central Asia could be confronted with resource-based conflicts or even, as some observers have suggested, with a "water war."[1] Water is *the* scarce commodity in a region that is rich in oil, gas, and mineral resources. Most of the water comes from two rivers, the Syr Darya and the Amu Darya. These feed the Aral Sea, previously the fourth largest inland fresh (actually brackish) water reservoir in the world. These rivers and their tributaries, together, form the Aral Sea basin. Since the 1960s, the Aral Sea has shrunk rapidly in surface area and in volume of water, representing "one of the world's worst ecological disasters." Increased demand for water for irrigation and hydroelectric power by the competing newly independent states, both upstream and downstream, is a potential source of interstate and even interethnic conflict. The latter could occur in the densely populated Ferghana Valley, where various countries such as Kyrgyzstan, Tajikistan, and Uzbekistan share common borders.

Introduction

During the past decade of transition, the newly independent Central Asian States (CAS) have been unable to tackle the root causes of the desiccation of the Aral Sea. This seemingly irreversible process has continued as the total irrigated agricultural acreage has expanded and

---

[*] "Water Wars" is the title of a series of short video documentaries produced by the BBC dealing with conflicts over water in the Colorado River Valley, the Jordan Valley, the Volga River, and the Aral Sea Basin (1991).

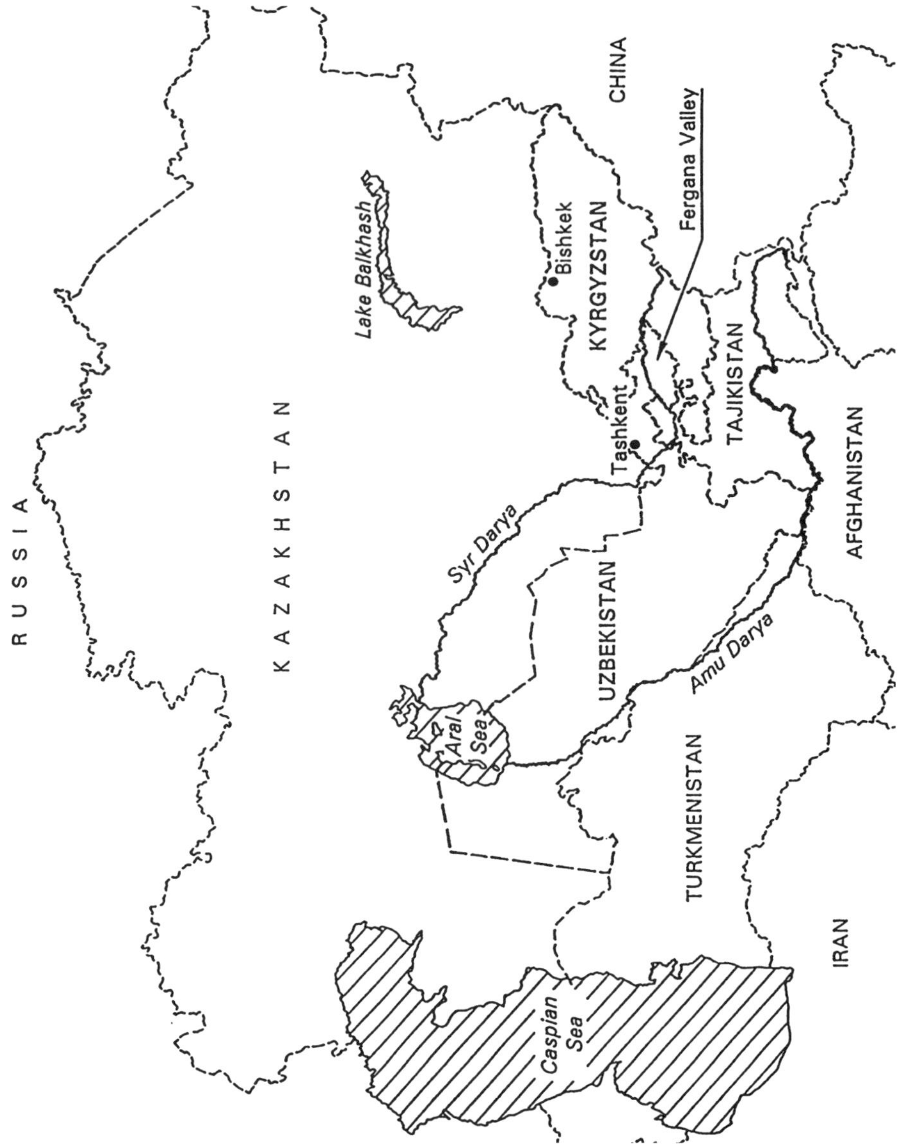

Figure 11.1. Central Asia and the Aral Sea Basin.

hydroelectric power generation has increased. The upstream "supplier" countries, such as Kyrgyzstan, Tajikistan, and Southeastern Kazakhstan, and the downstream "user" countries Turkmenistan, Uzbekistan, and Southwestern Kazakhstan, are increasingly finding themselves competing for *the* scarce resource of the region. Hydroelectric power is particularly important for Kyrgyzstan and Tajikistan, which have no hydrocarbons.

Competition for water comes mainly from agriculture, which is dominated by the cotton crop.

Cotton is a crucial foreign currency earner and a major provider of employment, especially in Turkmenistan, Uzbekistan, Tajikistan, and Southwestern Kazakhstan. Indiscriminate use of water for cotton since the early 1960s has led to the drying out of the Aral Sea and is causing severe environmental problems, such as climate change, soil and water salinity and air, soil and water pollution. The rapidly growing population in the downstream countries, increasing impoverishment in rural areas and the "economic nationalism" that the authoritarian regimes of Central Asia tend to pursue at the expense of regional cooperation are further ingredients for possible tension, social instability, and conflict.

This article will situate these developments in a political economy framework, departing from the Soviet legacy of forced cotton production, which will be analyzed in the second section. It emphasizes that water is an increasingly scarce resource that is under pressure from a variety of economic interests, including agricultural production and hydroelectric power generation. Environmental interests in the form of biodiversity, improved livelihoods of the Central Asian population, and the "voice" of the Aral Sea itself are underrepresented and are losing out.

The third section examines the environmental degradation of the Aral Sea basin in more detail. The drying out of the Aral Sea is having far-reaching consequences for the climate and biodiversity of the surrounding regions, while desert winds are transporting sand and salt over long distances, depositing millions of tons of (often polluted) salts on agricultural land all over the basin area. Due to inadequate and badly maintained drainage systems, water logging is widespread and soil salinity is an increasing environmental problem. The worsening ecology of the region makes living in many areas—such as Karakalpakstan in Uzbekistan and Kyzlorda in Kazakhstan, where poverty and environmental degradation are linked in a vicious downward spiral—quite inhospitable.

The fourth section discusses the institutional framework in which water is managed in the region. A transition has taken place in which the centralized allocation of water by the Ministry of Land Reclamation and Water Resources (*Minvodkhoz*) in Moscow via the Ministries of Water of the five Soviet republics has now been replaced by a new situation following the collapse of the Soviet Union. Since then, upstream and downstream countries must jointly allocate water resources. In the early 1990s, new institutions have been created to oversee this process, and each year agreements are negotiated at presidential level specifying the volume of water that is allocated to each country. At micro-level, there has been little change, except in Kyrgyzstan and Kazakhstan, where Water

Users' Associations (WUA) have been introduced in various regions. In Uzbekistan and Turkmenistan, the main "water user" countries, water is still centrally allocated and managed in the absence of reforms at local level. The principle of "use it or lose it" which was the outcome of centrally planned water allocation, is still in force as there are no sanctions against misuse or incentives for conserving water (Lerman, Garcia-Garcia, and Wichelns 1996: 170).

The final section argues that the environmental situation in the Aral Sea basin is critical. Continued water consumption at the current level and with a low efficiency ratio will lead to the further spread of soil salinity and the irreversible deterioration of the Aral Sea over the next decade. The power of water in a divided Central Asia, therefore may well lead to conflicts and tensions over an increasingly scarce resource within the context of a deteriorating environment in which water is critical. Although many in the region believe that water is "God-given," the current environmental disaster is man-made, and only a reduced and more efficient use of *the* scarce resource, supervised by well-designed micro and macro-institutions and subject to interregional cooperation, will be able to turn the tide.

The Soviet Legacy of Water Management

Most of the area covered by former Soviet Central Asia consists of steppes and deserts. Ever since ancient times, settlements and agricultural activities have only emerged in the traditional oases, which are fed by rivers or underground water reservoirs. During the Soviet period, and especially during and after the forced collectivization of the early 1930s, much of the existing sustainable cropping patterns (using grains, cotton, and fruits) were altered and traditional water management was destroyed and replaced by large-scale surface irrigation systems.

Cotton had already been grown for a very long time, and irrigated areas in the Central Asian plains, such as the Ferghana Valley, were found to have comparative advantages in producing this "white gold." However, since the 1940s and especially since the early 1960s, a cotton quasi-monoculture was introduced on the orders of Moscow (Spoor 1993). Central Asia, thus, increasingly became a peripheral region within the Soviet Union, producing raw materials for the center. Very little cotton processing was carried out in these Soviet republics, and most of the harvest was transported to the central and western parts of the country, where it was used as input for the textile industries. Cotton became a crucial commodity in the political economy of these republics, especially Uzbekistan, which developed into one of the largest cotton producing countries in the world. The power, and, ultimately, the fate of the political elites of some the SSRs (Soviet Socialist Republics) became dependent on the success or failure of the cotton sector. Corruption, the over and underreporting of the cotton output and the forced organization of labor during peak periods became structural features of the economies of the Uzbek, Turkmen, and Tajik SSRs in particular.

Water was desperately needed to rapidly increase the cotton output of the Central Asian region, since hardly anything can grow without irrigation in the desert climate. Water was—or at least seemed to be—available in sufficient quantities, since the Aral Sea basin is blessed with two main river systems, the Amu Darya and the Syr Darya. The Amu Darya, which flows along the south side of the basin, is the larger of the two, with an average annual flow of 73.6 km^3, and a variation of between 47 and 108 km^3. Around 19 km^3 of this volume is generated in Afghanistan, which falls outside the scope of this article, but should play a role in the institutional development of water management in the Aral Sea basin. The famous Karakum canal that runs into Turkmenistan over a distance of more than 1,100 km substantially taps the Amu Darya. The Syr Darya River, which originates from the Naryn and Karadarya rivers that flow through the Ferghana Valley and then turns northwest into Kazakhstan, has an average annual flow of 38.8 km^3, with a variation of between 21 and 54 km^3. Both rivers emerge from the mountain areas of Kazakhstan, Kyrgyzstan, and Tajikistan, and are largely consumed in the downstream areas of Turkmenistan and Uzbekistan. This "differential access" is at the root of tensions concerning the use of this precious resource in the region.

The expansion of the cotton acreage caused an increasing volume of water to be diverted to agricultural irrigation. In the region's largest cotton producer, Uzbekistan, the expansion of cotton cultivation was nothing short of spectacular. Starting from an acreage of 441,600 hectares in 1913, the cotton acreage grew from 1,022,600 hectares in 1940 to 1,427,900 hectares in 1960, and has since increased to even as much as 2,103,000 hectares in 1987 (Spoor 1993: 148).

Cotton became "king" in Uzbekistan, and to some extent also in Turkmenistan and Tajikistan (and to a lesser extent in southwestern Kazakhstan and Kyrgyzstan), and water was the essential ingredient in the success of this forced cultivation policy. The efficiency of water use is very low: canals are unlined, leakage is extremely high, and much of the water does not even reach the fields. As a consequence, progressively less water was available to replenish the Aral Sea, for which approximately 50 km^3 was needed annually to maintain 1960s levels. Very soon, only marginal quantities of water were still reaching the shores of what had once been the world's fourth largest brackish inland water reservoir, and as a result, the Sea started shrinking rapidly in size and volume (see Table 1). In just 30 years (1960-1990), the Aral Sea shrank in surface area to only half its

Table 11.1

The Chronology of Desiccation of the Aral Sea (1960-2010)

Year		Average level (m)	Average area (km^2)	Average volume (km^3)	Average salinity (g/l)
1960		53.4	66,900	1,090	10
1971		51.1	60,200	925	11
1976		48.3	55,700	763	14
1980		45.4	–	602	–
1985		41.5	45,713	468	–
1988		40.1	–	358	–
1990			36,500	330	
	large sea	38.6	33,500	310	~30
	small sea	39.5	3,000	20	~30
1993		37.1	33,642	300	
	large sea	36.9	30,953	279	~37
	small sea	39.9	2,689	21	~30
1998		34.8	28,687	181	~45
1999*			25,600	187	
	large sea	33.4	22,800	168	59
	small sea	39.4	2,700	19	18
2000*			24,003	173	
	large sea	32.5	21,200	149	67
	small sea	38.6	2,700	17	18
2010	(Scenario)	32.4	21,058	~124	~70

Sources: Spoor (1998); Adjusted data for the years 1985, 1998, and the future scenario for 2010, came from the German Aerospace Center (DLR) (*http://www.dfd.dir.de/app/land/aralsee/chronology.html*). The data for 1999 and 2000 are provided by the World Bank office, Tashkent, Uzbekistan.

Note: *The latest data for 2000 are slightly worse than the estimates made by Micklin (1992: 275).

original size (from 66,900 to 36,500 km^2), and its volume went down to a third (from 1,090 to 310 km^3). By the year 2000, this volume was less than a quarter of what it had been in 1960. The latest data on the Aral Sea are close to, and even slightly worse than, the estimates for 2000 made by Micklin (1993: 275) in the early 1990s. The scenario that was calculated for 2010 suggests a somewhat slower process due to the smaller water surface. However, a recent satellite photograph from Uzbek Hydromet Services (July 26, 2002) is close to the scenario for 2010 given in Table 1.

The shoreline of the Aral Sea has withdrawn in some places by more than 100 km, which means that towns such as Muynak (Uzbekistan) and Aralsk (Kazakhstan), which were built with sea-side promenades, are now in the middle of the desert. Due to the continued evaporation and the insufficient inflow of river water, the Aral Sea is not only disappearing and

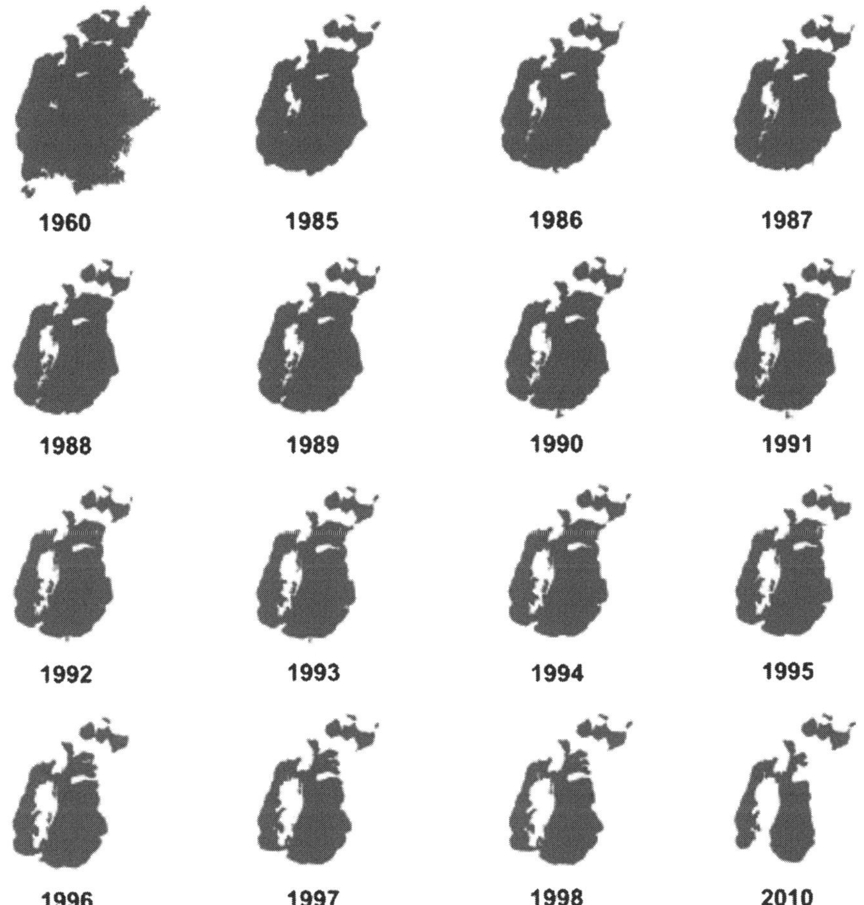

Figure 11.2. Images of a Shrinking Aral Sea (1960-2010). Source: http://www.dfd.div.de/uppl/and/aralsee/chrology.html.

splitting into three different smaller areas (northern, western, and eastern), but it is also becoming a saline sea in which most of the fish population has since died out. Figure 2, which is based on satellite photos and other data, shows the shrinkage of the Aral Sea. Based on current data, it also projects a scenario for 2010, which takes into account the fact that this shrinkage will slow down as less water evaporates due to increased salinity and the substantially smaller water surface.

The data presented above shows that the drying out of the Aral Sea did not stop or decelerate during the decade of transition. Clearly, the newly independent countries of former Soviet Central Asia were primarily concerned with their own survival; employment and the generation of foreign currency for Uzbekistan, Turkmenistan, Tajikistan, and the

Table 11.2

Irrigated Land and Water Use in Central Asia (1990-1999)

	Actual Water Use			Irrigated Areas in the Basin (×1,000 ha)		
	1990	1994	1999	1990	1994	1999
Kazakhstan	11.9	10.9	8.2	702	786	786
Kyrgyzstan	5.2	5.1	3.3	434	430	424
Tajikistan	13.3	13.3	12.5	709	719	927
Turkmenistan	24.4	23.8	18.1	1,329	1,744	1,744
Uzbekistan	63.3	58.6	62.8	4,222	4,286	4,277
Total	118.1	111.7	104.9	7,466	7,965	8,110

Sources: Spoor (1998) for 1990 and 1994; World Bank (2001: 19) for 1999.
Note: The two data sets do not coincide for the earlier years. Moreover, the ICWC reports that all countries, especially the downstream ones, withdrew more water than was allocated to them.

southwestern regions of Kazakhstan and Kyrgyzstan very much depended on cotton. Politically there was no choice, even if the governments had wanted to embark on a more sustainable resource management path, other than to keep most of the cotton production intact, while even expanding irrigated areas, mostly for grain production (Table 2).

As can be seen from Table 2, the irrigated area in the five Central Asian countries increased in the space of a decade from 7.5 to 8.1 million hectares. The expansion of the irrigated acreage in Turkmenistan, even though water consumption remained unchanged, can be explained by the introduction of new grain-producing areas that use less water per hectare. The tendency to expand irrigated areas, which can be seen in the 1990s, is expected to continue. According to a recent report by the International Crisis Group (ICG), Turkmenistan intends to increase its irrigated acreage by 450,000 hectares over the next few years, Kyrgyzstan by 230,000 hectares, and Tajikistan by 500,000 hectares. Not only will this expansion put more pressure on water resources but also a rising ground water table will also occasionally cause problems in the adjacent provinces of neighboring countries (ICG 2002: 3-4).

While water allocation has gradually been decreasing, it was reported that actual water consumption has been going up. The last two years, 2000-2001 (not shown in the table), were years of extreme drought, with a strongly reduced water availability and consumption, causing repeated loss of crops and increased poverty in areas such as Karakalpakstan, Kashkadarya, and Khorezm in Uzbekistan and Dashkhovuz in Turkmenistan. These areas received only very small shares of their water allocations, with disastrous consequences for their agricultural sectors. In 2001, the farmed acreage in Karakalpakstan fell by 44%, and there was an 80% reduction in grain output (ICG 2002: 22).

The twin competing demands for water by agriculture and hydroelectric power have not been contained over the past decade. Water use for agriculture has remained more or less constant, although there has been a shift in the crop mix of the downstream countries, with a slight reduction in the cotton acreage and a strong expansion in wheat, targeted towards self-sufficiency in wheat production. In 1990, the overall cropping pattern for the five Central Asian countries was 40% for cotton and 7% for wheat, while in 2000 this had shifted to 35% for cotton and 30% for wheat (World Bank 2001: 18). This shift involves no reduction in water demand; since wheat is grown using less water per hectare than cotton, the positive impact of the shift is cancelled out by increased water leaching to combat soil salinity and the further deterioration of irrigation systems in the 1990s due to lack of investment. The consumption of water is highly inefficient. Cotton requires around 13,000 m^3/ha, which is substantially higher than in other cotton-producing countries. Land has been used for cotton growing for several decades in a row, with no crop rotation or fallow periods, since irrigated land is extremely limited and pressure to comply with centrally planned output was high. Furthermore, on-farm and off-farm drainage systems are usually weak or nonexistent, which means that water logging is an ongoing problem, thereby increasing soil salinity (see next section).

Environmental Degradation in the Aral Sea Basin

There is a close link between environmental degradation and water, since water is often the "cause and cure" of many environmental problems. In Uzbekistan, for example, the complexity of the water system can be seen from its sheer size. The availability of water resources for sectors within the Uzbek national economy (mainly agriculture, but also industry and human consumption) depends on the operational reliability of a complex water management system. This consists of 5 regional and 53 national reservoirs, primary and inter-farm canals with a total length of 28,000 km, a drainage infrastructure and 1,465 pumping plants with 4,942 pumps supplying water to 2.3 million hectares of irrigated land. Surface irrigation systems are used to irrigate cotton on 2 million hectares of land. Many of the water management systems, however, are old and well beyond their service lifetime.

There are many environmental problems in the Aral Sea Basin, some of which are serious in themselves, some of which affect the current state of the Aral Sea, and others that are an indirect spin-off of the drying out of the Sea. Clearly, from the start of this analysis, water is not actually in short supply, but its highly inefficient use causes shortages during various parts of the year, especially in regions within the downstream countries.

In the 1970s and early 1980s, a mega-plan was drawn up to divert the north Russian Ob and Irtysh rivers through the steppes of Kazakhstan and into the Central Asian heartland over a distance of 1,500 km. This idea originated in Moscow and was supported by the Central Asian countries, where the local elites saw the advent of Siberian water as a panacea for their emerging problems. However, the plan was shelved after Gorbachev came to power. Not only was there less support for providing Russian water to "Islamic" Central Asia, but it would also have caused environmental disaster elsewhere, as pointed out by the increasingly vocal Russian environment movements. Interestingly enough, the option to divert Siberian waters to Central Asia never completely disappeared from the minds of the policymakers. Even in UNESCO (2000), the plan is still being mentioned as a possible way of redressing water levels in the Aral Sea.[1]

The environmental problems in the Aral Sea basin are as follows. First, a growing proportion of irrigated land in Central Asia is now more or less saline. The major cause of this should be sought in the lack of crop rotation, since in most places, cotton has been a monoculture for many decades. Furthermore, inadequate and archaic drainage systems cannot handle the serious problems of water logging and the upward flow of minerals. It can therefore be argued that there is a salt crisis as well as a water crisis in Central Asia. Soil salinity tends to reduce agricultural yields and to increase water consumption, since farmers get into the habit of water leaching to wash the soil, which consumes large quantities of water at the start of the season.

Salinity is even more severe in the downstream areas of the basin, since the rivers and drainage wash down salt canals, and there is hardly any natural drainage in these relatively flat areas. Table 3 shows that soil salinity increases from south to north in the basin. The upstream countries Kyrgyzstan and Tajikistan have low rates of salinization, while severe soil salinity is seen in the lower reaches of the Amu Darya (Khorezm, Karakalpakstan, and Kashkadarya in Uzbekistan, and Turkmenistan), the Syr Darya (Southern Kazakhstan), and the Zerafshan rivers (Bukhara

[1] In March 2002, there appears to have been some joint Russian-Uzbek interest in reviving the plans. For the Uzbek government, it would provide easy access to additional water resources, while for the Russian government it could mean a "new lever of influence." However, it is questionable, taking into account the enormous investment costs involved and the possible environmental impact, whether these reports should be taken seriously (ICG 2002: 27). A "design and science" lobby with the aim of milking donors and producing feasibility reports has also driven the mega-project.

Table 11.3

Soil Salinity of Irrigated Lands (1999 Level)

Basin/Country	1999 Irrigated Area (×1,000 ha)	Irrigated area by FAO soil salinity class** (×1,000 ha)				
		None 0-2	Slight 2-4	Moderate (dS/m) 4-8	Severe 8-15	Very Severe >15
Syr Darya Basin						
Kazakhstan*	786	157	330	199	84	16
Kyrgyzstan	424	302	110	7	4	1
Uzbekistan	1,876	797	618	332	115	14
Total	3,086	1,257	1,058	537	203	32
Amu Darya Basin						
Tajikistan	747	467	219	44	14	2
Turkmenistan	1,714	53	376	847	389	49
Uzbekistan	2,372	650	867	592	228	35
Total	4,832	1,170	1,462	1,482	632	86
Aral Sea Basin	7,919	2,427	2,520	2,020	834	118

Source: World Bank (2001: 106).
Note: *1994 values; **Average root salinity in dS/m (*deci-Simens per meter*). The total irrigated areas of various countries given by the FAO are not precisely identical to those reported by the World Bank, especially for Tajikistan. The differences (191,000 ha in total) remain unexplained.

in Uzbekistan). The regional disparities are quite wide, from 90-94% of the land in the Karakalpakstan, Khorezm and Bukhara provinces of Uzbekistan is salinized compared to 60-70% in Kashkadarya province and only 5% in Samarkand province.[2]

There has also been a marked increase in soil salinity in the downstream "user" countries in the 1990s during the first decade of transition (see Figure 3). This increase has been estimated as 30% for Uzbekistan, 24% for Turkmenistan, and 18% for Kazakhstan, while in Kyrgyzstan and Tajikistan, soil salinity has diminished (World Bank 2002: 10-15).

Soil salinity might seem to be "merely" a technical problem, yet it has major social and economic consequences. Salinity can negatively affect crop yields and, hence, the income of farm households. It is known that only tolerant plants will grow satisfactorily on soils with moderate salinity. On severely saline soils, only a few highly tolerant plants will flourish. If

[2] The severe environmental problems in a relatively new area such as Kashkadarya, where much of the agricultural land was taken into production in the 1970s, were visible during a field visit by the first author of this article. He was part of a UNDP mission to Uzbekistan on Macroeconomic Policy and Poverty Reduction in September 2002.

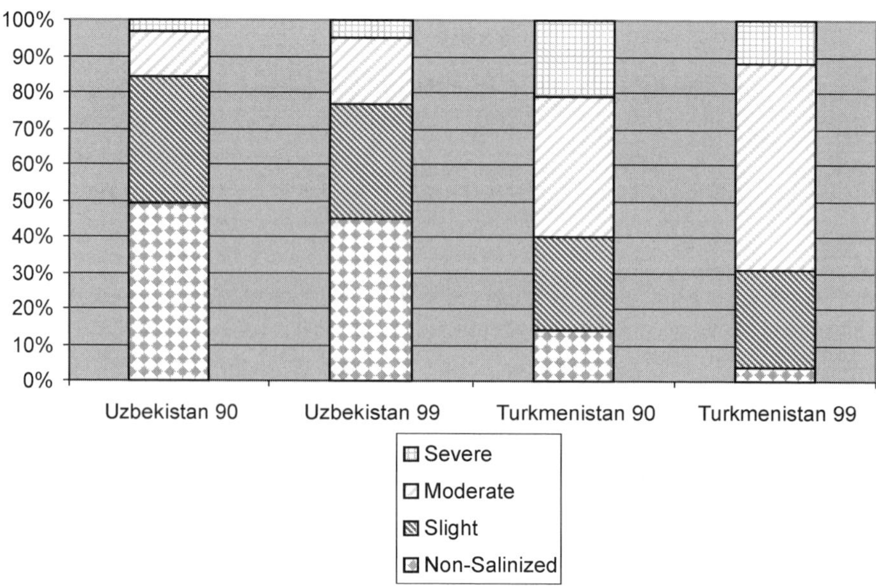

Figure 11.3. Land salinity Uzbekistan and Turkmenistan.

soil salinity is above a certain threshold value, yield losses can easily range between 10-50 percent.

Apart from increased soil salinity, the downstream river water is also increasingly saline, which affects agricultural yields and the quality of water in the aquifers. Average salinity levels are 0.45-0.60 g/l in the upper reaches of the two rivers. In the southern Amu Darya, water salinity increases in the middle and lower reaches (0.60 g/l in Termez, on the border between Uzbekistan and Turkmenistan, and 1 g/l near the Aral Sea). In the northern Syr Darya, it is even slightly higher: 1.1 g/l at the outlet of the Ferghana Valley, rising to 1.4 g/l further on (World Bank 2001). Most of the salt in the river comes from the drainage systems that discharge irrigation water back into the river, while the rest is deposited in desert "sinks." The total amount of salt transported in the two rivers has increased from 55-60 million tons in the mid-1960s to 135-40 million tons in the 1990s (*Ibid*). Finally, water pollution is caused not just by salt but also by nutrients. The intensive use of fertilizers and pesticides in cotton production, which has diminished during the 1990s, mainly for financial reasons, caused severe chemical pollution of the rivers in the basin with high concentrations of several toxic substances (Spoor 1998).

Second, the smaller quantities of water that actually flow as far as the deltas of both main rivers and the aforementioned increased water salinity in those areas have had devastating consequences for biodiversity. Part of the rich flora and fauna of these wetlands, which were also the breeding grounds for many birds and fish in the basin system, has disappeared. In the Amu Darya delta, the unique *tugai* forests have suffered enormously (Spoor 1998). UNESCO has estimated that around 30,000 ha of lakes and bogs have almost entirely dried out in the same delta. Much of the fish population in the Aral Sea itself has died out, with of course dramatic consequences, of cause, for the populations of the surrounding towns, which were largely dependent on catching fish. Again, loss of biodiversity is far from being an abstract issue in that it has had very negative consequences on employment, income generation, and health.

Third, as noted in Figure 2, there is a rapidly expanding area of exposed seabed in the Aral Sea. This is found on the shores as the Sea has shrunk, but also on the land that separates the deep western and shallow eastern sea, which consists largely of salt. With desert storms blowing during approximately three months per year, large quantities of salt are being deposited on surrounding agricultural lands. Because of desertification, windstorms move an increasing amount of salt in Central Asia, especially near the Aral Sea. An estimated 1.5-6.5 tons of particles per hectare, of which 260-1000 kg/ha is toxic salts, is transferred annually from the dried bed of the Aral Sea (an estimated 1.5 billion tons of salt covering 3.5 million ha) to an expanding area. Wind erosion also carries salts in areas such as the Central Ferghana Steppe. Bukhara Province in Uzbekistan receives a total of 300-400 kg/ha of salt-laden aerosols annually, of which 40-50% comes from the dried bed of the Aral Sea some 300 km distant away (World Bank 2002). This "salt pollution" not only has a negative impact on agricultural production, but also on human health. The number of cases of respiratory disease is relatively high, especially in the downstream areas close to the Aral Sea; this was referred to in the aforementioned BBC documentary as a form of "environmental AIDS."

Fourth, the shrinking of the Aral Sea has contributed to climate change in the surrounding areas. The planting season has shortened, the number of frost-free days has decreased, and summer temperatures (in the desert) are slightly higher. Previously, the huge size of the Aral Sea helped to regulate temperatures, and its drying out has had a negative impact on this process.

There is a clear relationship between the environmental degradation of soil and water and the increased incidence of poverty, especially downstream of the basin, as mentioned earlier. The environmental stock/capita (Z/N, where Z = environmental stock, including water and land resources and N = the total population in a particular region) has dropped due to the deterioration of the first variable and an increase in the population in the downstream areas (Spoor 1998). As demand for water increases and supply either remains constant or diminishes due to competing use (hydroelectric power, for example, which has become crucial for Kyrgyzstan), tensions can mount between countries, as recently occurred between Kazakhstan and Kyrgyzstan, and previously between Kyrgyzstan and Uzbekistan with regard to water management in the Ferghana Valley.

Water Management, Institutions, and Reforms

Centralized and regionally focused Soviet water management of the basin was abandoned in the wake of the collapse of the USSR. As UNESCO (2000: 19) concluded in its *Water-Related Vision for the Aral Sea Basin for the year 2025*:

> Regional co-operation was needed to restore a basin-wide mechanism and perspective in water and salt management. Following the independence of the Central Asian republics in 1991, Soviet central authority over basin development gave way to that of five sovereign governments acknowledging distinct interests. Management of water resources came to be undertaken according to national perspectives. If the interests of water users were addressed somewhat inefficiently, the interests of the Sea, deltas and wetlands were nearly orphaned.

Indeed, with national interests prevailing, the voice of the Aral Sea, the anonymous sixth player in the field of Central Asia, became even weaker. However, new national and regional organizations appeared during the first decade of transition, which would fill the institutional vacuum that remained after 1991, at least at macro-level. In February 1992, soon after the independence of the five Central Asian states, a joint agreement was reached establishing an Interstate Commission for Water Coordination (ICWC), which became responsible for the water allocation for the five former Soviet states in the Aral Sea basin. Even so, there were still substantial weaknesses in the agreement, such as the failure to address the problem of water quality or the potential conflict situations that might arise.

A subsequent agreement was signed in March 1993, establishing regional organizations such as the Interstate Council on the Aral Sea (ICAS), an advisory body for the five regional governments, which had an Executive Committee and a Secretariat. An International Fund for the Aral Sea (IFAS) was also established to finance the activities of ICAS, and in 1994, a Sustainable Development Commission (SDC) was formed, which focused on environmental protection and socioeconomic development (UNESCO 2000: 51). A few years later, ICAS and IFAS were merged to form a new IFAS, supported by a high-level board of deputy prime ministers.

The water management of the two main rivers at basin level is undertaken by two Water Basin Associations (*Basseynoe Vodnoe Ob'edinenie*, the Amu Darya BVO and the Syr Darya BVO). These organizations, which had existed since the 1980s, were given the complex task of managing the same water resources in a basin that was now covered by five newly independent countries (and one that is not represented, namely, Afghanistan). The Amu Darya BVO has under its mandate the water resource systems of the Pyandj Vaksh, Kafirnigan, and Amu Darya rivers from their source to the Aral Sea, including distribution facilities, pumping systems, canals, communication infrastructure, and power supply installations. It has offices in Kurgan-Tyube (Tajikistan), Turkmenabat (Turkmenistan), Urgench (Uzbekistan), and Tahkiatash (Karakalpakstan in Uzbekistan). The Syr Darya BVO manages the flow of the Naryn, Karadarya, Chirchik, and Syr Darya rivers up to the Chardara reservoir. It has offices in Tashkent, Charvak, Gulistan, Chirchik, and Uchkurgan (Uzbekistan). Finally, there is a separate Aral Syr Darya BVO, which is a purely Kazakh agency with offices in Kyzlorda and Shymkent (both in Southwestern Kazakhstan). The BVOs do not control or manage drainage as this falls under the responsibility of the national water authorities, while local drainage and desert sinks are managed by local institutions.

Apart from the complex and differentiated jurisdiction of the BVOs, their main problem is that agreements do not have the status of international law, and that they themselves are not even recognized by national legislatures, which means that they lack authority over the national use of resources (Horsman 2001: 73). Shortage of funding has also hampered their operational capacity. Funding obligations are linked to water allocation shares, but it seems that only Uzbekistan and Turkmenistan have complied regularly in recent years. Most of the financial contributions are used for direct operational costs, and insufficient quantities are left for capital repairs or replacement investments. During the 1990s, there was therefore a steady deterioration in the water management infrastructure (Hogan 2000; World Bank 2001: 22).

IFAS was supposed to be financed by annual budget allocations of 1% of each individual country's GNP. However, although these allocations were later reduced to 0.3% for Kazakhstan, Turkmenistan and Uzbekistan, and to 0.1% for Kyrgyzstan and Tajikistan, the countries have been very slow with their payments, thereby limiting IFAS' financial capacity (Horsman 2001: 73). Moreover, during the Almaty Summit of Heads of State in February 1998, it was agreed that the funds set aside to tackle the Aral Sea crisis would be utilized on their individual territories and they would not transfer funds to the central IFAS account.

Throughout the decade, the five Central Asian presidents have at various times promised closer cooperation and a future sustainable management of water (and environmental) resources in the region, such as in Nukus

(Karakalpakstan) in 1995, and also most recently in Dushanbe (Tajikistan) in October 2002. Nevertheless, tensions have remained between the countries, especially between those upstream and downstream, and also between the two main user countries, Turkmenistan and Uzbekistan.[3] At the end of the first decade of transition, the five Central Asian presidents signed the Ashgabad Declaration (April 9, 1999), in which they clearly acknowledged the need "to work out joint measures for the realization of a regional strategy and concrete actions for the rational use of water resources of the region, based on an ecosystem approach and integrated principle of water management" (UNESCO 2000: 53).

The Ashgabad Declaration reflected an important foundation of regional cooperation. However, in reality, the political economy is inwardly focused, and national interests weigh more heavily than transboundary ones. In an interview in Kazakhstan, the then ICAS/IFAS chairman Almabek Nurushev asked:

> Who will have the bravery to tell the farmers: 'reduce production and perish'? It will take quite some time to have rational production systems, where instead of cotton and rice, in some places the farms will produce wine and other products. Nevertheless, currently all states want to be independent in the production of grains, although nature defines the production of which commodities can be grown in each place. In fact, it is too hot during the summer in Turkmenistan and Uzbekistan to produce grains. At the same time, cotton is the foreign exchange earner. The question is a very important one, and has to be faced in the very near future. (Spoor 1998: 427)

Clearly, it is precisely this scenario that unfolded in the following years, with no concomitant reduction in overall water use. Institutional arrangements for water management have not changed in the downstream countries Turkmenistan and Uzbekistan. Irrigation is carried out mainly by gravity methods. The water supply is organized through planned allocations, mostly to the existing and predominantly *shirkat* farms, the successors of the former *sovkhozy* and *kolkhozy*. Only very small water charges have been introduced, and these are nowhere near enough to finance the costs of operation and maintenance. One example of these symbolic payments was given by Wegerich (2000: 5), who noted that a Water Users' Association in the Syr Darya *oblast* paid a water tax of

[3] Since both countries are highly dependent on the Amu Darya (Turkmenistan is entirely dependent), they compete for water consumption for the same economic activity, namely, cotton. There are many places where tensions can arise, such as in the Kashkadarya pumping cascade, which takes water from Turkmenistan through a series of pumps to southwest Uzbekistan. The World Bank has a project in place to renovate the cascade, but a final decision to implement has not yet been reached between the two countries. Another contentious issue is a large drainage canal leading from Urgench (Uzbekistan) into Southwestern Turkmenistan.

0.11 Soum/m^3, while, according to Uzbek experts, the real price of water was 0.9 Soum/m^3.

A new externally funded project, "Integrated Water Resources Management in the Ferghana Valley," will introduce WAUs to pilot areas, using the experience gained by these local institutions in Kyrgyzstan and Kazakhstan over the past five years (Dukhovny 2002). However, it should be noted that agricultural reform in Turkmenistan and Uzbekistan is far less advanced than it is in neighboring countries, which makes locally or even privately managed water systems more problematic. Early experiments with WAUs in Uzbekistan were complicated by the fact that much of the farm produce (cotton and wheat) was still covered by the obligatory state order system. As a result, farms and even "private farmers" (Spoor 2003) still have little room for maneuver when it comes to deciding the allocation of inputs and the choice of crop mix, and this makes it almost impossible to provide incentives for water savings.

The current irrigation and drainage infrastructure is in a fairly shaky state, since for many years now hardly any investments have been made. Many of the irrigation canals are unlined (i.e., 34,200 of the total 47,700 km), causing a high level of seepage. A recent survey of farms in Uzbekistan found that 60% of the water supplied to them did not reach the fields, and it characterized the deficiencies in management, leakage, and similar losses as very significant (World Bank 2001: 21). On-farm drainage systems in the Basin are in an even worse state, and drainage canals are filled with weeds and silt due to insufficient cleaning. Finally, one aspect that is often forgotten is the high cost of irrigation in the newly developed agricultural areas, where cascades of water pumps sometimes have to bring water up to levels of 100-200 meters. The pumps are often old, and electricity costs have gradually risen.

As indicated above, much still needs to be done to improve cross-border water legislation and the legal acceptance of regional agencies making decisions concerning water allocations and conflict management. However, fundamental institutional change also needs to be based on an appropriate national legal framework, a clear definition of property and user rights, the introduction of the "polluter pays" and "beneficiary pays" principles, and water pricing.

In Uzbekistan, the latter will be introduced on a two-phase basis. Payments for irrigation in the agricultural sector will be introduced in two stages. During the first stage (2002-2004), producers who are not within the state order system (of cotton and wheat) will begin paying. By the end of this period, 30-35% of the costs of irrigation will be covered, while during the following stage (2003-2005) all producers (and therefore consumers of water) will have to be covered. By the end of the period, only 15-25% will still be compensated.

Economic mechanisms for conserving water in different sectors of the economy will take the form of fines for the excessive use of irrigation water (compared with the allocated volume) and the establishment of special incentive funds (using a portion of the water fees) to encourage the reduction of water use per hectare. However, the effectiveness of these measures will depend largely on the progress in agrarian reform, since the *shirkat* farms are currently still entangled in systems of obligatory procurement for cotton and wheat with low administrative prices, political interference, and "missing markets."

Conclusion: Cooperation or Conflict over Water?

There is no doubt that the seriousness of the Aral Sea environmental disaster has now started to penetrate the minds of the Central Asian authorities. However, this is clearly not enough, given that their response is confined to strictly defined national interests. The planned reduction of water quota allocated to the countries in the Basin (as agreed in 1993) has not really materialized, since irrigated areas are still being expanded and water efficiency has worsened rather than improved. As analyzed above, there is also what can be termed the "salinity trap," namely, that agricultural enterprises in areas with increased salinity will start leaching farmland, thereby using more water than before and entering into a vicious circle of environmental degradation. Moreover, current water allocation quotas are not much lower than they were before independence and the volume of water reaching the Aral Sea shores is still negligible—and in the dry years 2000 and 2001, it was in fact nil. The drying out of the Sea therefore seems to have entered into a stage that will be irreversible if no dramatic changes are made at national and regional level, and especially at the macro- and micro- and the transboundary basin level.

A continuation of the water shortages, the increased soil salinity, the drying out of the Aral Sea and its disastrous environmental spin-offs (salt storms, climate change, diminished biodiversity, worsening human health conditions) can also fuel tensions between (and within) countries. In the mid-1990s, some analysts were already warning of possible resource-based conflicts. And while new institutions and organizations have replaced the central plan directives from Moscow, the potential for future conflict remains:

> In Central Asia, regional tensions may be enhanced by current water allocation practices. In recent years, Central Asia has experienced an increase in irredentist activities and inter-ethnic conflicts. Competition over natural resources may intensify such irredentist sentiments, with some

> viewing escalating future inter-ethnic confrontation in Central Asia as being driven in part by water allocation problems. (Smith 1995: 353)

Horsman (2001) notes that there are various examples of recent conflicts between the Central Asian states. For example, in 1998 Kyrgyzstan concluded agreements with Kazakhstan and Uzbekistan to release water for cotton irrigation in these (downstream) countries instead of keeping more for the generation of hydroelectric power. This water was traded for energy supplies (coal, gas, and mazout, a concentrated oil product). However, in 1997-98 there were fierce conflicts over these "water-energy swaps," which led to harsh words between governments and threats to cut off supplies. Following insufficient (or non-timely) "payments" in these barter agreements from the Kazakh and Uzbek side, the Kyrgyz decided to keep more water in the main Toktogul Reservoir during the summer. This caused water shortages in the downstream areas during the peak period in the irrigation season.

In the winter, more water was released due to increased electricity production, causing winter flooding in the western part of the Ferghana Valley and further downstream along the Syr Darya River. The vicious circle was then completed when the Uzbek government retaliated by cutting the gas supply during the winter of 1999-2000. Finally, in July 2000, there was a serious shortage of water in Southern Kazakhstan (Shymkent and Kyzlorda, where most of the cotton is grown) when Kyrgyzstan cut supplies (because Kazakhstan was not keeping its side of the water-energy swap agreement), and Uzbekistan used more water to combat a severe drought. Although the energy-water swaps were a first step towards multilateral management and agreement, it seems that much of the bilateral accords failed due to default. Lack of trust is also typical of the fate of these agreements, which follow the perverted logic of a prisoner's dilemma. The costs of individual behavior (speculating on the noncompliance of the partner or opponent) are higher than in the case of transparent cooperation.

There are also disputes concerning the Amu Darya, notably between Uzbekistan and Turkmenistan. These disputes primarily involve new plans and infrastructure being developed in one country and being viewed with suspicion in the other. In 2000, Turkmenistan began the construction of a huge desert sink known as Golden Century Lake. The Turkmen maintain that it will be filled only with drainage water, but the Uzbek suspect that in the future, the lake will also take more water from the Amu Darya, thereby reducing the Uzbek allocation, while more water to this lake will also mean that there will be less available for the Aral Sea. The project also appears to have an ethnic or irredentist connotation, since it has been reported that a large number of ethnic Uzbeks in Turkmenistan will be resettled in the Karakum desert (ICG 2002: 25-26).

Another problem is that when the civil war ended, Tajikistan claimed that it had a very small share of the water in the Amu Darya and that it intended to increase this allocation. It wants to expand its irrigated acreage and needs water to do so. Tajikistan is also seeking international finance to complete the Rogun Dam on the Vakhs River. This would bring it into conflict with Uzbekistan since a new large reservoir would put Tajikistan completely in control of the water supply to Uzbekistan.[4] In view of the balance of power in the region, the latter will never allow this to happen. Finally, now that Afghanistan is entering a new era of peace and reconstruction, it will need water to develop its own agricultural sector. In the near future, it is therefore likely to demand new allocations of water, mainly from the Panj River. Once again, it is Uzbekistan that will suffer if Afghanistan uses more water (ICG 2002: 27), which could, therefore, give rise to new tensions.

In spite of the bellicose language that is bandied around, and even the threat of military intervention,[5] Horsman (2001) does not believe that there will be an armed conflict since, in all these cases, the governments have reached a compromise and negotiated a solution. Even so, these water resource-based tensions seem to occur increasingly more often. Closer international cooperation between the Central Asian countries is crucial for diminishing these tensions, especially if each country is prepared to relinquish some its national authority to transboundary-based regional water and environmental agencies, which can then operate under international water and environmental law. International assistance also remains crucial, as noted by Hogan (2000):

> Central Asia has not ignited in the wide-scale resource war that some experts predicted. Early intervention—and large side payments—by international donors may stave off conflict in the near term. Nevertheless, as long as the region's leaders insist on making unilateral decisions that affect their neighbors, water will still remain a potential source of conflict in Central Asia.

---

[4] Tajikistan already controls around 40% of the flow of the Amu Darya, through the Nurek Reservoir. It now wants to build a hydroelectric power station near this dam (at Sangtuda). The Uzbek government is not against the plan, but it is highly critical of the Rogun Reservoir plan. It may simply be that the costs are prohibitive and international donors—who are already wary of large dam projects—see the project as potentially sensitive.

[5] He noted (2001: 76) that it was reported in 1996 that Uzbekistan had drawn up tactical plans to take the Toktogul Dam by force. While this was never confirmed, it is not impossible, since part of its water supply passes through the reservoir on Kyrgyz territory, and this is seen as a matter of national security.

The tendency to conclude agreements and resolve conflicts in a bilateral rather than a multilateral setting, especially by Turkmenistan and Uzbekistan, is not a good sign. However, the improvements made in the functioning of the regionally operating BVOs and other agencies, with the assistance of multilateral and bilateral donors, are moderately promising. The main bottleneck remains the fact that all the Central Asian governments, especially those of the downstream countries, remain very strongly focused on their own national interests and are reluctant to make compromises in this seemingly zero sum game.[6] Current debates are still dominated by technical solutions and ignore the fact that only institutional and political change can contain the environmental crisis and potential conflicts.[7]

References

DUKHOVNY, V.A.
2002 "Report on Project Inception Phase." Project "Integrated Water Resources Management in the Ferghana Valley." Tashkent (ICWC), January.

GLANTZ, M.H. (ED.)
1999 *Creeping Environmental Problems and Sustainable Development in the Aral Sea Basin.* Cambridge: Cambridge University Press.

LERMAN, Z., GARCIA-GARCIA, J. AND WICHELNS, D.
1996 "Land and Water Policies in Uzbekistan." *Post-Soviet Geography and Economics* 37(3): 145-74.

HOGAN, B.
2000 "Central Asian States Wrangle over Water." November 7. Retrieved November 20, 2002 (*http://www.eurasianet.org*).

HORSMAN, S.
2001 "Water in Central Asia: Regional Cooperation or Conflict?" in Allison, R. and Jonson, L. (eds.), *Central Asian Security: The New International Context,* pp. 69-94. London: Brookings and Royal Institute of International Affairs.

ICG
2002 "Central Asia: Water and Conflict." Asia Report No. 34, May. Osh/Brussels: International Crisis Group.

KRUTOV, A.
1999 "Environmental Changes in the Uzbek Part of the Aral Sea Basin," in Glantz, M.H. (ed.), *Creeping Environmental Problems and Sustainable Development in the Aral Sea Basin.* Cambridge: Cambridge University Press.

MICKLIN, P.P.
1992 "The Aral Sea Crisis: Introduction to the Special Issue." *Post-Soviet Geography* 33(5): 269-82.

SMITH, D.R.
1995 "Environmental Security and Shared Water Resources in Post-Soviet Central Asia." *Post-Soviet Geography* 36(6): 351-70.

SPOOR, M. (ED.)
2003 *Transition, Institutions and the Rural Sector.* Lanham and Oxford, Rowman & Littlefield, Lexington Books.

[6] It is seemingly a zero sum game since the available water supply could substantially increase following an improvement in water efficiency (through water savings, for example).

[7] Unfortunately, the debate was recently fueled by a technocratic and not very realistic UNESCO (2000) report, which presented a *Water Related Vision for the Aral Sea for the year 2025.*

1998 "The Aral Sea Basin Crisis: Transition and Environment in Former Soviet Central Asia." *Development and Change* 29(3): 409-35.

1993 "Transition to Market Economies in Former Central Asia: A Comparative Study of Uzbekistan and Kyrgyzstan." *The European Journal of Development Research* 5(2): 142-58.

UNESCO

2000 *Water Related Vision for the Aral Sea for the Year 2025.* Paris: UNESCO, in cooperation with the Scientific Advisory Board for the Aral Sea Basin (SABAS).

WEGERICH, K.

2000 "Water User Associations in Uzbekistan and Kyrgyzstan." *Studies on Conditions for Sustainable Development.* IWMI (International Water Management Institute).

WEGERICH, K. AND K. BEKTEMIROV

2001 "Food Self-Sufficiency or Food Sovereignty for Uzbekistan: An Analysis in the Context of Water Scarcity." *OCEES Research Paper No. 23.* Oxford: Oxford Centre for the Environment, Ethics & Society.

WORLD BANK

1996 "Developing a Regional Water Management Strategy: Issues and Work Plan." *SBP Technical Paper Series.* Washington: World Bank.

2002 *Irrigation in Central Asia: Where to Rehabilitate and Why.* Main Report, Vol. 1. Washington: World Bank.

2001 Water and Environmental Management Project. 2001. Sub-component A1. National and Regional Water and Salt Management Plans. Regional Report No. 2. Phase III Report—Regional Needs and Constraints. Supporting Volume. Tashkent: World Bank.

XII. A Transnational Policy for Conflict Reduction and Prevention in the South Caucasus

ROBERT M. CUTLER

ABSTRACT

This article examines conflicts in the South Caucasus with a view towards means for their interdependent resolution. It begins by reviewing briefly in succession situations in Georgia: Abkhazia, Ajaria, Javakhetia, and Tskhinvali (South Ossetia). A comparative qualitative analysis then follows that is heuristic rather than definitive. The situation in Mountainous Karabagh is juxtaposed to this, and complicating factors are identified. On that basis, a policy initiative for the South Caucasus is described, building upon the extensive considerations previously elaborated on in a report by the Centre for European Policy Studies. A focus on nongovernmental actors in particular leads to reflections on how to create potential transgovernmental and transsocietal sociopolitical coalitions for conflict reduction and prevention. Specifically, possibilities are considered for moving toward an institution such as a transnational Assembly for Regions and Peoples of the South Caucasus. Issues of institutional design are considered and assessed on the basis of existing comparative work on international parliamentary formations.

Ethnic Disputes and Conflict Resolution

Tishkov (1997) has decisively demolished the applicability of many general Western theories of ethnic conflict to the former Soviet area. He indicts Western scholars also for an over-dependence upon quantitative data from official sources, without attention to how those data were collected or aggregated. Most notably, he has empirically demonstrated, by reference to Soviet census methodology, the fallacies of relying upon categories of nationality derived from Soviet census data (see Lijphart 1980, for a related, general warning about drawing conclusions about intersubjective phenomena from subjective individual-level data). As Director of the Institute of Ethnology in Moscow since the Gorbachev era and Minister of Nationalities under Yeltsin in the mid-1990s, Tishkov arrives at propositions that he intends to be policy-relevant. He does this by establishing categories of variables under the rubrics "out of conflict," "from tension to violence," and "governing ethnicity in the non-violent stage" (cited in Cutler 2000).

His conclusions inform the further empirical examination here undertaken. Affirming Tishkov's criticisms of Western approaches, this research adopts a qualitative methodology, using the "most similar systems" method (Meckstroth 1975; Skocpol and Somers 1980). The four intra-Georgian conflicts enumerated below provide a series of pair-wise comparisons. Indicative binary variables that operationalize concepts from Tishkov's concluding chapters include the presence or absence of a local strongman, of an irredentist ethnos or a militant Diaspora, of the feasibility of a federal solution, and of a majoritarian ethnic group on the ground. The "coding" of these variables and the necessarily heuristic analysis of the ensemble follow a brief narrative background to the conflict situations themselves.

Abkhazia

In 1988, an organization called the Abkhazian Forum proclaimed Abkhazia independent from Georgia, provoking military clashes. In 1990 the Supreme Soviet of Georgia overruled a formal declaration of independence adopted a few days earlier by the Supreme Soviet of Abkhazia. In 1992, the Russian Federation mediated the first unsuccessful ceasefire agreement. The Abkhaz rebellion festered through the fall and winter of 1992-1993, during which time Eduard Shevardnadze won a landslide presidential victory in Georgia. In August, the UN Observer mission in Georgia (UNOMIG) was created. In mid-September, after UN monitors began to arrive, the ceasefire was massively violated to the advantage of the Abkhaz, with strong evidence of complicity by Russian military staff. Ethnic cleansing of the non-Abkhaz populations of Abkhazia during and after the fighting created nearly 300,000 internally displaced persons in Georgia.

In December 1993, a "Memorandum of Understanding between Georgia and Abkhazia" was agreed on in Geneva, followed in April 1994 by a "Declaration on Measures for a Political Settlement of the Georgian-Abkhaz Conflict." (The latter is the only official document that discusses possible constitutional arrangements and power sharing.) Under the terms of international arrangements agreed to by the parties, Russia has the authority in Abkhazia to convene meetings with the conflicting sides and to motivate the activities of a variety of multilateral forums under the aegis of the Commonwealth of Independent States and also the United Nations, the latter including Friends of the UN Secretary-General for Georgia (FOG), which comprises France, Germany, Russia, the United Kingdom, and the United States.

The Sukhumi authorities have never backed off from their demand that Georgia recognize Abkhazia's independence as a precondition for any formal negotiations. They contend that these should culminate in an Abkhazian sovereignty within an equal federation with Georgia. Georgia, for its part, has refused to consider any settlement other than an Abkhazia within Georgia. Georgia remains willing to grant Abkhazia a large degree of autonomy within a federal Georgia, exceeding even the degree of Tatarstan's autonomy within the Russian Federation, but the Abkhazian leadership will negotiate nothing other than the details of independence. In November 1997, under the UN's aegis, the Coordinating Council of the Georgian and Abkhaz Parties was created, with the participation by the OSCE and the Russian Federation. It comprises three working groups: military security, refugee problems, and economic cooperation and development. Since then a *modus vivendi*—but little real progress towards a political settlement—has been achieved.

Ajaria

Ajaria, an Autonomous Republic inside Georgia under the Soviet regime, retained de facto autonomy after 1991 even though Georgian independence was established as a unitary state without autonomous subunits. Aslan Abashidze has run Ajaria since the early 1990s as President of its Supreme Soviet and leader of its dominant political force, the Revival Party. For the October 31, 1999, parliamentary elections, opponents to Shevardnadze's rule throughout Georgia largely coalesced around Abashidze's party, although many of Shevardnadze's opponents were in Tbilisi. Shevardnadze's party, the Union of Citizens of Georgia, won a solid majority of the seats.

In mid-February 2000, Abashidze split the opposition coalition by filing papers to oppose Shevardnadze in the presidential election. Shevardnadze held talks with Abashidze, agreeing to a division of power between the regional and central authorities and also resolving a dispute over the region's contributions to the state budget. Two days before the election, Shevardnadze again visited Abashidze in Batumi; the next day Abashidze withdrew from the race, declined to endorse the candidacy of Shevardnadze's main remaining opponent, and retracted his previous threats to boycott the election. Within days after the elections, which gave Shevardnadze another term of office, the Parliament in Tbilisi amended the constitution to create the Ajarian Republic as a political entity, effectively federalizing the Georgian state.

There is no question of Ajarian secession from Georgia. Partly for this reason, Ajaria is actually one of the economically more prosperous regions of Georgia. The issue of the Russian military presence in Batumi is, for Abashidze, in principle separate from the internal Georgian question of determining the rights and responsibilities of the region vis-à-vis the central authorities in Tbilisi (Radvanyi and Beroutchchvili 1999).

Javakhetia

Javakhetia is divided into two districts called Akhalkalaki and Ninotsminda (formerly Bogdanovka), which are also the names of the district capitals that make up about 20% of the total population. Together the two districts cover about 850 square miles, with a population slightly over 100,000, of which over 90% is Armenian. Armenians settled in southern Georgia after 1828, when a treaty ceded the region from Turkey to Russia. However, Armenians make up only about one-third of the population of Meskhetia, mostly in the Akhaltstikhe district, so that they constitute about 40% of the population of the whole administrative region called Samtskhe-Javakhetia. This region is also called Meskheti-Javakhetia. Meskhetia will be remembered as the place where the "Meskhetian Turks" lived, all 90,000 of whom Stalin deported to Central Asia in one night during World War II.

The Armenian national movement in Javakhetia formed in response to events in Mountainous Karabagh. Both regions border Armenia proper, and Armenians are the overwhelming majority of the population in each. Volunteers from Akhalkalaki in Javakhetia went to fight in Karabagh from the time of the first armed clashes there. An anti-Armenian sentiment infused Georgia in the early 1990s as Armenians in Abkhazia initially supported that region's separatism. Although Javakhetia was effectively outside Tbilisi's control from the late 1980s through 1991, the self-constituted political-administrative apparatus of the region voluntarily dissolved itself once Shevardnadze came to power and named a prefect acceptable to the local population. The region accepted the Tbilisi regime; the Armenian organization "Javakhk" no longer existed per se in Javakhetia. Formed in response to Gamsakhurdia's "Georgianization" policies, the Javakhk mainstream and its representatives have, under Shevardnadze, sought only cultural autonomy. This is now guaranteed, as the large majority of schools are taught in Armenian, using textbooks published in Armenian that are provided to the region via an intergovernmental agreement with Tbilisi.

Javakhk was not a political party and its members have dispersed their activities among legally constituted parties. The most radical members of Javakhk have had ties with the Armenian "Dashnak" party and demand unification with Armenia. There is also a significant pro-Georgian faction, as well as a segment through which Russia exerts a certain influence.

Neither the Armenians in Javakhetia nor the government in Erevan seeks to detach Javakhetia from Georgia, although the insistence upon autonomy is a propaganda tool of some Armenian political parties. Leaders of the former Javakhk mainstream agree that tensions are rooted in social problems; Russian military bases in and around Akhalkalaki are strategically sensitive but, even more important for the local economy. Indeed, these bases are a source of employment for many Armenians, who have taken temporary Russian citizenship to qualify for the work. These bases are indeed the most important employers in the region. Shevardnadze has signed an agreement with Russia permitting them to remain, but the Georgian Parliament has not ratified the agreement. Over a thousand families depend on the main Russian base. For local residents, the bases represent a job, cheap products, and locally accepted Russian money. Ex-Javakh leaders feel that deeper ties with Armenia may help to resolve local problems. President Kocharian of Armenia has agreed that his country could indeed play a role in relieving the socioeconomic tension in the region, by providing electricity, building roads and even sending even school teachers.

However, the Armenian government has not supported the demands of some in Javakhetia for the region to obtain a separate administrative status within Georgia. Under conditions of the Turkish blockade on trade, Armenia's only overland egress is through Ajaria, which is adjacent to the Samtskhe-Javakhetia administrative region. Were Javakhetia to separate, Armenian imports and exports would have to traverse three rather than two Georgian provinces, adding administrative complication. The Armenian residents tend to regard their relations with the Russians as an integral part of the existing social order, and some even claim that the Russians are a deterrent against Turkey. They realize that this situation opens the way for the Russians to use them as a geopolitical pawn, but for the moment, they see no alternative, despite Shevardnadze's stated willingness to increase social programs and economic investment in the region. The realization of such programs is complicated by the fact that the Akhalkalaki region, one of the economically least developed in Georgia, is principally agricultural and has a sometimes difficult topography. Means of transportation and communications in Javakhetia (road, rail, etc.) are in general poor, as is infrastructure overall.

Tskhinvali (South Ossetia)

In 1989, the South Ossetian Autonomous Oblast within the Georgian Soviet Socialist Republic declared itself part of the Russian Soviet Federated Socialist Republic, where North Ossetia is to be found. In August 1990, it declared itself sovereign; four months later Georgia replied by abolishing South Ossetia's autonomous status within Georgia. This led first to armed confrontations and then, on November 28, 1991, to South Ossetia's declaration of independence. In April 1992, Georgia reestablished the South Ossetia as an Autonomous Oblast. In June of the same year, a cease-fire agreement, negotiated between the presidents of Russia and Georgia, stopped the eighteen-month war.

Ethnic unrest nevertheless escalated in mid-1992. Within a period of weeks, over 100,000 refugees fled to North Ossetia, in the Russian Federation, where ethnic Ingush refugees in the Prigorodnyi (literally "Suburban") region around the capital Vladikavkaz were demanding the reattachment of that region to Ingushetia, from which Stalin had severed it. The presence of so many refugees strained resources, led to disputes and unrest, and resulted in the appointment of a special prefect from Moscow to head an emergency administration. Ethnic Ossetes in North and South Ossetia alike began to call for reunification of their territory. In South Ossetia, Russia brokered an agreement providing for the deployment of a tripartite Russian, Georgian, and Ossetian force to guarantee civil peace and encourage residents to return there.

The Russian Federation continues to play a leading role in various multilateral forums under the aegis of the Organization for Security and Cooperation in Europe (OSCE). The OSCE provides political guidance to the Joint Control Commission (JCC), created by the 1994 agreement and originally charged with overseeing the trilateral, Georgian-Russian-South Ossetian peacekeeping force. The JCC later expanded its activities to include promotion of South Ossetia's economic reintegration into Georgia. In this connection, it has undertaken practical programs for cooperation among local officials. North Ossetia, which is part of the Russian Federation, participates autonomously in the activities of the JCC.

In 1995, the Parliament of Georgia adopted a new constitution that left open the question of Georgia's territorial and administrative structure in relation to South Ossetia (as well as Abkhazia). President Shevardnadze proposed a federal solution. Bilateral talks led to the agreement in Moscow, in July 1996, of a framework agreement officially titled the "Memorandum on Measures to Provide Security and Strengthen Mutual Trust Between the Sides in the Georgian-South Ossetian Conflict." Of importance also is that, in 1996, Georgia changed the official name of the region from South Ossetia to Tskhinvali, which is also the name of its administrative center. The 1996 Memorandum, however, provides for return of refugees, negotiations on political arrangements, and round-table meetings of mass media, civic organizations and intellectuals from both sides. A new administration took office in the region that was not connected with the immediately preceding conflict period. Working arrangements on practical everyday matters have followed since then.

Negotiations over the status of the region began in March 1997 in Moscow but have not made progress. Neither have proposals for an interim agreement been followed up. The South Ossetian side awaits the outcome in Abkhazia to define the widest limit of any possible autonomy they may subsequently negotiate. The fact that the region now has a government that is not implicated in the earlier conflict has been very important in readying the population to accept eventual Georgian jurisdiction. The approximately 30,000 refugees from the region now living in Georgia appear to consider improved economic conditions on par with security issues in determining to return. On April 8, 2001, South Ossetia held a referendum for changes to its constitution that were intended to increase presidential power. Voter turnout was roughly two-thirds, of whom two-thirds again approved the changes. Since the "Republic of South Ossetia" held the referendum on its own initiative without central Georgian participation, the EU and the OSCE condemned it and declared it illegal and void.

Analysis

Inspection of the Four Cases

Table 1 takes relevant variables from Tishkov's work and evaluates their presence or absence in each of the four situations discussed above. The situations are assessed as they stand at the end of 2002. This classification exercise is essentially ahistorical. It does not address how the "observations" may have evolved over time or the interdependence of the conflict situations. For example, ethnic Armenians were a majority in Javakhetia until 1995, when the region was merged into the larger Samtskhe-Javakheti administrative district. Also, it is possible to argue that in South Ossetia there was a local strongman earlier in the 1990s, although not today. In the instance of Abkhazia, some irredentism with ethnic groups in the neighboring region of the Russian federation was evident in the early 1990s.

Table 12.1

Abkhazia, Ajaria, Javakhetia, and Tskhinvali (South Ossetia)

	Abkhazia	Ajaria	Javakhetia	Tskhinvali (South Ossetia)
1) Local strongman?	Yes	Yes	No	No
2) Irredentism/Diaspora?	Yes	No	Yes	No
3) Federal solution?	No	Yes	No	No
4) Majoritarian?	Yes	No	No	No
5) "Success"?	No	Yes	Yes	No

The inspection of Table 1 from the "most similar systems" standpoint yields the following observations. In the Abkhazia-Ajaria comparison, the absence of irredentism/Diaspora promotes success, while the absence of an ethnic majority on the ground and the possibility of a federal solution may be contributing factors. In the Javakhetia-Tskhinvali comparison, the presence of irredentism/Diaspora promotes success. In the Abkhazia-Javakhetia comparison, the absence of a local strongman and the absence of an ethnic majority, separately or together, may promote success. In the Ajaria-Tskhinvali comparison, the presence of a local strongman promotes success, while the availability of a federal solution may be a contributing factor. From these observations, the following findings arise. The presence of a majoritarian ethnic group on the ground does not promote success. The possibility of a federal solution has promoted success for Javakhetia but not for Abkhazia or Tskhinvali. The presence of an irredentism/Diaspora has promoted success for Javakhetia but not for Abkhazia. The presence of a local strongman has promoted success for Ajaria but not for Abkhazia.

From a methodological standpoint, causal inferences are unwarranted. To start with, there are four cases and five variables. It is possible to eliminate sets of variables and do multiple pairwise comparisons with the identical pair of cases—that is, considering first only variables 1, 2, and 5; then 1, 3, and 5; and so forth—such that each "pairwise" comparison of cases in fact comprises six three-variable comparisons. Even then, however, all these tests would not be mutually independent. For such an hypothesis as "an ethnic majority never produces success," the requisite qualifying condition is "where the majority does not protect the minority"; and even here, there is only one case (on the utility of case studies for nomothetic research, see Lijphart 1971, 1975).

Clearly, there is a more complex dynamic at work than this simple categorical analysis is able to capture. Nevertheless, consideration of the cases on their idiosyncratic basis, eschewing the search for nomothetic laws, reveals that federalism is not a panacea for resolving problems of Georgia's territorial integrity, although it can help. South Ossetian elites have in the past greeted favorably the prospect of federal status within Georgia. The establishment of South Ossetia as a federal Georgian entity in this context would be a promising development. Unfortunately, this is less likely an outcome today than a few years ago. Even if such a federal precedent may help resolve the status of South Ossetia, it would not satisfy Abkhazian demands. In Abkhazia, the indigenous leadership has long rejected the idea of inclusion within the Georgian state. For example, the congratulations addressed to Shevardnadze by Abkhazia's leader Vladislav Ardzinba upon Shevardnadze's reelection were couched in protocol reserved for communications between heads of state. Neither traditional federal nor even confederate arrangements will solve the Abkhazia problem.

To establish Javakhetia as a federal entity, on the other hand, could create more problems than it solves. The districts comprising Javakhetia are part of a larger administrative region called Samtskhe-Javakhetia. The Virk party in Javakhetia, demanding autonomy, rejected a call by Armenian President Robert Kocharian to support Shevardnadze for re-election. Armenia and Georgia have taken steps to ameliorate the region's difficult economic situation. Although President Shevardnadze has identified the guarantee of Georgia's territorial integrity as the state's highest priority, even the emergence of Georgian federalism might not be enough. Georgia and the South Caucasus as a whole require a comprehensive international political initiative.

The Karabagh Case

Does the foregoing shed any light on the conflict in Mountainous Karabagh? The current conflict in Mountainous Karabagh broke out in the late 1980s when, under conditions of Gorbachev's glasnost, the Karabagh Armenians began political organizing to take their territory out of Azerbaijan's hands. They sought to unify the territory with Armenia, notwithstanding the absence of a common land border. In February 1988, the Supreme Soviet of Mountainous Karabagh voted for such a reunification with the Armenian Soviet Socialist Republic. Moscow thereupon abolished the local government there and instituted direct rule from Moscow through a Special Administration Committee. Karabagh forces responded by seizing the Azerbaijani town of Lachin, key to the "Lachin corridor" where a narrow winding mountain road (much improved since, thanks to funds from the Armenian Diaspora) connects Karabagh to Armenia. Then, turning northward, they seized and held Azerbaijan's Kelbajar district, which is not part of Mountainous Karabagh. This move abolished Karabagh as an enclave and attached it geographically to the main body of Armenia. It also turned Karabagh into an aggressor in the eyes of world public opinion. The situation on the ground has changed little since.

The OSCE Minsk Group is the focal point for multilateral consultations about Karabagh since the disintegration of the Soviet state. Its consultations were the basis for the decision by the OSCE to deploy peacekeeping forces in Karabagh, as Russia did not receive—either from the CIS Collective Security Committee or from the United Nations—the mandate it sought to play such a role.

At its 1995 Lisbon Summit, the OSCE passed a resolution calling for the "highest degree of autonomy" of Mountainous Karabagh within Azerbaijan, the territorial integrity of which was to be preserved. All

OSCE states except for Armenia accepted the resolution. In 1997 the Minsk Group proposed: (1) Armenian withdrawal from all of the occupied territories, (2) a buffer zone to be patrolled by an OSCE peacekeeping force, (3) an OSCE-administered lease of the "Lachin corridor" from Azerbaijan to Karabagh, (4) return of all ethnic-Azeri displaced persons to the occupied region, (5) lifting of all economic blockades, and (6) Karabaghi self-government within Azerbaijan. Azerbaijan accepted the document as a basis for negotiations, but Armenia and Karabagh responded that adequate guarantees of security were necessary to facilitate withdrawal from the occupied territories.

In 1998, the Minsk Group adopted the Russian proposal for a "common state," meaning a de facto independent Karabagh that could not secede unilaterally from Azerbaijan, and with which it had non-hierarchical relations. Armenia and Karabagh accepted this proposal while Azerbaijan rejected it. The presidents of Armenia and Azerbaijan were said to have been close in 1999 to an agreement somewhat resembling the "common state" although under another name, until the political situation within Armenia became stalemated, after an Erevan journalist entered the Armenian Parliament building while it was in session and assassinated many members of the national leadership (see, *inter alia*, Tavitian 2000).

The later "Key West" formula has not been publicly disclosed, but it seems to include withdrawal of Armenian troops from six of the seven Azerbaijani administrative districts that they occupy, the bundling together of Mountainous Karabagh with its Lachin land corridor as a self-governing region within Azerbaijan, and an internationally patrolled corridor through Armenia's Meghri district, to link Azerbaijan to its exclave Nakhichevan. The Armenian-Azerbaijani negotiations in Key West were said to have been "very fruitful," with a significant narrowing of differences between the sides and the development of basic principles of a new package deal. Despite the optimism emerging from the Key West summit, the momentum for a settlement has nevertheless been lost.

Categorically, following Table 1, Mountainous Karabagh is most identical with the Abkhazia case. The distinctive feature of the Karabagh case, however, which is not characteristic of any of those four cases, is that the leadership of the irredentist region has itself gained executive power in one of the existing states in the region. It is not without reason that many Armenians in Erevan refer to their state, since President Robert Kocharian's elevation to the presidency, as the "Republic of Greater Karabagh." Nevertheless, the fact of gaining influence over the state policy of Armenia would be promising if this meant that the leadership could take progressive steps towards settlement.

A First Attempt at Comprehensive Settlement

The only comprehensive initiative for a region-wide resolution of conflict situations in the South Caucasus remains the "Stability Pact for the South Caucasus," published by the Centre for European Policy Studies (Celac and Emerson 2000; Emerson, Tocci, and Prokhova 2000; Tocci 2001). In this vision, neither of two European stability pacts of the 1990s was a pure model for the Caucasus. The Balladur Stability Pact of 1994–95 was EU preventative diplomacy, designed to clear up frontier and minority problems among accession candidate countries, using this as a precondition for accession and therefore a strong incentive mechanism to settle the dispute. The Balkan Stability Pact of 1999 is a soft conference mechanism, dependent upon the earlier NATO intervention, which held the incentive of integration into the EU as a carrot. The Caucasus Stability Pact was to be devised differently. One suggested possibility, for example, was for a trilateral understanding, formal or informal, comprising the Russia, the EU, and the U.S., and requiring such to sketch its agenda. Such an understanding might include not only conflict resolution and prevention but also a South Caucasus Community eventually linked to an OSCE regional security system, as well as broader institutionalized Black Sea-Caucasus-Caspian economic and political cooperation.

The ambitiousness of the project is signaled, *inter alia*, by its proposal to establish without adequate preparation a regional parliament for the South Caucasus that bears a striking resemblance to the European Parliament, which has required over four decades to reach its present stage of evolution, departing moreover from cultural, economic, historical, and sociological conditions that differed radically from those of the South Caucasus today. It is not surprising that, despite the positive reception of the plan as a whole and its modest success in certain other areas, its parliamentary complement has remained a nonstarter. The whole Caucasus Stability Pact initiative has folded into a series of meetings and conferences now dubbed the Peaceful Caucasus Process.

The remainder of this article seeks to refine and redefine the parliamentary component of the ambitious Caucasus Stability Pact plan. A more theoretically informed approach to policy, akin to what Ruggie (1998) called "social constructivism in action," may be indicated. In the present instance, one might seek pragmatically to "unpack" the states into institutional and cognitive elements. The reason for making this distinction is to better induce change in state behavior through influence upon the formation of national interest and policies to realize the change. One might even refer to this as the creation of a "transnational interest." Relatively new inter-NGO networks in the South Caucasus have already begun this process. It needs to continue also at the national political levels. The question in both theory and practice then becomes the combination of research with initiatives for policy activity to implement the societally generated "transnational interests" within key sections of the respective state bureaucracies.

Towards the Parliamentarization of Transnational Conflict

Applied Theory as the Basis for a Real Solution

Such an approach is founded in a constructivism that recognizes the primacy of states in certain security issues and works with them (Boekle, Rittberger, and Wagner 1999). A domestic political system must be able to convert the elements of national power—particularly human demography and economic geography—into political resources. However, that "conversion process" develops at different rates in different places (Knorr 1970, 1975). Indeed, "overlay" (Buzan 1991) has made the operation of independent security dynamics in the region impossible. It is accurate to speak of a race between the governments' penetration of their own societies and their own penetration by the international system. States, while enjoying some relative autonomy (Moore 1966), remain an interface between demands and supports originating internationally and domestically. This dynamic is further complicated, and the acuteness of the situation continually exacerbated, by sociological phenomena such as diasporas, low-intensity ethnic conflict, and international migration, which significantly affect the construction, development, and definition of national interests of states and their construction of images of security (Coppieters 1996; see also Adler and Barnett 1998; Buzan, Waever and deWilde 1998).

It is instructive to synthesize the perspectives of social constructivism and organizational theory. Trondal (1999) has done this with an applied focus on the EU's committees and working groups, but the South Caucasus is a more conflictual region than that. Paradoxically, to promote a "security community" in the South Caucasus requires *de*-securitization of conflict issues in the region. How is this to be accomplished? Transgovernmental institutions (Slaughter 1997) are spaces of focus for emergent transnational advocacy coalitions. There they can communicate with one another and with disaggregated elements of formal state organizations. A focus on transgovernmental institutions ceases to restrict the stakeholders to states alone, yet does not exclude them altogether.

It follows that any sort of a regional South Caucasus parliamentary assembly should not seek from the start to legislate supranationally. Rather, it should more modestly act as a focal point for NGOs and inter-NGO networks. Such an assembly should build upon recently formed interparliamentary working commissions among the South Caucasus governments, which are organized by the Speakers of the three national parliaments under the aegis of the Parliamentary Assembly of the Council of Europe. This strategy promotes the resolution of the state security dilemma through a transnationalized "deepening" of security (Krause and Williams 1997), because such forums present the opportunity to adopt common goals and to identify the means to realize them. Finnemore (1996a, 1996b) provides the basis for this possibility in discussions of "sociological institutionalism."

When specified and operationalized in space and time, as it is in the CEPS Caucasus Stability Pact initiative, this institutional approach may be called "sociologized security." This approach may also be expressed in the language of regime theory. Trondal's (1999) synthesis of social constructivism and organization theory concludes on a threefold classification that characterizes social mechanisms as rational, cognitive, or integrative. This typology is likewise parallel to the Hasenclever, Mayer and Rittberger (1997) classification of schools of regime thought. Their power-based school is rational, in Trondal's terms, because it emphasizes maximizing relative gains. Their interest-based approach is integrative in Trondal's terms, because it emphasizes maximizing absolute gains. Their knowledge-based school is cognitive, in Trondal's terms, because it emphasizes sociological intersubjectivity. Using this shorthand, a pragmatic approach may be said to implant cognitive norms engineering rational behavior into integrative structures.

Recent research indicates that balance-of-power maneuvers such as we have seen in the Caucasus are a short-term, stopgap solution to providing international security. Current international relations theory, relying more heavily on sociological approaches (as summarized above), draws attention to the proliferation of contact among transnational forces, specifically as a mechanism that motivates states to overcome their perceived security dilemma. The South Caucasus is thus a crucial test case, both for theories holding that minimal levels of self-evolving cooperation are sufficient to coordinate international security, and for theories holding that multilateral frameworks for international public policy are necessary to manage geoeconomic conflict. The multiplicity of conflict situations and issues indeed makes the Caucasus a crucible for how examining transnational processes can encourage policymakers to trust one another, and how nonsecurity relations might contribute to an enhanced security environment.

Transnational Parliamentarization in Practice

How would one begin to establish, in the South Caucasus, even an unofficial or semiofficial transnational forum for discussions of security, economic, and political issues among individuals from the most important demographic and geopolitical formations in the region? The following answer does not explicitly consider the role such a forum might have, should have, or could have within the context of a broader stability and/or security pact for the South Caucasus. However, with proper organizational design and political engineering, it could become useful and facilitative if not occasionally catalytic.

It is, first of all, advisable to limit the initiative to the South Caucasus, because the North Caucasus is a complex unity where unexpected interdependencies crop up as unintended consequences of even well-intended moves. Not only would the inclusion of Chechnya be problematic—not least because of the difficulty of choosing among potential Chechen representatives—but also it would eventuate in the further necessity of including at least Dagestan as well. The problem here is that the ethnic variety and complexity within Dagestan is so great that the very number of communities to be represented would expand the size of the prospective forum beyond the point where it would be manageable in the first instance. Participation by various North Caucasus entities in the discussions and deliberations of such a forum is a matter for later consideration. It should not be excluded, in principle, for limited and well-defined purposes in specific instances. The participation of diverse "regions of the South Caucasus" should not be conditioned upon their separate corporate representation in a national parliament or their actual political existence as distinct administrative regions within a state.

The list of prospective non-state invitees nevertheless expands beyond Abkhazia and Mountainous Karabagh. Such a list could include Ajaria, Tskhinvali (South Ossetia), and perhaps Javakhetia. The list of politically Georgian entities, already quite long, must be stopped here, lest the Georgian ethnos itself begin to subdivide, leading to separate Svan, Mingrelian, and Kartvelian demands for representation, all of which could be given real and reasonable basis in both demography and geography. Nevertheless, some thought would need to be given to representation by the Meskhetian Turks, an ethnos deported by Stalin, whose return to Georgia has been mandated by the Council of Europe. Therefore, they too may be added to a provisional list. States might have some kind of presence at the forum, but representation of the respective national political executives should be tightly limited. The Azerbaijani exclave Nakhichevan is well represented in the executive of the country's national government and therefore does not require a separate voice as a region.

A provisional forum such as this may be convoked upon invitation from an external body such as one European organ or another. If the South Caucasus states themselves are to be responsible for the convocation, then such vexing and distracting questions about the status of the assemblage under international law would be raised too early. One need only imagine the debates, for example, as to whether invitation represents any kind of official or even unofficial recognition of one or another international-legal status of the invited entities. It makes sense, therefore, for a respected non-Caucasus diplomatic actor to issue, in the first instance, a convocation to *South Caucasus Congress for Stability and Cooperation.* Properly prepared, that convocation may then be supported by official, semiofficial, and unofficial instances from the South Caucasus proper. The goal of "pre-Congress" preparation should be to guarantee that no political authority in the region will frown on a formal initiative. Progress on this should not be hostage to progress on a more ambitious regional settlement. The form of this may be optimized by considering three matters together: the range of issues with which the assemblage may deal with initially and with which it may seek to deal with subsequently;, a comparative assessment of potential "models" for such an assemblage; and representation.

Concerning the range of issues. What should be involved is a transnational social-consultative forum that may later evolve towards a quasi-parliamentary body. It should not have actual international legislative authority, certainly not in any immediately foreseeable future. At the same time, such an assemblage should conserve its sociological and demographic basis, which could be also expanded in the more foreseeable future, incorporating even nonethnically-based actors and co-opting them into its deliberations. The pertinent issues would require careful circumscription. The competences and issue areas that animate respondents to the initial convocation will play a determining role. Some initial programmatic determination may be necessary at the outset. Speakers of the three national parliaments of the South Caucasus countries have met regularly in Western Europe following an initiative of French diplomacy, under the aegis of the Council of Europe. The Azerbaijan side has stated that any formal international parliamentary institution or official parliamentary assembly of the South Caucasus states would have to await definitive settlement of the Mountainous Karabagh conflict. Therefore, an initial failure that would discredit the overall exercise is to be avoided, therefore, any such issue or conflict should not be a first focus.

Concerning potential "models" for such an assemblage. The word "parliament" and its various forms are to be avoided as designations for this initiative, because these echo too loudly with the idea of official representation. States could easily balk at this. The adjective "regional" should be avoided because it threatens confusion between, first, the assemblage as a forum where the regions of the South Caucasus may have a voice and, second, the assemblage as a forum for the South Caucasus as a region for international politics. The latter of these two connotations opens up, too much and too soon, the Pandora's box about which non-Caucasus entities may have a voice on which issues and to what extent. It is a good nuance nevertheless to consider this assemblage as a forum "for regions" of the South Caucasus, rather than as one "of regions" of the South Caucasus, because a forum "of regions" implies questions about what a "region" is and who defines a "region." A forum "for regions," on the other hand, comfortably allows for participation by entities from outside the South Caucasus while not mandating such participation, but still maintaining a problématique focused on the South Caucasus itself.

Concerning representation. Parts of the South Caucasus that are not "regions" will be represented, at least partly, through states. The proposed assemblage differs slightly, in this respect, from the Assembly of European Regions associated with the Council of Europe. However, such state-based ethnic communities are best represented in the assemblage at a level other than the level of the national political executives. For this, there are two reasons: first, analogous political-executive authorities do not necessarily exist for all potential invitees; and second, this forum should remain, in the first instance, an assemblage of social forces. This does not mean that some of the representatives may not be chosen or seconded from the national parliaments. Still, the Assembly to be created by the initial Congress should be "for regions and peoples" and, therefore, better to include social formations that, in fact, will have a place within the bureaucratic and representative apparatus of the South Caucasus national-states themselves. If the convocation is issued by some voice from outside the South Caucasus altogether, then parties engaged in conflict are not the ones determining the invitation list, and multiple representations from various disputed territories become a possibility, either as direct participants or as observers. Still, it is not yet productive to think about allocation of "delegates" among various possible participants. Rather, it is more practical to sketch in slightly more details the first few steps of institutional development that such a nascent assemblage might follow. This sketch draws on Cutler's (2001) comparative study of the evolution of international parliamentary institutions.

Next Steps

It is not enough simply to convoke such a Congress as described. It is also necessary to consider in advance not only how such a assemblage may be convoked but also how it may develop and grow. It is natural to foresee at least two stages of development in addition to the preparatory work necessary for the initial convocation. As suggested above, the first convocation should not be issued to convoke a Congress of representatives of these various social and political forces. The three national parliaments, which are natural constituencies to be interested in such a proposal, are unlikely to accept even virtual participation without the consent of their national political executives. Therefore, the initiative would have to the outlined to those leaderships and their suggestions solicited.

Let us identify the "Congress" as the first meeting of those who may decide to establish on a more permanent basis an *Assembly for Regions and Peoples of the South Caucasus*. Such a Congress should, of course, not be a rubber-stamp of decisions in practice previously agreed on and awaiting some manner of formalized approval. By the same token, every precaution should be taken to avoid fiasco, at least by cataloguing the (at first limited number of) issues that may be raised at the Congress, even if these do not appear on a provisional agenda. This first meeting, the Congress, should aim at establishing the Assembly on a regular basis. However, it is possible that constructive developments may occur at the meeting itself, making a second Congress propitious. There should be an explicit *a priori* limit of two such Congresses. Any further deliberation will best following in a more institutionalized context.

For the Congress, it will be well to consider establishing an interim or provisional secretariat either within the South Caucasus itself or, temporarily and only provisionally, under the auspices of a European organization, awaiting the more definite preparation of institutional infrastructure. In principle, it could be better to have a "native" South Caucasus provisional secretariat, but choosing its geographic location would be a delicate task. It may well be desirable to give such a bureau significant institutional support by coordinating its activities with a European body on an ongoing basis. In this connection, it is worth considering later whether the precedent is pertinent of the Central European Initiative, which has a functionally specific secretariat in London, actually inside the European Bank for Reconstruction and Development, and which coordinates work with it.

Conclusion and Prospect

Based upon the actual situation in the South Caucasus, the present diplomatic conjuncture and existing comparative research into European and international representative assemblies and institutions, the schema emerges of convoking a Congress for a maximum of two meetings. In the meantime a secretariat is established for the purpose of circulating information and settling procedural norms. This will be also an organizational womb for the nascent institutional memory. From the labor, an Assembly emerges. The provisional name of the follow-on, standing Assembly should avoid the explicit rubrics "security" and/or "stability" because their habitual usage in the lexicographic field in diplomacy suggests giving privilege to the interests of states over these peoples. Standing international diplomatic habits will doubtless take good care of the states as political formations without the need for such special attention.

The Assembly will decide upon the pace at which it will create rules and enforce them, and will regulate those matters over which its participants agree to its competence. To some degree, these rules are functional requisites; to another degree, they will determine the Assembly's autonomy and capacity to set its own goals. The Assembly, therefore, will begin to evolve towards a transnational-deliberative institution in international affairs and world society. This schema will be further refined and elaborated as the entire South Caucasus diplomatic process develops in the near future. Fine-tuning will be necessary as the process gathers momentum, and only through a continuous process will it be possible to optimize the institutional infrastructure, first, of the Congress and, then, of the Assembly. This optimization does not imply a static institutionalization. Rather, it means an evolving organization that will sooner or later participate in the setting of its own agenda and procedures. It also means, later if not sooner, an organization that should not hesitate autonomously to launch self-originated activities, creating subsidiary or auxiliary bodies on its own initiative, and even spinning these off after they have acquired sufficient life and momentum of their own. The Benelux, Central American, and Baltic integration organizations would be useful precedents for comparison to the Caucasus situation, as each of them, like the Caucasus, comprises a handful of small countries having a tightly interwoven common history, began as a trade organization, and has subsequently acquired a political structure, including a regional parliamentary assembly.

References

ADLER, E. AND M. BARNETT (EDS.)
1998 *Security Communities*. Cambridge: Cambridge University Press.

BOEKLE, H., V. RITTBERGER AND W. WAGNER
1999 "Norms and Foreign Policy: Constructivist Foreign Policy Theory." *Tübinger Arbeitspapiere zur Internationalen Politik und Friedensforschung* No. 34a. Tübingen: University of Tübingen, Institute for Political Science.

BUZAN, B.
1991 *People, States and Fear: An Agenda for International Security Studies in the Post-Cold War Period*. London: Harvester Wheatsheaf.

CELAC, S. AND M. EMERSON (EDS.)
2000 *A Stability Pact for the Caucasus*, Working Document No. 145. Brussels: Centre for European Policy Studies.

COPPIETERS, B. (ED.)
1996 *Contested Borders in the Caucasus*. Brussels: VUB Press.

CUTLER, R.M.
2001 "The Emergence of International Parliamentary Institutions." Pp. 201-229 in Smith, G.S. and Wolfish, D. (eds.), *Who Is Afraid of the State?* Toronto: University of Toronto Press.

2000 "Policy Options for Resolving Post-Soviet Ethnic Conflict." *Central Asian Survey* 19: 451-468.

EMERSON, M., N. TOCCI AND E. PROKHOROVA

2000 *A Stability Pact for the Caucasus in Theory and Practice: A Supplementary Note.* Brussels: Centre For European Policy Studies.

FINNEMORE, M.

1996 *National Interests in International Society.* Ithaca, NY: Cornell University Press.

1996 "Norms, Culture, and World politics: Insights from Sociology's Institutionalism." *International Organization* 50: 325-347.

HASENCLEVER, A., P. MAYER AND V. RITTBERGER

1997 *Theories of International Regimes.* Cambridge: Cambridge University Press.

KRAUSE, K. AND M.C. WILLIAMS

1997 *Critical Security Studies.* Minneapolis: University of Minnesota Press.

LIJPHART, A.

1980 "The Structure of Inference." Pp. 37-56 in Almond, G. and Verba, S. (eds.), *The Civic Culture Revisited.* Boston: Little, Brown & Co.

1975 "The Comparable-Cases Strategy in Comparative Research." *Comparative Political Studies* 8(2): 158-177.

1971 "Comparative Politics and the Comparative Method." *American Political Science Review* 65: 682-693.

MECKSTROTH, T.

1975 "'Most Different Systems' and 'Most Similar Systems': A Study in the Logic of Comparative Inquiry." *Comparative Political Studies* 8(2): 133-157.

MOORE, B. JR.

1966 *Social Origins of Dictatorship and Democracy.* Boston: Beacon Press.

RADVANYI, J. AND N. BEROUTCHCHVILI

1999 "L'Adjarie, Atout et Point Sensible de la Géorgie." *Cahiers d'Études sur la Méditerranée Orientale et le Monde Turco-Iranien*, no. 27: 229-239.

RUGGIE, J.G.

1998 "Epistemology, Ontology, and the Study of International Regimes." Pp. 85-101 in Ruggie, J.G., *Constructing the World Polity: Essays on International Institutionalization.* London/New York: Routledge.

SKOCPOL, T. AND M. SOMERS

1980 "The Uses of Comparative History in Macrosocial Inquiry." *Comparative Studies in Society and History* 22: 174-197.

SLAUGHTER, A.

1997 "The Real New World Order." *Foreign Affairs* 76: 183-197.

TAVITIAN, N.

2000 "An Irresistible Force Meets an Immovable Object: The Minsk Group Negotiations on the Status of Nagorno Karabkah," WWS Case Study 1/00. Princeton: Princeton University, Woodrow Wilson School of Public and International Affairs.

TISHKOV, V.

1997 *Ethnicity, Nationalism and Conflict in and After the Soviet Union: The Mind Aflame.* London: Sage for PRIO and UNRISD.

TOCCI, N.

2001 "The Stability Pact Initiatives: Reactions and Perspectives." *Paper Presented to the Baku Conference on Europe and the South Caucasus* (11 June 2001). Brussels: Centre for European Policy Studies.

TRONDAL, J.

1999 "Unpacking Social Mechanisms: Comparing Social Constructivism and Organization Theory Perspectives." *ARENA Working Papers* 99/31. Oslo: University of Oslo, Advanced Research on the Europeanization of the Nation State Program.

DEWILDE, J. AND O. WAEVER

1998 *A Theory of International Security: Threats, Sectors, Regions.* Boulder, CO: Lynne Rienner.

XIII. International Challenges and Domestic Preferences in the Post-Soviet Political Transition of Azerbaijan

Ayça Ergun

Abstract

In the post-Soviet transition, the role of the international element (i.e., external actors-foreign countries and international governmental and non-governmental organizations) is one important aspect of the study of state-society relations. The international element highlights the fact that post-Soviet transition is not a question of internal politics only and is affected by redefinition of the relationships with the outside world. Thus, the interplay between international and domestic actors should not be ignored in conceptualizing the process of transition, but similarly it should not be used as a single tool to explain the process but rather as a complementary factor that aids in our understanding of its path. This article puts the post-Soviet political transformation in Azerbaijan into the context of both the regional and the global framework, exploring the interaction between the international element and domestic actors. The first part focuses on Azerbaijan's relations with Armenia, Iran, Russia, Turkey, and the West and explores the question of how Azerbaijan's perceptions about its external environment, with reference to these countries, can contribute or hinder political transformation. The second part of the article examines the degree of impact and promoting effect of the international element on political transformation, while focusing on the relationship between international organizations, the state, and the societal actors.

Introduction

The study of the international aspect of democratization is included in works on East and Central European [1] and Latin American transitions. [2] The regional and international environment is conceived as having a promoting effect on the democratizing countries although its impact is argued to vary in different phases of transition (Pridham 2000: 294). The impact of the international environment is also crucial for the analysis of post-soviet political transformation. The interaction between international and domestic actors shapes the path towards democratization and plays a remarkable role in the formation of state-society relationships.

The emergence of new independent states provided Western and neighboring countries with new areas of commerce, market, cultural and social exchange. For the newly independent republics making bilateral and multilateral agreements meant new foreign policymaking and redefining "friends and foes." Membership of international and regional organizations [3] became the "first priority ... to gain international recognition and to take their place in the international community" (Herzig 1999: 94). Moreover, the first contacts with the international element (i.e., foreign countries and international governmental and non-governmental organizations) were conceived of as the recognition of independence, which would help to secure statehood.

[1] See, for example, P. Kopecky and E. Barnfield (1999) "Charting the Decline of Civil Society: Explaining the Changing Roles and Conceptions of Civil Society in East and Central Europe," in J. Grugel (ed.) *Democracy Without Borders: Transnationalization and Conditionality in New Democracies*; G. Pridham (ed.) (1991), *Encouraging Democracy: The International Context of Regime Transition in Southern Europe*, London: University of Leicester Press; G. Pridham and T. Vanhanen (eds.) (1994), *Democratisation in Eastern Europe, Domestic and International Perspectives*, London: Routledge; L. Whitehead (1986), International Aspects of Democratization, in G. O'Donnell, P. Schmitter and L. Whitehead (eds.), *Transitions from Authoritarian Rule: Comparative Perspectives*, Baltimore: Johns Hopkins University Press; (1996), "Three International Dimensions of Democratization," in L. Whitehead (ed.), *The International Dimension of Democratization*, Oxford: Oxford University Press.

[2] See, for example, C.L. Freres (1999), "European Actors in Global Change: The Role of European Civil Societies in 'Democratization'," in J. Grugel (ed.), *Democracy Without Borders*; J. Grugel (1999), "European NGOs and democratization in Latin America: Policy Networks and Transnational Ethical Networks," in J. Grugel *Democracy Without Borders*; L. Taylor (1999), "Globalization and Civil Society-Continuities, Ambiguities and Realities in Latin America," *Indiana Journal of Global Legal Studies* (7).

[3] Among these are the United Nations, the Conference on Security and Cooperation in Europe, the International Monetary Fund, and the World Bank; see Herzig, *The New Caucasus*, p. 94. The regional organizations include the Commonwealth of Independent States (CIS), Black Sea Economic Cooperation Organization and the GUUAM.

In the former Soviet Union, the successor states have established their links with the international element in three respects: membership of these countries in international organizations, bilateral and multilateral relationships between governments, and support for the further development of democratization and formation of civil society. Membership of regional and international organizations such as the United Nations and the Council of Europe provides governments in the transitional countries with legitimacy in their new existence. It also means the approval of their independence and aids consolidation of their statehood. Bilateral and multilateral relationships result in new foreign policy formations and contribute to an increasing interaction between the West and the post-Soviet space. Through these relationships the international actors offer expertise and experience to the transitional states and propose guidelines. These include electoral monitoring, support for democratic reforms, promotion of human rights, advice in national law-making, training for state officials, political parties, and conflict resolution. However, to what extent the efforts shown by the international element to promote the transition to democracy is yet questionable.

All these interactions between international and domestic actors take place in a context of "triple transition" (Offe 1996). The attempt to establish a democratic polity within the newly defined territorial boundaries is also shaped by the simultaneous formation of a new national identity that serves as the basis of nationhood. Additionally, the achievement of independence brings the issue of building a state within the newly defined post-Soviet boundaries and the reorientation of domestic and foreign policy. Transition also involves a new reconstruction towards a market economy. This "triple transition" of state, economy, and community and the simultaneous attempts by the successor states to cope with the problems of all three different but interrelated processes make the nature of transition even more problematic.[4] The triple transition resulted in the intertwining of the three processes which reciprocally affect and interact with each other. Moreover, the triple transition is challenged by simultaneity and uncertainty. Simultaneity does not only mean the establishment of a democratic market-oriented nation-state but also implies the overcoming of the legacies of the past. The destruction of the ancient regime does not necessarily lead to the immediate construction of a new one, as construction takes more time than destruction. Thus, simultaneity should

[4] For this argument, see R. Allison (1994), "On the Gap Between Theories of Democracy and Theories of Democratization" *Democratization* 1(1), 8-26; C. Offe, *Varieties of Transition*, p. 34; S.M. Terry (1993), "Thinking About Post-Communist Transitions: How Different They are," *Slavic Review* 52(2), p. 334; H. Welsh (1994), "Political Transition Processes in Central and Eastern Europe," *Comparative Politics*, p. 380.

be understood with reference to the new formations within the state, nation, and polity as well as in terms of the competing factors of the legacy of the past and the challenges introduced by the regime change. Furthermore, the simultaneous attempts and the instability caused by these "triple" transformations, institutions, social actors, and the context itself remain highly fluid, and this results in uncertainty.[5]

Similarly, the post-Soviet political transformation of Azerbaijan is a triple transition that is shaped both by domestic and international actors. The question here is whether and to what extent the international factor plays a significant, promoting, and supportive role. The penetration of the international element in the post-Soviet context creates an unbalanced impact on political transformation when the state and society components of democratization are considered. Although Caucasian and Central Asian states have established relationships with the international element, they fall short of the standards of democratic achievement to varying degrees.[6] In terms of the regional context, in contrast to East and Central Europe and the Baltic States, none of Azerbaijan neighbors have implemented or encouraged reform for further democratization or aspired to democratize, with the exception of Turkey. The international element, however, promotes the sphere of civil society via teaching, guidance, training, and funding. This leads to a much higher degree of sensitivity in civil society regarding the undemocratic practices exercised by their governments. Consequently, the international element interacts with countries where anti-democratic practices are persistent and where civil societies have increasingly acquired the tools to promote democratization.

---

[5] See, for example, R. Bova (1992), "Political Dynamics of the Post-Communist Transition: A Comparative Perspective," in Bermeo, N. (ed.), *Liberalization and Democratization, Change in the Soviet Union and Eastern Europe* (also in *World Politics*, October 1991, 44, pp. 113-38); P.C. Schmitter and T.L. Karl (1994), "The Conceptual Travels of Transitologist and Consolidatologists: How Far to the East Should They Attempt to go?" *Slavic Review* 53(1); Offe, *Varieties of Transition*; R. Rose, W. Mishler, C. Haerpfer (1998), *Democracy and Its Alternatives*, London: Polity Press, p. 45.

[6] For transition and leadership patterns in Central Asia, see D. Carlisle (1995), "Islam Karimov and Uzbekistan: Back to the Future," in T.J. Colton and R.C. Tucker (eds.), *Patterns in Post-Soviet Leadership*, Boulder: Westview Press; M. Ochs (1997), "Turkmenistan: A Quest for Stability and Control," in K. Dawisha and B. Parrott (eds.), *Conflict, Cleavage and Change in Central Asia and the Caucasus*; M.B. Olcott (1995), "Nursultan Nazarbaev and the Balancing Act of State-Building in Kazakhstan," in T.J. Colton and R.C. Tucker (eds.), *Patterns in Post-Soviet Leadership*; R. Taras (ed.) (1997), *Post-Communist Presidents*, Cambridge: Cambridge University Press, See Dawisha and Parrott (eds.) (1997), *Conflict, Cleavage and Change in Central Asia and the Caucasus*; T.J. Colton and R.C. Tucker (eds.) (1995), *Patterns in Post-Soviet Leadership*, Boulder: Westview Press. A. Altstadt (1997), "Azerbaijan's Struggle Toward Democracy," in K. Dawisha and B. Parrott (eds.), *Conflict, Cleavage and Change in Central Asia and the Caucasus*.

In this article, I concentrate on three aspects of the international dimension of democratization in the post-communist transition of Azerbaijan. First of all, the post-Soviet lands offered a vast source of economic goods and have the potential to become important markets for the Western countries. The former Soviet countries' resources (mainly oil and gas) attracted foreign investment and paved the way to the flourishing of economic relationships. Thus, the international community is concerned not only with the democratic achievements and the promotion of democratic practices in the post-communist countries but also with the building of new economic relations, the opening up of new markets which consequently contribute to the global political economy (Strange 1992). This indeed necessitates a stable regime in order to secure economic interest. For this reason, the international community seems content with stability even if it is achieved by old authoritarians—*recent* democrats. In other words, the stability generated by authoritarian or semi-authoritarian leaders in Azerbaijan and Central Asia is assumed to be a guarantee for the West's economic interests. For the successor states, economic issues also shape their priorities and preferences in foreign affairs and determine whether they favor the West, Russia, Turkey, and Iran.

Second, the international governmental (IGOs) and non-governmental organizations (INGOs) showed a particular concern for the democratization of the successor countries of the former Soviet Union. The impact of Western democratic countries and organizations is highly valued by both governments and societal actors (political parties and local non-governmental organizations) and deemed to promote democratization. In this respect, the international factor is welcomed by both sides not only for assistance and collaboration but also for promoting and preserving the "democratic [or democratizing] image of the country." Interaction with international actors also highlights the willingness of the successor states to be an integral part of the Western world. Foreign observers are invited to observe elections, international non-governmental organizations have opened branches in these countries, and governments are quite open to interaction with their international counterparts.

Lastly, IGOs and INGOs conduct activities in the region for the formation of a civil society. In the organizational sphere, the INGOs constitute the main sources of funding for local NGOs. They also provide teaching, technical assistance, and training. Offering the Western experience and expertise through international funding provides the basis for *construction above and from abroad* and the international community constitutes one of the major players in this construction. However, this should not imply that the democratization process is heavily defined, directed, or guided from outside. Rather, the international element highlights the fact that democratization is not a question of internal politics only and is affected by redefinition of the relationships with the outside world. Thus, it should not be ignored in conceptualizing the process of post-Soviet political transformation, but similarly it should not be used as a single tool to explain the process but rather as a complementary factor that aids in our understanding of its path.

Azerbaijan is a former Soviet Socialist Republic located between the mountains of the South Caucasus and the Caspian Sea. Neighboring countries are Armenia, Georgia, Iran, Russia, and Turkey. It is a small country with a population of eight million. Azerbaijan's population is not homogenous. The majority of the people of Azerbaijan are considered as "Turkic-speaking people" (Swietochowski 1995) or "ethnically and linguistically Turkic" (Wimbush 1979). The majority of the population is Shi'a Muslims (93.4%). Religious minorities are Russian Orthodox, Armenian Orthodox, and Jews.

The country has rich oil and gas resources attracting foreign investment and capital since the collapse of the Soviet Union. Its process of transition include three government changes since independence;[7] attempts at coup d'état; a long period of instability; interethnic and interstate conflicts in the Karabagh region; occupation of almost twenty per cent of its territory; and the problems caused by refugees in the subsequent years of independence.

The introduction of Soviet *glasnost* (openness) and *perestroika* (restructuring) brought change to the country. This change was initially marked by the formation of literary unions in order to enhance glasnost and contribute to the debates surrounding *perestroika*. At this time, the people of Azerbaijan (hitherto divided first between two empires, Russian and Persian, and then between two states, the Soviet Union and Iran)[8] revealed their forgotten and/or hidden past and started to seek a niche for their sovereignty, independence, nation, and finally democracy.

Stimulated by the Karabagh Armenians' demands over the Nagorno-Karabagh Autonomous Region of the Azerbaijan Soviet Socialist Republic, the main concerns of the literary unions such as the Organisation for the Defense of Azerbaijan's Sovereignty, Varlik (Wealth), Birlik (Unity), Chenlibel Literary Union and Baku Intellectuals Club shifted slightly from cultural, intellectual and literary issues to more politically oriented

[7] The former first secretary of the CPAz, Ayaz Muttalibov, became the first president of Azerbaijan in September 1991. He was replaced by the leader of the Popular Front of Azerbaijan, Ebulfez Elchibey, in June 1992. After the overthrow of the Popular Front Government, Heydar Aliyev became the president of the country in October 1993.

[8] Azerbaijani territory was divided between Russian and Persian Empires by the Turkmenchay Treaty signed in 1828.

and nationalist ones. This paved the way for the formation of a mass movement engaged in open political action in the form of rallies, meetings, and strikes. They first protested against Soviet rule and then against two "brother" nationalities, the Armenians and Russians. The reluctance of the Communist Party of Azerbaijan (CPAz) to implement both *glasnost* and *perestroika* as well as to deal with the Karabagh problem accelerated the formation of a nationalist independence movement. Aiming first at sovereignty and then independence, the People's Front Movement of Azerbaijan (hereafter referred to as the PFA) was established. It first became a nationwide organization challenging the party elite and later a candidate to rule the independent Republic of Azerbaijan.

On August 30, 1991, following the abortive Soviet coup, Azerbaijan became an independent state. After a short period of rule by the former first secretary of the CPAz, Ayaz Muttalibov, the PFA elite gained control and a new phase of transition to democracy started under the presidency of the PFA's leader, Ebulfez Elchibey. The PFA's coming to power symbolized the emergence of a binary divide in Azerbaijani political transformation, that is, the government versus opposition dichotomy. The fact that the PFA was formed in opposition to the Soviet regime and the leaders who still represented this regime created a strict dichotomy between those who were in government and those that formed the opposition. More importantly, in Azerbaijan, the binary divide shaped the three interdependent processes of democratization, nation-building and state-building as well as state-society relations in the years of political transformation.

The People's Front Government's (hereafter referred to as the PFG) eleven months in power signaled new prospects for democratization, new definitions of national identity and nationhood as well as attempts at a new reconstruction of the independent state. In other words, it introduced the patterns of change such as new institution-building, national law-making, the formation of political parties, the establishment of social organizations, the shift to a market economy, rule of law, and a free press. However, arising from the Movement, the PFA did not have a solid basis to consolidate its rule. The fact that all these patterns of change were introduced into the political order of the old regime meant that the *new* forces were surrounded by the *old*. The legacies of the previous rule were represented by patterns of continuity. These patterns of continuity include the heritage of rule by the old Communist *nomenklatura* (i.e., old state cadres), their authoritarian and semi-authoritarian practices, and the lack of democratic institutions prior to the Soviet period. While trying to reshape the existing structures with new institution-building, democratic principles, and the replacement of the cadres, the PFG was also challenged by the old elite whose legacies were not as yet insignificant. In this sense, the problems caused by the simultaneity were related not only to the reconstruction of regime, nation, and state but also to the struggle between *old* and *new*.

Moreover, in an environment of transition, new rival groups emerged to challenge its rule. Suret Huseyinov, formerly a cotton entrepreneur, formed a military base in the city of Genje. He first demanded the resignations of the Prime Minister, Penah Huseyinov, and the parliament speaker, Isa Gamber, and then that of the President. The PFG, unable to control the revolt, started to negotiate with Huseyinov and also "invited" Heydar Aliyev—formerly KGB boss, former first secretary of the CPAz between 1969-82, a member of the Politburo and head of the Nakhchivan Autonomous Region—to collaborate in eliminating the threat posed by Husseyinov's forces.

The "invitation" to Aliyev was the real proof of how dominant the legacies of the past were. This was not only a "tactical mistake" of the PFG to collaborate with the "old man" but also a natural consequence of the remaining, but dormant old cadres in the parliament and in state bureaucracy, perhaps more important, old loyalties.

Continuity came to the surface with the overthrow of the PFG and its replacement by Heydar Aliyev. Aliyev's coming to power in 1993 indicated the victory of past legacies. After his coming to power, Azerbaijan's political transformation was troubled due to anti-democratic practices in elections, repression of the opposition, the consolidation of one-man rule, bribery, corruption, and regionalism. The change of regime and its replacement by Aliyev can be defined as a step back towards the past. Nevertheless, it did not lead to a restoration of Soviet-type authoritarianism. The main reason was that the patterns of change were already incorporated into the transition and the former PFG elite constituted the opposition against Aliyev's rule.

In Aliyev's first term in office, elections paved the way for the formation of the executive and the parliament based on the principle of the popular will. The new constitution was adopted in 1995. It determined the type of the regime as presidential. However, one-man rule facilitated the dominance of the executive over parliament, which was the product of anti-democratic elections. The elections that were held on November 12, 1995, promised to be a significant turning point since they were the first parliamentary elections of the new independent republic. This meant that, for the first time, Azerbaijani citizens went directly to the ballot box to elect their representatives. In doing so, the popular will was to become the foundation of state-building. However, the 1995 elections, rather than contributing to the path towards democratization, proved to be a continuation of anti-democratic practices both during the election campaign and during the counting of ballots. The reports produced by international and local observers discussed in detail how malpractices were carried out.[9] According to the election law, every candidate had to collect 2000 signatures from the voters.[10] During the election campaign, the lists of signatures collected by the opposition parties were stolen and destroyed. The election council, without any reasonable explanation, did not accept the lists of the candidates known to be in opposition. Every party leader had the right to give a seven-minute speech on television, but the speeches were prerecorded and those considered unsuitable by the government were cut or not broadcast at all. As the state had control of the publishing houses, the newspapers and publications by the opposition parties were also censored or their publications delayed before the election (Ergun 1998: 67). Three out of eight of the participating political parties gained seats in the new parliament. With 62.66% of the votes, Aliyev's party, the New Azerbaijan, constituted the majority in the parliament (*Azerbaycan*, November 12, 1995: 1). As a result, the parliament acted under the supervision and control of the presidential apparatus.

In the case of Azerbaijan, the democratization process was started by the overthrow of an authoritarian regime as a result of the independence movement. Yet, the transition to democracy remained incomplete due to the impact of the patterns of continuity derived from the prior regime type.[11] The impact of prior regime type was reflected in repressions of opposition, censorship over media,[12] dominance of old party cadres in state bureaucracy, a personalized bureaucracy which facilitates corruption, and the myth of "strong man." These patterns of change did not produce a consolidated democracy. Thus, in the whole process of political transformation, Azerbaijan is a case of an "unconsolidated democracy" or a "failed democratization."

In this failed democratization, the struggle between continuity and change persists. Aliyev's government symbolizes continuity whereas op-

[9] See, for example, *Azerbaycan Demokrasinin nkiaf Fondu Parlament Seçkileri Hakknda Raport* (Report on Parliamentary Elections by Foundation of the Development of Democracy) (1995), p. 4; OSCE/UN of the OSCE UN Joint Electoral Observation Mission in Azerbaijan on Azerbaijan's. 12 November Parliamentary Elections and Constitutional Referendum (1996), *http://www.osce.org/odihr/documents/reports/az/azer1-1.pdf*, p. 12; Ayça Ergun (1998), *The Process of Democratization and Political Elite in Azerbaijan*, unpublished MSc. Thesis, pp. 65-68.

[10] For the full text of law, 5 October 1995, *Azerbaycan.*

[11] For the discussion of the impact of prior regime type, see Linz, J.J. and Stepan (1996), *Problems of Democratic Transition and Consolidation*, Baltimore: The Johns Hopkins University Press.

[12] On August 8 1998, the government censorship body, the Department for Protection of State Secrets in publications and other media, was abolished.

position and most of the local NGOs represent change. The case proves how the path of democratization has been shaped by the coexistence of prior regime type and newly introduced societal initiatives. During the course of change, new political parties were established; the number of social organizations increased. This has led to agreements between political parties and social organizations under umbrella groups. Even though the government elite remained reluctant to introduce change, societal actors became highly active, demanding change, the democratization of society, and the widening of representation.

In this challenging and complicated process of democratization, the relations with Western and neighboring countries play a significant role in the post-Soviet political transformation of Azerbaijan. The country is of crucial importance not only for its oil resources but also for further economic cooperation and security-building in the South Caucasus and the Caspian regions. For Azerbaijan, two main issues shape its priorities and preferences in foreign affairs and determine whether it favors the West, mainly the USA, Turkey or Russia, and Iran: economic growth and the Karabagh problem. The flourishing of the economy through oil revenues, foreign investment, and equal distribution of wealth, together with the resolution of the Karabagh conflict, will lead to further economic cooperation and stability not only for individual countries but also for the region. In this respect, the West symbolizes new economic ties, development and strategic interests for Azerbaijan. Iran, Russia, and Turkey, on the other hand, have historical, political, and cultural ties with Azerbaijan. The relations with these three countries have significant impact on nation-building and state-building aspects of political transformation.

The purpose here is not to provide an analysis of the foreign policy of Azerbaijan nor of foreign countries' relations with Azerbaijan.[13] Rather, the question of how Azerbaijan's perceptions of its external environment affect the post-Soviet political transformation will be explored.

---

[13] On foreign policy of Azerbaijan, L. Alieva (2000), "Reshaping Eurasia: Foreign Policy Strategies and Leadership Assets in the Post-Soviet South Caucasus;" T. Dragadze (1996), "Azerbaijan and the Azerbaijanis," in G. Smith (ed.), *The Nationalities Question in The Soviet Union*, London: Longman; pp. 269-290; E. Herzig (1999), *The New Caucasus*; S. Hunter (1996), *Transcaucasus In Transition Nation Building and Conflict*, Washington D.C.: The Center For Strategic and International Studies; S. Hunter (1997). "Azerbaijan: Searching for new Neighbors," in I. Bremmer and R. Taras (eds.), *New States, New Politics: Building the Post-Soviet Nations*, Cambridge: Cambridge University Press; T. Swietochowski (1994) "Azerbaijan's Triangular Relationship: The Land Between Russia, Turkey and Iran," in A. Banuazizi and M. Weiner (eds.), *The New Geopolitics of Central Asia and Its Borderlands* London: I.B. Tauris, pp. 118-35; and (1999), "Azerbaijan: Perspectives from the Crossroads," *Central Asian Survey* 18(4), pp. 419-434.

Azerbaijan and Its Neighbors

After the achievement of independence, protecting newly acquired sovereignty became the focal point of the Azerbaijani state-building. The construction of new statehood required new institution-building, national law-making and formulation of domestic and foreign policy agendas. In the post-independence period of Azerbaijan, there is a common understanding between government and opposition on statehood and state-building. The prerequisites defining statehood are first, independence, which signifies the sense of sovereignty in internal and external affairs; second, territorial integrity, which is still under threat, as 20% of the territory is under Armenian occupation; third, international recognition, which renders the status of statehood guaranteed via protection by international law. Membership of regional and international organizations was viewed as consolidating the existence of the independent statehood as well as its "being an integral part of the world." Both Elchibey's and Aliyev's governments declared the new statehood project as being democratic, secular, and legal in accordance with international standards.

The post-Soviet political transformation is also about new nation-building. The post-communist countries are considered as "incomplete" and "unrealized" nation-states, with the process of nation-building conceived of as a dynamic rather than a "static condition" (Brubaker 1996: 63). These countries are still "nations in the making" and "nationalizing states" (ibid. 9) as well as states-in-making. States that require a nation-building project are in a far more difficult position than those confident of national identity. Independence generated ideas of the nation-state and citizenship. Both nation and state-building inherited factors from the period of Soviet rule and from the preindependence movements. The task was the construction of a national identity that was an alternative to the Soviet policies of Sovietization[14] and Russification.[15] Similarly, in Azerbaijan, the post-Soviet nation-building is a process in which the boundaries of the "national" have been redrawn, the content of "being national" has been rediscovered and redefined and in some sense reformulated in accordance with the ideas of independence and sovereignty.

In this project of democratic and secular statehood and nationhood, Azerbaijan's neighbors are of particular importance. The People's Front Government (PFG) in general and President Elchibey in particular represented a rupture from Soviet rule. The PFG was characterized as

[14] What I refer to as Sovietization is the aim of creating the "Soviet man," regardless of national and religious identities.

[15] What I refer to as Russification is the widespread use of the Russian language, both in education and daily life.

pro-Turkey, pro-West, anti-Iran and anti-Russia (Hunter 1996; 1997; Swietochowski 1999) and defined its nation- and state-building project accordingly.

Armenia's and Russia's impact on state-building and nation-building can be seen in the emphasis on the territorial integrity and on the protection of independence. For the PFG, both countries constituted threats. The main point of reference is the Karabagh problem. Initially a secessionist movement, the conflict became an interstate war that resulted in an estimate of 25,000 dead (Macfarlane and Minear 1997: 1). The number of estimated casualties is 7,000-8,000 from the Armenian side and 20,000-22,000 from the Azerbaijani side (Panossian 2001).

In Azerbaijan, Karabagh's significance is tied to all three processes of political transformation. First and foremost, the occupation of the Azerbaijani territory by Armenians permeated the redefinition of national identity by redefining "others" and fostered national independence movement. At the beginning of the conflict, Armenians and Russians were considered as being the main "others." It has been suggested that without the help of the Russians the Armenians would not have been able to invade the territory (Elchibey 1998: 47).

The Karabagh problem is also an integral part of state-building since the territorial integrity of the country is violated. It is also an integral part of the democratization process, as both defeat and success in handling the issue was and is always at the center stage of opposition policies in the struggle against government. Besides being the main stimulus of the independence movement, Karabagh is the sacred and untouchable signifier of territorial integrity, which is the basis for statehood. Azerbaijanis regard Nagorno Karabagh as an integral part of their history and territory, where Armenians "unjustly" occupying 20% of Azerbaijani territory, resulting in approximately one million refugees and internally displaced people. Unless territorial integrity can be achieved, the project of statehood and nationhood will remain incomplete.

New state-building required new foreign policymaking. The agenda of state-building and the definition of "friends and foes" were also interrelated to nation-building. With the PFA's coming to power, the issue of national identity that was the stimulus of the demand for independence and democracy became a tool for state-building and shaped foreign policy orientations. The conduct of foreign policy towards these four states inspired the PFG's agendas of state-building and nation-building.

Under the presidency of Ebulfez Elchibey, Azerbaijan declined membership of Commonwealth of Independent States (CIS), arguing that it would create another version of Russian dominance. It was the first republic in the Soviet Union who freed its territory of Soviet military bases and did not allow Russian border troops and military bases in the country (Aliyeva 2000: 3). During the PFG years, Russia was considered as a main threat for the independence and security of statehood since the country symbolizes the past, hegemony, and dominance as well as the authoritarian system.

As for the relations with Turkey, the country has a special prestige, especially in the PFG's years. The fact that Turkey was the first country to recognize Azerbaijan's independence and did support Azerbaijan on the Karabagh problem created very close ties of friendship between two countries. Because of the cultural, ethnic, and linguistic similarities, Turkey is considered as the best ally and ultimate friend of Azerbaijan by both the Elchibey and Aliyev governments. President Aliyev frequently described Azerbaijan and Turkey as being one nation but two states. Heydar Aliyev's foreign policy can be considered as the partial continuation of the PFG's principles, close friendship with Turkey, as a gateway to the West and the development of economic ties. The most concrete example is the realization of the Baku-Tbilisi-Ceyhan main export pipeline, which will transport Azerbaijani oil to the Western markets through Turkey. However, Aliyev was quite cautious not to disturb Russia and Iran and recognized these two countries as regional powers and encouraged collaboration rather than antagonisms. He conducted a balanced policy, developed peaceful relationships and Azerbaijan became a member of the CIS.

The post-Soviet political transformation of Azerbaijan is marked by strong emphasis on secularism. Both the Elchibey and Aliyev governments have declared their will to remain a secular country and kept a distance while arranging their relationship with neighboring Iran, a Shi'ite country whose impact in promoting religiosity could be potentially significant. In this respect, Iran does not provide any inspirations for nation and state-building projects, nor does it constitute a model. Although both countries are Shi'a Muslims and share historical and cultural heritage (Hunter 1996; Swietochowski 1994a; 1994b), close collaboration and approachment are not achieved. The PFG also had hostile relationships with Iran. The fact that the Azerbaijani territory was divided into two parts—one part remaining in Russia and the other in Iran by the Turkmanchay Treaty signed in 1828—widely affected the nationalist inspirations of the PFA elite. This has been revived by the PFG concerns for the Azerbaijanis living in Iran. The issue of "divided Azerbaijan" became especially significant during the early years of independence. The PFA asked for reestablishment of cultural links with "compatriots" living in Iran (PFA's Programme 1989) and gave a particular emphasis to the development of relationships between Iranian (Southern) Azerbaijan and the Republic of Azerbaijan, demanding the protection of the cultural rights of Azerbaijanis in Iran. The idea of a Unified Azerbaijan—*Butov Azerbaycan*—was launched by Elchibey, aiming to unify both people and territory.[16] The relations between "nationalist" and "dissident" Iranian Azerbaijani intellectuals were increased. Aliyev's government, however, took cautious measures against this policy, arguing that such ideas could harm the Azerbaijani-Iranian relationship and that the status of ethnic Azerbaijanis was an internal affair of Iran and avoided intervening.

A close collaboration between Azerbaijan and Iran has not been possible due to three main reasons. First, during and after the Karabagh war, Iran provided military support and food supply to Armenia. Iran's support for Armenia created mistrust and was severely criticized by the PFG. Second, Iran was criticized for encouraging pro-Islamic activities in Azerbaijan, such as opening up religious schools, bookshops, and cultural centers, allegedly supporting the Islam Party and allegedly encouraging fundamentalism (Huseyinli 2001: 169-170). Last, the status of the Caspian Sea and the equitable of distribution of its natural resources have been a major area of controversy among the Caspian states. On July 2001, the incursions of the Iranian navy into the Azerbaijani territorial waters have been considered as hostile acts of Iran, amounting to the violation of Azerbaijani territorial integrity. The Azerbaijani government perceived this as an attempt of intimidation by the Iranian government with regards to Azerbaijani policy on the maritime delimitation of the Caspian Sea. Iran's ambitions over the status of the Caspian Sea, known for its rich oil resources, strengthened the hostility.

Azerbaijan perceives the countries of the West as being more complicated than any other states. The West symbolized economic development, integration to the global markets, and strategic alliances. The global political economy is argued to promote the shift from "military or authoritarian governments to democracy and from protectionism and import substituting industrialisation towards open border and export promotion" (Strange 1992: 2). For the West, Azerbaijan is significant for its oil and gas resources and offers "economic, geopolitical and transportation significance of the regions of South Caucasus and Caspian Sea" (Swietochowski 1999: 432). The signing of the "contract of the century" in September 1994, which gave "a consor-

[16] For a detailed discussion of Southern Azerbaijan, see, E. Elchibey *Butov Azerbaycan* and the web page of Movement for Unified Azerbaijan, *www.bab.az*. The idea has been considered as a utopia. Elchibey said, "I will never allow any other countries to become our lords. When I said that the Soviet Union would collapse, no one believed me. They considered me insane. But it happened. Now I say that Iran will collapse and all Azerbaijanis will be united. The day is coming, you will see," (Elchibey's speech to university students, 18.12.1999, Caspian University, and interview with E. Elchibey, July 1998, Baku).

tium of Western oil companies the right to extract oil from Azerbaijan" (Hunter 1997: 469), signaled the hopes for future economic development. Prospects for economic development are dependent on foreign investment. One politician, however, argued that this was an unequal relationship: "We do not have a mutual interest with the West. We gave them the oil but could not get political support yet."[17] Similarly, Leila Aliyeva argued that "the international environment was not particularly favourable for the development of post-Soviet states," especially with regard to conflict resolution. For the territorial integrity of the Caucasian states she identifies the "wait and see policy" of the West as having made "the involvement of the OSCE and UN very cautious, belated and discrete" (2000: 2).

In the process of peace talks, Azerbaijan has had ongoing contacts with international bodies, such as the UN, the OSCE-Minsk Group, and the Council of Europe, alongside bilateral meetings with the diplomats and state officials of the USA, Russia, Turkey, and France. With the involvement of the Minsk Group in the peace talks on the Karabagh conflict, the West in general appears to be another "other" because of its ineffectiveness and its reluctance to act in favor of Azerbaijan. Azerbaijan argues that it is subject to discrimination even though international law supports its right to territorial integrity.[18] The government and opposition have blamed the international community for underestimating the importance of these principles. This contradicts the fact that both government and opposition have welcomed international agencies in the political and economic spheres. The only exception where the international element has been criticised both by government and opposition is the Karabagh problem. Due to the unsolvable nature of the problem and the burden of refugees for the last ten years, international actors were charged with reluctance and inefficiency to mediate between conflicting parties and to reach a resolution.

Aliyev's government had to legitimize its rule in terms of democratic practices, rule of law, and respect for human rights, although legitimization is far from convincing. The UN, the OSCE, the Council of Europe, and international non-governmental organizations are particularly concerned with various aspects of democratization in Azerbaijan. However, the degree of their impact is still debatable. The West affects Azerbaijan in

---

[17] Author's interview with E.F., July 2000, Baku.

[18] The main reference points are the UN Charter, UN Security Council Resolution 822 adopted on 30 April 1993, and the UN Security Council resolution 853 adopted on 29 July 1993, as well as, the Helsinki Final Act of the Conference on Security and Cooperation in Europe adopted on 1 August 1975, and the Paris Charter adopted in 1990.

terms of democratization in the sense that it introduces new concepts through advice, guidance, and monitoring. It is a source of inspiration for democratization of the country and provides international standards for democratization. It also provides guidelines for legislative reforms, economic reconstructing, institution-building and the formation of civil society. One of the party leaders argued that "The improvements in Azerbaijan are thanks to the Azerbaijani people's resistance with its opposition and free press and the Western demands to democratize the country. Aliyev understood this and he involuntarily allied himself to the universal rules of the game." [19] However, to what extent the propagation of "universal rules of the game" has a democracy-promoting impact is as yet questionable.

Azerbaijan and International Organizations

The collapse of the Soviet Union provided the international governmental and non-governmental organizations with new areas "to democratize" or "to promote democracy and human rights." Through international governmental and non-governmental organizations, the international community proposes guidelines for democratization and the formation of civil society in the former Soviet Union. [20] Within the framework provided by the internationally defined aims of "the right to democracy," "agenda for democratization," and "promoting and consolidating democracy," the question is to what extent the current relationship between the international and domestic actors impacts on the nature of democratization in Azerbaijan. Firstly, that relationship provides material resources and job opportunities through funding. Secondly, the international discourse of democracy has been incorporated into countries that have different political, cultural, and social backgrounds. Third, new concepts are introduced into

[19] Author's interview J.K., November 1999, Baku.

[20] One prominent example of the Western perception for democratization is the "Agenda for Democratization" (Boutros-Ghali, 1996). In the text, the expected process of democratization, its favorable conditions, its requirements, and the role of international governmental and non-governmental organizations were all outlined. Two other documents complemented the international element's vision for democratization: one is the UN Commission on Human Rights Resolution 1999/57, entitled "Promotion of the Right to Democracy," (UN Doc.E/CN.4/1999/SR.57); the other is the UN Commission on Human Rights Resolution 2000/47, entitled "Promoting and Consolidating Democracy," (UNDoc E/CN4/2000/SR.58). The significance of these resolutions lies in the fact that they are strongly supported by other international NGOs who have a consultative status in the Commission and take part in its meetings. The INGOs aimed at promoting aforementioned issues in the conduct of their activities in the former Soviet Union.

the domestic political vocabulary. Last, the relationship between domestic and international actors contributes to the pace of social and political change.

The IGOs and the INGOs have participated in the political transformation of the post-Soviet republics. In the South Caucasus, they acted as mediators and observers of peace-keeping in the Caucasian conflicts (Herzig 1999; 51), observed elections, monitored human rights violations, provided consultation in the institution-building, and promoted the formation of "civil society" via training, partnership, assistance, and funding (Herzig 1999: 36; Macfarlane 1997; 399). They are also supporting democratic reforms, respect of human rights, advice in national law-making, training for state officials and political parties. However, the impact of the international element in the post-Soviet transition of Azerbaijan is more complex than it first seems.

In most parts of the Soviet space, the idea of democracy and the attempt to further democratization had been already incorporated into the independence movements. Domestic struggles for democratic political transformation are internationalized because of the presence of IGOs and INGOs. These organizations opened up their regional offices, developed projects, and initiated exchange programes in the post-Soviet states. Particularly, the INGOs—among which are the SOROS Foundation, the ISAR, the National Democratic Institute, the National Republican Institute, the National Endowment for Democracy, the National Republican Institute, the Westminster Foundation, the Frederich Ebert Foundation and the Naumann Foundation offer funding to and form partnership with the local non-governmental organizations.

In Azerbaijan, the INGOs constitute the main sources of funding for local NGOs. Neither the state nor private sector provides any financial assistance. Membership fees do not provide substantial financial contribution since they are very small and are not usually paid. Thus, funding from abroad is vital for local organizations' existence. They also provide technical assistance, training, and exchange programs for local organizations in order to "teach" what civil society is and what its functions and operations are. Through these relationships, the international factor offers its expertise and experience to the newly democratizing states and proposes guidelines. In Azerbaijan, the integration of local NGOs with their international partners, provides them legitimacy, efficient working and productivity. Via projects, they produce monthly reports about their respective areas. Moreover, the financial support not only aids the realization of projects but is the source of income that facilitates most of the NGOs functioning as "job creation schemes for the professional classes" (Herzig 1999: 38). However, this funding-based relationship also creates tensions among local NGOs and between international and domestic ones. On the one hand, internationally funded projects make the local NGOs dependent on foreign assistance. This dependency prevents the continuity of their activities. The end of a project can leave an NGO without resources and their activities may cease. Moreover, there is a monopoly of grant-taking and international recognition. The previous experiences of local NGOs in getting financial assistance increase their chances of receiving subsequent projects and collaboration with other international organizations. Additionally, since local NGOs compete in order to get funding, the chances for collaboration among them remain limited. This competition also constitutes an obstacle for further development of civil society. On the other hand, the representatives of the INGOs complained that they were seen only as "money givers" and noted that training without financial benefit was not well-attended.[21] This is due to the fact that, for most of the local NGO people, their organizations are a source of income and a place of "self-realization of their vast potential."[22] They partly overcome economic difficulties through engaging in civil society activity.

The relationship between international and domestic organizations also influences the relationship between local organizations and the state. The relationship with the international element forms a milieu of interaction and becomes a support mechanism for local NGOs, a place from which they can expect support against government and a tool that puts pressure on the government and an opportunity to legitimize their demands.

Local NGOs, while trying to protect and defend "ordinary people's rights," need to be supported and protected by international organizations. As most of the local NGOS are funded by international NGOs, via projects, they have regular contacts with these organizations by e-mails, mail, reports, and newsletters. The international organizations become aware of the local conditions by getting firsthand information. They realize their supervision and monitoring roles via local organizations that provide them with reports on the country's conditions about the issues that they deal with. Due to these contacts, local NGOs consider them as a source of support and protection against government. If they are unable to apply any pressure directly on government, they ask their "international" partners to take the necessary steps. These steps can include making government accept their proposals concerning the laws and regulations related to the organizational sphere, supporting and asking for help for registration, releasing political prisoners, and asking for a more fair trial for prisoners.

[21] Author's interviews with the INGOs, operating in Azerbaijan, July 1999, Baku.

[22] Author's interviews with the local NGOs in Azerbaijan, August 1999, Baku.

This mutual relationship paves the way for the government to make new arrangements—even limited—in order to "protect its democratic and civic image."

Government favors the international element in order to promote its "democratizing image." The international non-governmental organizations do not face direct obstruction of their operations by the government. On the contrary, the representatives of IGOs and INGOs are highly welcomed both by the president Aliyev and governmental cadres. To promote the country's image of democratization and to show government "willingness to be an integral part of the Western world," the concepts of human rights, democracy, and the rule of law are frequently used by the representatives of the government and ruling cadres. By 1998, as a result of both the growing number of local NGOs and the impact and the advice and guidance of the international element of democratization, the state-building project of Aliyev's government was rearranged towards the inclusion of "civil society" or the "citizens' community" in the political transformation. New units and departments were established within the presidential apparatus and parliament. The UN Resource and Training Center for NGOs initiated the formation of an umbrella group of local NGOs, the NGO Forum, on the basis of an agreement with the government. By late 1999, when the parliament was working on a new draft law on social organizations, in order to rearrange their legal status, the representatives of some local NGOs were invited to parliamentary meetings and presented their proposals. The new arrangements in institution-building meant two things. On the one hand, government collaborated with the international community by supporting "civil society formation" in Azerbaijan. On the other hand, the international element realized its duty in encouraging civil society in order to promote democratization.

Evidently the interaction with the international community has also introduced new concepts, such as "civil society," "human rights," and "citizens' initiative" into political vocabulary and discourse. The introduction of these concepts has also raised issues of teaching and training. In this respect, in a transitional state like Azerbaijan, *the international* becomes *the national.* The penetration of the internationally used concepts leads to domestication of their usage and creates an internationalized vocabulary of politics. Local NGOs shape their discourses accordingly and internalize this new vocabulary. The interactive relationship between international and domestic societal actors makes the latter learn what the issues are and what kind of terminology is dominant in the international environment. Not only do the terms such as "citizen's initiative" and "citizenship community" become integral parts of daily language, but also original

English words such as "gender," "conflict management," "monitoring," "manager" and "training" are used without translation into the Azerbaijani language.

The interaction between domestic and international actors contributes to societal change, especially in the formation of civil society in Azerbaijan. The international element supports local organizations and introduces the international vocabulary of democratization into the domain of civil society. It provides technical assistance, funding, training, and guidance for the local NGOs, via which it teaches "how to be a democratic country." It also introduces the ways in which local NGOs can act within the international context of civil society by consultations, projects, and conferences.

Conclusion

The penetration of the international element into the post-Soviet transformation of Azerbaijan helps to define emerging state-society relationships. Although international support is officially highly appreciated, respect of the internationally recognized values remains cosmetic on the part of the government. The country still falls short of democratic standards. There are two reasons for this fact. Since the international element has to operate within the country, its relationship has to be compatible with the regime's demands. As they have to work with the current regime within and outside the country, these transnational actors do not wish to disturb or to challenge directly the government. In this respect, democratization seems to be less an international issue and more of a domestic one. Moreover, the international element is still subject to the limits defined by the regime and it is not that powerful to affect the government. Thus, the regime has considerable weight over international element at least domestically. Therefore, the outcome of the triple relationship between international element, government and local organizations is far from democratizing and/or promoting democratization when the domestic actors paricularly governments are resistant to it.

References

ALIEVA, L.

2000 "Reshaping Eurasia: Foreign Policy Strategies and Leadership Assets in the post-Soviet Caucasus." Retrieved April 19, 2001 (*http://ist-socrates.berkeley.edu/bsp/publications/alieva.pdf*).

ALLISON, R.

1994 "On the Gap Between Theories of Democracy and Theories of Democratization." *Democratization* 1(1): 8-26.

ALTSTADT, A.
1997 "Azerbaijan's Struggle Toward Democracy," in Dawisha, K. and Parrott, B. (eds.), *Conflict, Cleavage and Change in Central Asia and the Caucasus*. Cambridge: Cambridge University Press.
AZERBAYCAN DEMOKRASININ NKIAF FONDU PARLAMENT SEÇKILERI HAKKNDA RAPORT
1995 (Report on Parliamentary Elections by Foundation of the Development of Democracy). Baku.
BOVA, R.
1991 "Political Dynamics of the Post-Communist Transition: A Comparative Perspective." *World Politics* 44: 113-38.
BOUTROS-GHALI, BOUTROS.
1996 *An Agenda For Democratization*. New York: United Nations Publications.
BRUBAKER, R.
1996 *Nationalism Reframed*. Cambridge: Cambridge University Press.
CARLISLE, D.
1995 "Islam Karimov and Uzbekistan: Back to the Future," in Colton, T.J. and Tucker, R.C. (eds.), *Patterns in Post-Soviet Leadership*. Boulder: Westview Press.
DRAGADZE, T.
1996 "Azerbaijan and the Azerbaijanis," in Smith, G. (ed.), *The Nationalities Question in The Soviet Union* (2nd edition). London: Longman.
ELCIBEY, E.
1999 Butov Azerbaycan Yolunda: On the Road of Unified Azerbaijan. Ankara: Ecdad Publications.
ERGUN, A.
1998 *Process of Democratization and Political Elites in Azerbaijan*. Unpublished M.Sc. Thesis.
FRERES, C.L.
1999 "European Actors in Global Change: The Role of European Civil Societies in Democratization," in Grugel, J. (ed.), *Democracy Without Borders: Transnationalization and Conditionality in New Democracies*. London: Routledge.
GRUGEL, J.
1999 "European NGOs and democratization in Latin America: Policy Networks and Transnational Ethical Networks," in Grugel, J., *Democracy Without Borders: Transnationalization and Conditionality in New Democracies*. London: Routledge.
HERZIG, E.
1999 *The New Caucasus*. London: RIIA.
HUNTER, S.T.
1996 *Transcaucasus in Transition Nation Building and Conflict*. Washington D.C.: The Center For Strategic and International Studies.
1997 "Azerbaijan: Searching for New Neighbors," in Bremmer, I. and Taras, R. (eds.), *New States, New Politics: Building the Post-Soviet Nations*. Cambridge: Cambridge University Press.
HUSEYINLI, G.
2001 "Azerbaycanda Siyasal Partiler ve Siyasal Iliskiler." *Avrasya Dosyasi* 7(1).
KOPECKY, P. AND E. BARNFIELD
1999 "Charting the Decline of Civil Society: Explaining the Changing Roles and Conceptions of Civil Society in East and Central Europe," in Grugel, J. (ed.), *Democracy Without Borders: Transnationalization and Conditionality in New Democracies*. London: Routledge.
LINZ, J.J. AND A. STEPAN
1996 *Problems of Democratic Transition and Consolidation*. Baltimore: The Johns Hopkins University Press.
MACFARLANE, N.S. AND L. MINEAR
1997 Humanitarian Action and Politics: The Case of Nagorno-Karabakh. *Occasional Paper 25*. Providence: Watson Institute.
MACFARLANE, N.S.
1997 "Democratization, Nationalism and Regional Security in the Southern Caucasus." *Government and Opposition* 32(3): 399-420.

OCHS, M.
"Turkmenistan: A Quest for Stability and Control," in Dawisha, K. and Parrott, B. (eds.), *Conflict, Cleavage and Change in Central Asia and the Caucasus*. Cambridge: Cambridge University Press.

OFFE, C.
1996 *Varieties of Transition*. Cambridge: Polity Press.

OLCOTT, M.B.
1995 "Nursultan Nazarbaev and the Balancing Act of State-Building in Kazakhstan," in Colton, T.J. and Tucker, R.C. (eds.), *Patterns in Post-Soviet Leadership*. Boulder: Westview Press.
OSCE/UN of the OSCE UN Joint Electoral Observation Mission in Azerbaijan on Azerbaijan's 12 November Parliamentary Elections and Constitutional Referendum.1996. Retrieved November 1, 2002 (*http://www.osce.org/odihr/documents/reports/az/azer1-1.pdf*).

PRIDHAM, G. (ED.)
1991 *Encouraging Democracy: The International Context of Regime Transition in Southern Europe*. London: University of Leicester Press.

PANOSSIAN, R.
2001 "The Irony of Nagorno-Karabakh: Formal Institutions versus Informal Politics." *Regional and Federal Studies*.

PRIDHAM. G. AND T. VANHANEN (EDS.)
1994 *Democratisation in Eastern Europe, Domestic and International Perspectives*. London: Routledge.

PRIDHAM, G.
2000 *The Dynamics of Democratization*. London: Continuum.

ROSE, R., W. MISHLER, AND C. HAERPFER
1998 *Democracy and Its Alternatives*. London: Polity Press.

SCHMITTER, P.C. AND T.L. KARL
1994 "The Conceptual Travels of Transitologist and Consolidatologists: How Far to the East Should They Attempt to Go?" *Slavic Review* 53(1).

STRANGE, S.
1992 "States, Firms and Diplomacy." *International Affairs* 68(1).

SWIETCHOWSKI, T.
1999 "Azerbaijan: Perspectives from the Crossroads." *Central Asian Survey* 18(4): 419-434.
1995 *Russia and Azerbaijan: A Borderland in Transition*. New York: Columbia University Press.
1994 "Azerbaijan's Triangular Relationship: The Land Between Russia, Turkey and Iran," in Banuazizi, A. and Weiner, M. (eds.), *The New Geopolitics of Central Asia and Its Borderlands*. London: I.B. Tauris.
1994 "Islam and the Growth of National Identity in Soviet Azerbaijan," in Allworth, E. (ed.), *Muslim Communities Reemerge*. North Carolina: Duke University Press.
1994 "Azerbaijan: A Borderland at the Crossroads of History," in Starr, F.S. (ed.), *The Legacy of History in Russia and the New States of Eurasia*. Armonk, NY: M.E. Sharpe.

TARAS, R. (ED.)
1997 *Post-Communist Presidents*. Cambridge: Cambridge University Press.

TAYLOR, L.
1999 "Globalization and Civil Society-Continuities, Ambiguities and Realities in Latin America." *Indiana Journal of Global Legal Studies* (7).

TERRY, S.M.
1993 "Thinking About Post-Communist Transitions: How Different They Are." *Slavic Review* 52(2).

WHITEHEAD, L.
1986 "International Aspects of Democratization," in O'Donnell, G., Schmitter, P. and Whitehead, L. (eds.), *Transitions from Authoritarian Rule: Comparative Perspectives*. Baltimore: Johns Hopkins University Press.
1996 "Three International Dimensions of Democratization," in Whitehead, L. (ed.), *The International Dimension of Democratization*. Oxford: Oxford University Press.

Wimbush, E.S.
1979 "Divided Azerbaijan: Nation Building, Assimilation, and Mobilization Between Three States," in Mccagg, W., and Silver, B.D. (eds.), *Soviet Asian Ethnic Frontiers*. New York: Pergamon Press.

Index